# Attractors for Degenerate Parabolic Type Equations

Mathematical Surveys and Monographs
Volume 192

# Attractors for Degenerate Parabolic Type Equations

Messoud Efendiev

**American Mathematical Society**
Providence, Rhode Island

**Real Sociedad Matemática Española**
Madrid, Spain

2010 *Mathematics Subject Classification.* Primary 35K55, 35K99, 35K65, 35L05, 37L30.

---

For additional information and updates on this book, visit
**www.ams.org/bookpages/surv-192**

---

**Library of Congress Cataloging-in-Publication Data**

Efendiev, Messoud, author.
Attractors for degenerate parabolic type equations / Messoud Efendiev.
pages cm — (Mathematical surveys and monographs ; volume 192)
Includes bibliographical references and index.
ISBN 978-1-4704-0985-2 (alk. paper)
1. Differential equations, Parabolic. 2. Degenerate differential equations. I. Title.

QA377.E37 2013
515′.3534—dc23

2013016419

---



∞ The paper used in this book is acid-free and falls within the guidelines
established to ensure permanence and durability.
Visit the AMS home page at http://www.ams.org/

10 9 8 7 6 5 4 3 2 1 18 17 16 15 14 13

# Contents

# Preface

This book deals with the long-time behavior of solutions of degenerate parabolic dissipative equations arising in the study of biological, ecological, and physical problems. It is well known that the long-time behavior of many dissipative systems generated by evolution partial differential equations (PDEs) of mathematical physics can be described in terms of the so-called global attractors. By definition, a global attractor is a compact invariant set in the phase space which attracts the images of all bounded subsets under the temporal evolution.

In particular, in the case of dissipative PDEs in bounded domains, this attractor usually has finite Hausdorff and fractal dimension; see [9], [93], [32], and the references therein. Hence, if the global attractor exists, its defining property guarantees that the dynamical system (DS) reduced to the attractor $\mathcal{A}$ contains all of the nontrivial dynamics of the original system and the reduced phase space $\mathcal{A}$ is really "thinner" than the initial phase space $X$. (We recall that in infinite-dimensional spaces, a compact set cannot contain, for instance, balls and as a result should be nowhere dense.) One of the important questions in this theory is, *In what sense is the dynamics reduced to the attractor finite or infinite dimensional?*

Usually for regular (nondegenerate) dissipative autonomous PDEs in a bounded spatial domain $\Omega$, the Kolmogorov $\varepsilon$-entropy of their attractors has asymptotics such as

$$C_1|\Omega|\log_2\frac{1}{\varepsilon}\le H_\varepsilon(\mathcal{A},X)\le C_2|\Omega|\log_2\frac{1}{\varepsilon}.$$

Consequently, in spite of the infinite dimensionality of the initial phase space, the reduced dynamics on the attractor is (in a sense) finite dimensional and can be studied by the methods of the classical (finite-dimensional) theory of dynamical systems.

In contrast, infinite-dimensional global/uniform attractors are typical for dissipative PDEs in *unbounded* domains and/or for *nonautonomous* equations. In order to study such attractors, one usually uses the concept of Kolmogorov $\varepsilon$-entropy and its asymptotics in various functional spaces; see the recently published book [32] for the systematic study and appropriate details.

However, we note that the above results have been obtained mainly for evolution PDEs with more or less regular structure (e.g., uniform parabolic). In contrast to this, very little is known about the long-time dynamics of degenerate parabolic equations, such as porous media equations, $p$-Laplacian and doubly nonlinear equations, as well as degenerate diffusion with chemotaxis and ODE-PDE coupling systems (and their degenerate extensions), etc., which also play a significant role in modern mathematical physics, biology, and ecology. In this book we aim to fill this gap. Therefore the main goal of the present book is to give a detailed and systematic study of the well-posedness and the dynamics of the associated semigroup generated by the degenerate parabolic equations mentioned above in terms of their global and exponential attractors (e.g., existence, convergence of the dynamics, and the rate of convergence) as well as studying fractal dimension and Kolmogorov entropy of corresponding attractors. Our analysis and results in this book show that there are new effects related to the attractor of such degenerate equations which cannot be observed in the case of nondegenerate equations in bounded domains.

The book consists of eleven chapters. In Chapter 1 for the convenience of the reader we give some details of several well-known facts which are used in the sequel. In particular, we recall asymptotics of the $\varepsilon$-Kolmogorov entropy in various functional spaces, $L^q$ regularity and interior regularity of solutions for nondegenerate parabolic equations, classical embedding theorems as well as embedding theorems in weighted Sobolev spaces with degenerate weights. Moreover, Chapter 1 contains properties of Nemytskii (superposition) operators in Sobolev spaces and Hölder spaces which we use in the analysis of the next chapters.

Chapter 2 is concerned with the long-time behavior of solutions of evolution equations in terms of the global and regular attractors, including existence of attractors and properties of the attractor. Moreover we derive an estimate of time derivatives for nonautonomous perturbations of regular attractors. This is a cornerstone for developing a new method of proving stabilization to equilibria for solutions of an ODE-PDE coupling problem studied in Chapter 11, because, as we will see in Chapter 11, when the ODE part of the coupling is nonmonotone, the equilibria set of ODE-PDE coupling is not compact in any reasonable topology, and, as a result, the standard Lojasiewicz technique fails for stabilization of trajectories to the equilibria.

Chapter 3 is devoted to the systematic study of exponential attractors both for autonomous and for nonautonomous dynamical systems. We deal with existence theorems as well as perturbation theory of the exponential attractors and give some recent development on pull-back exponential attractors.

In Chapters 4–7 we are concerned with the well-posedness (global in time solutions) as well as long-time dynamics (finite and infinite dimensional) of porous medium and $p$-Laplacian equations both in homogeneous and heterogeneous media. In these chapters we present some new features related to the attractors of such equations that one cannot observe in nondegenerate cases, namely,

(a) the infinite dimensionality of the attractor,
(b) the polynomial asymptotics of its $\varepsilon$-Kolmogorov entropy,
(c) the difference in the asymptotics of the $\varepsilon$-Kolmogorov entropy depending on the choice of the underlying phase spaces.

These are the first examples in the mathematical literature of infinite-dimensional attractors admitting polynomial asymptotics of their $\varepsilon$-Kolmogorov entropy. It is worth noting that although infinite-dimensional global attractors are typical for nondegenerate equations in unbounded domains, even in that case the asymptotics of their Kolmogorov $\varepsilon$-entropy were always logarithmic in nature (such as $\left(\log_2 \frac{1}{\varepsilon}\right)^{n+1}$; see the book [32] for a systematic study of this issue).

We emphasize that, in our analysis in Chapters 4–7, to obtain properties (a)–(c) we cannot rely on the techniques that apply to nondegenerate parabolic equations. Indeed the usual method for obtaining lower bounds of the Kolmogorov entropy of attractors (as a result of its dimension) is based on the instability index of hyperbolic equilibria (see [9], [93], [32], and the references therein), which in turn requires differentiability of the associated semigroup with respect to the initial data. However, this method is not applicable for degenerate parabolic equations, since the associated semigroups (in contrast to nondegenerate parabolic equations) are usually not differentiable. That is why we were forced in Chapters 4–7 to develop alternative methods for proving properties (a)–(c) based on the existence of a localized solution and a scaling technique, which is closely related to the degenerate nature of the problem considered.

In Chapter 8 we give a detailed study of some classes of doubly nonlinear degenerate equations (we allow polynomial degeneration with respect to $\partial_t u$). We emphasize that the structure of a doubly degenerate equation considered in Chapter 8 does not fit the assumptions of the general fully nonlinear theory (see, e.g., [68] and [64]), so the highly developed classical theory in these books is not formally applicable. Thus we are forced to develop a new method, which in turn shows that a class of equations considered in Chapter 8 (under some assumptions on the data of the problem under consideration) possesses very good regularity properties and in particular has classical solutions. We believe that this phenomenon has a general nature and we clarify the difficulties related to finding stronger solutions of more general doubly nonlinear equations. Moreover, we obtain the uniqueness of solutions, which in fact was also a very delicate problem, because the simplest ODE example with polynomial degeneracy in $\partial_t u$ shows that the uniqueness of solutions fails. We also study the long-time behavior of solutions of doubly nonlinear degenerate problems in terms of the associated global and exponential attractors.

In Chapters 9 and 10 we consider both autonomous and nonautonomous chemotaxis systems with degenerate diffusion. Such classes of equations arise in the study of the role of chemotaxis for biofilm formation. We prove both global existence in time and uniqueness of solutions when the underlying domain is three dimensional. Such a well-posedness is done under certain "balance conditions" on the order of the porous medium degeneracy and the growth of the chemotactic functions.

The main aim of Chapter 11 is to study the long-time behavior of solutions of a class of degenerate parabolic systems describing the development of a forest ecosystem. From the mathematical point of view, the problem considered is a coupled system of second-order ODEs with a linear PDE (heat-like equation). Heuristically, it is clear that the dynamics of such coupled dissipative systems should depend drastically on the monotonicity properties of the ODE component. In this chapter we justify this in a mathematically rigorous way in the example of the ODE-PDE coupled system.

We finally note that the methods developed in Chapters 4–11 in order to study the long-time dynamics of certain classes of degenerate parabolic equations of different kinds seem to have a general nature and can be applied to other classes of degenerate equations, both autonomous and nonautonomous.

I would like to thank many friends and colleagues who gave me suggestions, advice, and support. In particular, I wish to thank Professors H. Berestycki, N. Kenmochi, K. H. Hoffmann, F. Hamel, R. Lasser, J. Mazon, A. Miranville, H. Matano, E. Nakaguchi, M. Otani, L. A. Peletier, R. Temam, J. R. L. Webb, W. L. Wendland, A. Yagi, S. Zelik. Futhermore, I am greatly indebted to my colleagues at the Helmholtz Zentrum München and Technische Universität München, the Alexander von Humboldt Foundation, as well as the AMS-RSME book series Mathematical Surveys and Monographs for their efficient handling of the publication.

Last but not least, I wish to thank my family for continuously encouraging me during the writing of this book.

CHAPTER 1

# Auxiliary materials

## 1.1. Functional spaces and embedding theorems

We shall use the following notation. We shall denote by $\mathbb{R}, \mathbb{C}, \mathbb{Z}$, and $\mathbb{N}$ the sets of real, complex, integer, and natural numbers, respectively; $\mathbb{Z}_+ = \{x \in \mathbb{Z} | x \geq 0\}$ is the set of nonnegative integers. $\mathbb{R}^n$ is the standard real vector space of dimension $n$. We denote by $D_i$ the operator of partial differentiation with respect to $x_i$,

$$D_i u = \frac{\partial u}{\partial x_i} \quad (i = 1, \ldots, n).$$

As usual, we use multi-index notation to denote higher-order partial derivatives;

$$D^\gamma = D_1^{\gamma_1} \cdots D_n^{\gamma_n}, \quad |\gamma| = \gamma_1 + \cdots + \gamma_n,$$

is a partial derivatives of order $|\gamma|$ for a given $\gamma = (\gamma_1, \ldots, \gamma_n)$, $\gamma_i \in \mathbb{Z}_+$.

Let $u : \Omega \subset \mathbb{R}^n$ be a real function defined on a bounded domain $\Omega$. The space of continuous functions over the bounded domain $\bar{\Omega}$ is denoted by $C(\bar{\Omega})$; the norm in $C(\bar{\Omega})$ is defined in a standard way:

$$\|u\|_{C(\bar{\Omega})} = \sup\{|u(x)| \mid x \in \bar{\Omega}\}.$$

The space $C^m(\Omega)$ consists of all real functions on $\Omega$ which have continuous partial derivatives up to order $m$. By definition, $u$ belongs to $C^m(\bar{\Omega})$ iff (abbreviation for if and only if) $u \in C^m(\Omega)$ and $u$ and all its partial derivatives up to order $m$ can be extended continuously to $\bar{\Omega}$. Let $0 < \gamma < 1$ and $k \in \mathbb{Z}_+$. By definition $C^{k,\gamma}(\bar{\Omega})$ denotes the Hölder space of functions $u : \Omega \to \mathbb{R}$ such that $D^\alpha u : \Omega \to \mathbb{R}$ exists and is uniformly continuous when $|\alpha| = k$ and such that

$$|u|_{k,\gamma} \equiv \sup \left\{ \frac{|D^\alpha u(x) - D^\alpha u(y)|}{|x - y|^\gamma} \mid x, y \in \Omega,\ x \neq y,\ |\alpha| \leq k \right\} \tag{1.1}$$

is finite. For $u \in C^{k,\gamma}(\bar{\Omega})$, we set

$$\|u\|_{k,\gamma} = |u|_{k,\gamma} + \sum_{|\alpha| \leq k} \max\{|D^\alpha u(x)| \mid x \in \bar{\Omega}\}.$$

We also have

$$C^{k,\gamma}(\partial\Omega) = \{\varphi : \partial\Omega \to \mathbb{R} \mid \text{ there exists } u \in C^{k,\gamma}(\bar{\Omega}) \text{ with } u|_{\partial\Omega} = \varphi\},$$

and for $\varphi \in C^{k,\gamma}(\partial\Omega)$ we set

$$\|\varphi\|_{k,\gamma} = \inf \left\{ \|u\|_{k,\gamma} |\ u|_{\partial\Omega} = \varphi; u \in C^{k,\gamma}(\bar{\Omega}) \right\}.$$

In cases when it is clear from the context where the function under consideration is defined, we shall sometimes simply write $u \in C^k$ instead of, for example, $u \in C^k(\mathbb{R}^n)$. In several examples we shall use the spaces of functions that are $2\pi$-periodic in every variable $x_i$ $(i = 1, \ldots, n)$. We shall consider such functions as being defined on the $n$-dimensional torus $T^n = \mathbb{R}^n/(2\pi\mathbb{Z})^n$. We denote by $L^p(\Omega)$, $1 \leq p \leq \infty$, the space of measurable functions with the finite norm

$$\|u\|_{0,p} = \|u\|_{L^p} = \left(\int_\Omega |u(x)|^p dx\right)^{1/p}. \tag{1.2}$$

We denote by $L^\infty(\Omega)$ the space of almost everywhere bounded functions,

$$\|u\|_{0,\infty} = \|u\|_{L^\infty} = \operatorname{vrai}\sup\{|u(x)| \mid x \in \Omega\}$$

(for continuous functions this norm coincides with the norm of $C(\bar{\Omega})$). The norm in the Sobolev space $W^{l,p}(\Omega)$, $l \in \mathbb{Z}_+$, $1 \leq p < \infty$, is defined by the formula

$$\|u\|_{l,p} = \left(\sum_{|\alpha|\leq l} \|D^\alpha u\|^p_{L^p}\right)^{1/p}. \tag{1.3}$$

In the case $p = 2$ this Sobolev space is a Hilbert space and is denoted $H^l(\Omega)$, $H^l(\Omega) = W^{l,2}(\Omega)$. The scalar product in $H^l(\Omega)$ is defined by the formula

$$(u, v)_l = \sum_{|\alpha|\leq l} \int_\Omega D^\alpha u(x) \cdot D^\alpha v(x) dx. \tag{1.4}$$

The space $W^{l,p}(\Omega)$ is the completion of $C^l(\Omega)$ with respect to the norm (1.3). The norms $C^{k,\gamma}(T^n)$ and $W^{l,p}(T^n)$ are defined by (1.1) and (1.3) with $\Omega = (0, 2\pi)$. The scalar product and the norm in $H^l(T^n)$, which are equivalent to those defined by (1.4), are defined in terms of Fourier coefficients,

$$(u, v)_l = \sum \tilde{u}(\xi) \cdot \overline{\tilde{v}(\xi)} \cdot (1 + |\xi|^2)^l; \ \|u\|^2_l = \langle u, u\rangle_l, \tag{1.5}$$

where the summation is over $\xi \in \mathbb{Z}^n$; the bar denotes complex conjugation; $\tilde{u}(\xi)$ and $\tilde{v}(\xi)$ are the Fourier coefficients,

$$\tilde{u}(\xi) = (2\pi)^{-n} \int_\Omega u(x) e^{-ix\cdot\xi}\, dx.$$

Here the integral is over $[0, 2\pi]^n$;

$$x \cdot \xi = x_1\xi_1 + \cdots + x_n\xi_n.$$

The formula (1.5) defines the norm in $H^l(T^n)$ for $l \in \mathbb{R}$ as well as $l \in \mathbb{Z}_+$. We denote by $C^\infty(\bar{\Omega})$ the space $\bigcap_{k\geq 0} C^k(\bar{\Omega})$ and by $C_0^\infty(\Omega)$ the set of functions from $C^\infty(\bar{\Omega})$ which vanish on a neighborhood of the boundary $\partial\Omega$. We shall also use spaces of functions which vanish on $\partial\Omega$. In this case we shall denote the corresponding space as follows:

$$C^{k,\gamma}(\bar{\Omega}) \cap \{u|_{\partial\Omega} = 0\}, \quad W^{l,p}(\Omega) \cap \{u|_{\partial\Omega} = 0\}.$$

We denote the completion of $C_0^\infty(\Omega)$ with respect to the norm of $H^l(\Omega)$ by $H_0^l(\Omega)$ and with respect to the norm of $W^{1,p}(\Omega)$ by $W_0^{1,p}(\Omega)$. It is well known that

$$H_0^1(\Omega) = H^1(\Omega) \cap \{u|_{\partial\Omega} = 0\} \quad \text{and} \quad W_0^{1,p}(\Omega) = W^{1,p} \cap \{u|_{\partial\Omega} = 0\}.$$

The Sobolev spaces $H^{\rho}(\Omega)$ with noninteger $\rho \geq 0$, $\rho = k + \beta$, $k \in \mathbb{Z}$, $0 \leq \beta < 1$ are endowed with the norm

$$\|u\|_{\rho}^{2} = \|u\|_{k}^{2} + \int_{|y| \leq \delta} \|u(x+y) - u(x)\|_{k}^{2} \cdot |y|^{-n-2\beta} dy$$

($u(x)$ is extended over a $\delta$-neighborhood of the boundary; see [78]).

By $S(\mathbb{R}^n)$ we denote the class of rapidly decreasing (at $\infty$) functions $u(x) \in C^{\infty}(\mathbb{R}^n)$, with

$$(1 + |x|)^{k} \, |D^{\alpha} u(x)| \leq C_{k,\alpha}$$

for each $\alpha = (\alpha_1, \ldots, \alpha_n) \in \mathbb{Z}_{+}^{n}$ and $k \in \mathbb{Z}_{+}$, where $C_{k,\alpha}$ are constants.

Recall that an operator $j : X \to Y$ between Banach spaces with $X \subseteq Y$ is an embedding iff $j(x) = x$ for all $x \in X$. The operator $j$ is continuous iff $\|x\|_Y \leq$ constant $\|x\|_X$ for all $x \in X$. Further, $j$ is compact iff $j$ is continuous, and every bounded set in $X$ is relatively compact in $Y$. If the embedding $X \hookrightarrow Y$ is compact, then each bounded sequence $\{x_n\}$ in $X$ has a subsequence $\{x_{n'}\}$ which is convergent in $Y$.

We will widely use the Sobolev embedding theorems formulated below.

THEOREM 1.1. *Let $\Omega$ be a bounded domain in $\mathbb{R}^n$, with smooth boundary $\partial\Omega$ and $0 \leq k \leq m-1$. (See [2], [78], [97].) Then*

$$W^{m,p}(\Omega) \hookrightarrow W^{k,q}(\Omega), \quad \text{if} \quad \frac{1}{q} \geq \frac{1}{p} - \frac{m-k}{n} > 0, \tag{1.6}$$

$$W^{m,p}(\Omega) \hookrightarrow W^{k,q}(\Omega), \quad \text{if} \quad q < \infty, \; \frac{1}{p} = \frac{m-k}{n}, \tag{1.7}$$

$$W^{m,p}(\Omega) \hookrightarrow C^{k,\delta}(\bar{\Omega}), \quad \text{if} \quad \frac{n}{p} < m - (k + \delta), \; 0 < \delta < 1.$$

*The last two embeddings are compact; the first embedding is compact if $\frac{1}{q} > \frac{1}{p} - \frac{m-k}{n}$.*

THEOREM 1.2. *Let $0 \leq \beta < \alpha \leq 1$ or $\alpha, \beta \in \mathbb{Z}$ with $0 \leq \beta < \alpha$ (see [79], [97]). Then*

(1) *the embedding*

$$C^{\alpha}(\bar{\Omega}) \hookrightarrow C^{\beta}(\bar{\Omega}) \text{ is compact} \tag{1.8}$$

*and*

(2) *for $k + \beta < m + \alpha$, with $0 \leq \alpha, \beta \leq 1$, $m \geq k \geq 0$, the embeddings*

$$C^{m,\alpha}(\bar{\Omega}) \hookrightarrow C^{k,\beta}(\bar{\Omega}) \tag{1.9}$$

*are compact.*

Lemma 1.1 will be used in Chapter 8. Since it is not known by the broader audience, we present the proof below.

Let $\Omega$ be a bounded domain in $\mathbb{R}^3$ and let $a \in L^{\infty}(\Omega)$ be a nonnegative function. Set $\Omega_0 := \{x \in \Omega : a(x) = 0\}$ and consider the spaces

$$V := L^2([0,T], L^2(\Omega \backslash \Omega_0))$$

and

$$V_1 := \left\{ u \in L^2([0,T], W^{1,2}(\Omega)) : \int_0^T (a(\cdot)\partial_t u(s), \partial_t u(s))\, ds < \infty \right\}.$$

LEMMA 1.1. *The embedding $V_1 \subset V$ is compact for $n \le 3$.*

PROOF. We set

$$\Omega_\delta^+ := \{x \in \Omega, \quad a(x) > \delta\}, \quad \delta \ge 0.$$

Then, obviously, $\Omega_0^+ = \Omega \backslash \Omega_0$. Moreover, due to the continuity of the Lebesgue measure, we have

$$\lim_{\delta \to 0} \operatorname{mes}\{\Omega_0^+ \backslash \Omega_\delta^+\} = 0.$$

On the other hand, due to the Hölder inequality and the embedding $W^{1,2} \subset L^6$ (since $n \le 3$), we have

$$\|\chi_X(\cdot) v\|_{L^2(\Omega)} \le \operatorname{mes}\{X\}^{1/3} \|v\|_{L^6(X)} \le C \operatorname{mes}\{X\}^{1/3} \|v\|_{W^{1,2}(\Omega)}$$

for any set $X \subset \Omega$ and, consequently,

$$\|\chi_X(\cdot) v\|_{L^2([0,T]\times\Omega)} \le C \operatorname{mes}\{X\}^{1/3} \|v\|_{V_1}$$

where the constant $C$ is independent of $v \in V_1$. Thus, for verifying the compactness of the embedding $V_1 \subset V$, it is sufficient to verify the compactness of the embedding

$$V_1 \subset V^\delta, \quad V^\delta := L^2([0,T] \times \Omega_\delta^+),$$

for any *positive* $\delta$.

Now let $\delta > 0$ be fixed. Then, according to the Arzela-Ascoli theorem, we need to verify that there exists a function $\mu : \mathbb{R}_+ \to \mathbb{R}_+$, $\lim_{z \to 0+} \mu(z) = 0$, such that

$$\int_0^T \int_\Omega \chi_{\Omega_\delta^+}(x) |u(t+s,x) - u(t,x)|^2 \, dx\, dt \le \mu(|s|), \quad s \in \mathbb{R}, \tag{1.10}$$

and

$$\int_0^T \int_\Omega |\chi_{\Omega_\delta^+}(x+h) u(t,x+h) - \chi_{\Omega_\delta^+}(x) u(t,x)|^2 \, dx\, dt \le \mu(|h|), \quad h \in \mathbb{R}^3, \tag{1.11}$$

uniformly with respect to all $u$ belonging to the unit ball in $V_1$ (in these estimates function $u$ is assumed to be extended by zero for $(t,x) \notin (0,T) \times \Omega$).

Let us first verify (1.10). Let $s > 0$ (the case $s < 0$ can be considered analogously). Then, using the obvious formula

$$u(t+s,x) - u(t,x) = s \int_0^1 \partial_t u(t + \kappa s, x)\, d\kappa$$

together with the fact that $a(x) > \delta$ if $x \in \Omega_\delta^+$, we have

$$\begin{aligned} &\int_0^{T-s} \int_\Omega \chi_{\Omega_\delta^+}(x) |u(t+s,x) - u(t,x)|^2 \, dx\, dt \\ &\quad \le s \int_0^T \int_\Omega \chi_{\Omega_\delta^+}(x) |\partial_t u(t,x)|^2 \, dx\, dt \\ &\quad \le \delta^{-1} s \int_0^T \int_\Omega a(x) |\partial_t u(t,x)|^2 \, dx\, dt \le \delta^{-1} s \|v\|_{V_1}^2. \end{aligned} \tag{1.12}$$

On the other hand, using the fact that

$$\|u(\cdot,x)\|_{L^\infty([0,T])} \le C(\|\partial_t u(\cdot,x)\|_{L^2([0,T])} + \|u(\cdot,x)\|_{L^2([0,T])}),$$

we obtain

$$\begin{aligned}
&\int_{T-s}^{T}\int_\Omega \chi_{\Omega_\delta^+}(x)|u(t+s,x)-u(t,x)|^2\,dx\,dt\\
&\qquad\le Cs\int_\Omega \chi_{\Omega_\delta^+}(x)(\|\partial_t u(\cdot,x)\|^2_{L^2([0,T])} + \|u(\cdot,x)\|^2_{L^2([0,T])})\,dx\,dt\\
&\qquad\le C\delta^{-1}s\int_0^T\int_\Omega a(x)(|\partial_t u(t,x)|^2+|u(t,x)|^2)\,dx\,dt \le C\delta^{-1}\|u\|^2_{V_1}. \qquad (1.13)
\end{aligned}$$

Estimates (1.12) and (1.13) show that (1.10) holds with $\mu(z) := 2C\delta^{-1}z$. Let us now verify (1.11). Indeed, due to the estimate

$$\begin{aligned}
&|\chi_{\Omega_\delta^+}(x+h)u(t,x+h) - \chi_{\Omega_\delta^+}(x)u(t,x)|\\
&\qquad\le |\chi_{\Omega_\delta^+}(x+h) - \chi_{\Omega_\delta^+}(x)|\cdot|u(t,x)| + |u(t,x+h)-u(t,x)|
\end{aligned}$$

and the embedding $W^{1,2}\subset L^6$, we have

$$\begin{aligned}
&\int_0^T\int_\Omega |\chi_{\Omega_\delta^+}(x+h)u(t,x+h) - \chi_{\Omega_\delta^+}(x)u(t,x)|^2\,dx\,dt\\
&\qquad\le CT\|u\|^2_{V_1}\Big(\int_\Omega |\chi_{\Omega_\delta^+}(x+h) - \chi_{\Omega_\delta^+}(x)|^3\,dx\Big)^{2/3}\\
&\qquad\quad+\int_0^T\int_\Omega |u(t,x+h)-u(t,x)|^2\,dx\,dt. \qquad (1.14)
\end{aligned}$$

The first term on the right-hand side of (1.14) tends to zero since $\chi_{\Omega_\delta^+}\in L^\infty(\Omega)\subset L^3(\Omega)$ and the second one tends to zero uniformly with respect to $u$ analogously to (1.10). Thus, estimates (1.10) and (1.11) are verified and Lemma 1.1 is proven. □

## 1.2. Kolmogorov ε-entropy and its asymptotics in functional spaces

We start with the definition of Kolmogorov $\varepsilon$-entropy, via which we define fractal dimension of the compact set in the metric space. We will use these two concepts in the sequel.

DEFINITION 1.1. Let $K$ be a (pre)compact set in a metric space $M$. Then, due to Hausdorff criteria, it can be covered by a finite number of $\varepsilon$-balls in $M$. Let $N_\varepsilon(K,M)$ be the minimal number of $\varepsilon$-balls that cover $K$. Then, we can call Kolmogorov $\varepsilon$-entropy of $K$ the logarithm of this number:

$$\mathbb{H}_\varepsilon(K,M) := \log_2 N_\varepsilon(K,M).$$

We now give several examples of typical asymptotics for the $\varepsilon$-entropy.

EXAMPLE 1.1. We assume that $K=[0,1]^n$ and $M=\mathbb{R}^n$ (more generally, $K$ is an $n$-dimensional compact Lipschitz manifold of the metric space $M$). Then

$$\mathbb{H}_\varepsilon(K,M) = (n+\overline{o}(1))\log_2\frac{1}{\varepsilon} \quad \text{as } \varepsilon\to 0.$$

Here $\overline{o}(1)$ denotes, as usual, a function of $\varepsilon$ that converges to 0 as $\varepsilon$ goes to 0. This example justifies the definition of the fractal dimension.

DEFINITION 1.2. The fractal dimension $\dim_F(K,M)$ is defined as

$$\dim_F(K,M) := \limsup_{\varepsilon\to 0} \frac{\mathbb{H}_\varepsilon(K,M)}{\log_2 1/\varepsilon}.$$

Hence, for a compact $n$-dimensional Lipschitz manifold $K$ in a metric space $M$, $\dim_F(K,M) = n$.

The following example shows that, for sets that are not manifolds, the fractal dimension may be a noninteger.

EXAMPLE 1.2. Let $K$ be a standard ternary Cantor set in $M = [0,1]$. Then $\dim_F(K,M) = \frac{\ln 2}{\ln 3} < 1$.

PROOF. Let $K$ be the Cantor set obtained from the segment [0,1] by the sequential removal of the center thirds. First we remove all the points between 1/3 and 2/3. Then the center thirds $(1/9, 2/9)$ and $(7/9, 8/9)$ of the remaining segments $[0,1/3]$ and $[2/3,1]$ are deleted. After that, the center parts $(1/27, 2/27)$, $(7/27, 8/27)$, $(19/27, 20/27)$, and $(25/27, 26/27)$ of the four remaining segments $[0,1/9]$, $[2/9,1/3]$, $[2/3,7/9]$, and $[8/9,1]$, respectively, are deleted. If we continue this process to infinity, it will lead to the standard Cantor set $K$. Next we calculate its fractal dimension. We emphasize that $K = \bigcap_{M=0}^{\infty} \theta_m$, where $\theta_0 = [0,1]$, $\theta_1 = [0,1/3] \cup [2/3,1]$, $\theta_2 = [0,1/9] \cup [2/9,1/3] \cup [2/3,7/9] \cup [8/9,1]$, and so on. Each of the sets $\theta_m$ can be considered as a union of $2^m$ segments of length $3^{-m}$. In particular, the cardinality of the covering of the set $K$ with segments of length $3^{-m}$ is equal to $2^m$. Consequently $\dim_F(K,[0,1]) = \lim_{m\to\infty} \frac{\ln 2^m}{\ln(3^m)} = \frac{\ln 2}{\ln 3}$. □

It is not difficult to show the following:

(1) If $K_1 \subseteq K_2$, then $\dim_F(K_1,M) \le \dim_F(K_2,M)$.

(2) $\dim_F(K_1 \cup K_2, M) \le \max\{\dim_F(K_1,M); \dim_F(K_2,M)\}$.

(3) $\dim_F(K_1 \times K_2, M\times M) \le \dim_F(K_1,M) + \dim_F(K_2,M)$.

(4) Let $g$ be a Lipschitzian mapping of one metric space $M_1$ into another $M_2$. Then $\dim_F(g(K),M_2) \le \dim_F(K,M_1)$.

The next example gives the typical behavior of the entropy in classes of functions with finite smoothness.

EXAMPLE 1.3. Let $V$ be a smooth bounded domain of $\mathbb{R}^n$. Let $K$ be the unit ball in the Sobolev space $W^{l_1,p_1}(V)$ and let $M$ be another Sobolev space $W^{l_2,p_2}(V)$ such that the embedding $W^{l_1,p_1} \subset W^{l_2,p_2}$ is compact, i.e.,

$$l_1 > l_2 \ge 0, \quad \frac{l_1}{n} - \frac{1}{p_1} > \frac{l_2}{n} - \frac{1}{p_2}.$$

Then, the entropy $\mathbb{H}_\varepsilon(K,M)$ has the following asymptotics (see [94]):

$$C_1\left(\frac{1}{\varepsilon}\right)^{n/(l_1-l_2)} \le \mathbb{H}_\varepsilon(K,M) \le C_2\left(\frac{1}{\varepsilon}\right)^{n/(l_1-l_2)}.$$

Finally, the last example shows the typical behavior of the entropy in classes of analytic functions.

EXAMPLE 1.4. Let $V_1 \subset V_2$ be two bounded domains of $\mathbb{C}^n$. We assume that $K$ is the set of all analytic functions $\phi$ in $V_2$ such that $\|\phi\|_{C(V_2)} \le 1$ and that $M = C(V_1)$. Then

$$C_1 \left(\log_2 1/\varepsilon\right)^{n+1} \le \mathbb{H}_\varepsilon(K\big|_{V_1}, M) \le C_2 \left(\log_2 1/\varepsilon\right)^{n+1}$$

(see [61]).

## 1.3. Interior regularity estimates for linear parabolic equations

We are now going to discuss the interior regularity estimates for linear parabolic equations. To this end, we first construct a special cut-off function.

PROPOSITION 1.1. *Let $V \subset B(R, 0, \mathbb{R}^l)$ be a bounded set in $\mathbb{R}^l$ and let $V_\delta := \mathcal{O}_\delta(V)$ be its $\delta$-neighborhood. Then, there exists a cut-off function $\phi \in C^\infty(\mathbb{R})$, $\phi(x) \in [0, 1]$, such that, for every $\beta \in (0, 1)$ and every $k \in \mathbb{Z}_+$,*

$$\begin{cases} (1) \ \ \phi(x) = 1, \quad \text{for } x \in V \text{ and } \phi(x) = 0 \ \text{ for } x \notin V_\delta, \\ (2) \ \ |D_x^k \phi(x)| \le C_{k,\beta}[\phi(x)]^{1-\beta}, \ \ x \in \mathbb{R}^l, \end{cases} \tag{1.15}$$

*where the constant $C_{k,\beta} = C(k, \beta, \delta, R)$ is independent of $x$ and the concrete choice of $V$ and $D_x^k$ means the collection of all $x$-derivatives of order $k$.*

PROOF. Indeed, let us introduce the standard bump function in $\mathbb{R}^l$:

$$\psi_r(x) := \begin{cases} e^{-\frac{1}{r^2 - |x|^2}}, & |x| < r, \\ 0, \ \ |x| \ge r. \end{cases} \tag{1.16}$$

Then, this function obviously satisfies estimate (1.15)(2).

Let us now fix a covering of the $\mathbb{R}^l$ by the balls of radius $\delta/2$ and let $W_\delta := \delta/2\mathbb{Z}^l$ be the centers of that covering. Let us now also construct the partition of unity associated with that covering and (1.16) via

$$\phi_q(x) := \frac{\psi_{\delta/2}(x - q)}{\sum_{p \in W_\delta} \psi_{\delta/2}(x - p)}, \quad q \in W_\delta. \tag{1.17}$$

Obviously, $\{\phi_q(x)\}_{q \in W_\delta}$ is a partition of unity associated with the above covering and, moreover, these functions satisfy (1.15)(2) *uniformly* with respect to $q \in W_\delta$.

Let us now define the required cut-off function $\phi(x) = \phi_V(x)$ by the following expression:

$$\phi_V(x) := \sum_{q \in W_{\delta/2} \cap V_{\delta/2}} \phi_q(x). \tag{1.18}$$

Indeed, since $\operatorname{supp} \phi_q \subset B(\delta/2, q, \mathbb{R}^l)$ and the sum of all such functions equals one identically, the function $\phi_V$ thus defined satisfies (1.15)(1). Moreover, since the number of points

$$\#(W_{\delta/2} \cap V_{\delta/2}) \le \#(W_{\delta/2} \cap B(R + \delta, 0, \mathbb{R}^l)) \le N(\delta, R)$$

is finite and uniformly bounded with respect to $V \subset B(R, 0, \mathbb{R}^l)$ and the functions $\phi_q(x)$ satisfy (1.15)(2) uniformly with respect to $q \in W_{\delta/2}$, then the function $\phi_V(x)$ also satisfies this inequality uniformly with respect to $V \subset B(R, 0, \mathbb{R}^l)$. Proposition 1.1 is proven. □

We now recall the classical $L^q$-regularity estimate for second-order parabolic equations on the following model example:

$$\begin{cases} \partial_t w = a(t,x)\Delta_x w + b(t,x)w + h, \\ w\big|_{\partial\Omega} = 0, \quad w\big|_{t=0} = 0. \end{cases} \tag{1.19}$$

PROPOSITION 1.2. *Let $\Omega$ be a smooth domain and let $a \in C^\alpha(\Omega_T)$ (with $\alpha > 0$ and $\Omega_T := [0,T] \times \Omega$) satisfy*

$$0 < C_1 \le a(t,x) \le C_2, \quad (t,x) \in \Omega_T, \tag{1.20}$$

*for some positive $C_i$. Also let $h \in L^q(\Omega_T)$ for some $1 < q < \infty$, $q \neq 3/2$. Assume finally that*

$$b \in L^r(\Omega_T) \tag{1.21}$$

*for a sufficiently large $r$ depending on $q$ ($r > \max\{q, \frac{n+2}{2}\}$). Then, problem (1.19) possesses a unique solution $w \in W^{(1,2),q}(\Omega_T)$ and the following estimate holds:*

$$\|w\|_{W^{(1,2),q}(\Omega_T)} \le C\|h\|_{L^q(\Omega_T)}, \tag{1.22}$$

*where the constant $C$ depends on $q$, $\Omega$, $\|a\|_{C^\alpha}$, $\|b\|_{L^r}$ and on the constants $C_i$ from (1.20) but is independent of the concrete choice of $a$, $b$, and $h$.*

The proof of proposition (in the more general setting) can be found in [68, see Chapter IV, §9, Th. 9.1]. In particular, the assertion of the proposition is proven there without the assumption of Hölder continuity of $a$, and the constant $C$ in (1.22) depends on the modulus of continuity of the function $a$. However, for our purposes it is more convenient to control this modulus of continuity by the Hölder norm.

We are now able to verify the $L^q$-interior regularity estimate for equation (1.19) (which is analogous to estimate (10.12) of [68, see Chapter IV, §10, page 355]).

PROPOSITION 1.3. *Let the above assumptions hold and let $V$ be an arbitrary open set in $\Omega$. Then, for every $0 < t_0 < T$, $\delta > 0$, and $q > 2$, the solution $w$ satisfies*

$$\|w\|_{W^{(1,2),q}([t_0,T]\times V)} \le C(\|h\|_{L^q([0,T]\times V_\delta)} + \|w\|_{L^1([0,T]\times V_\delta)}), \tag{1.23}$$

*where $V_\delta := \mathcal{O}_\delta(V) \cap \Omega$ and the constant $C$ is independent of $w$ and of the concrete choice of $a$, $b$, and $h$.*

PROOF. According to Proposition 1.1 there is a cut-off function $\phi \in C^\infty(\mathbb{R}^{n+1})$ such that

$$\begin{cases} (1) \ \ \phi(t,x) \equiv 1, \quad \text{for } (t,x) \in [t_0,T] \times V, \\ (2) \ \ \phi(t,x) \equiv 0, \quad \text{for } (t,x) \notin [3t_0/4,T] \times V_{\delta/2}, \\ (3) \ \ |D^k_{(t,x)}\phi(t,x)| \le C_{k,\delta,\beta}[\phi(t,x)]^{1-\beta}, \end{cases} \tag{1.24}$$

where $\beta > 0$ is arbitrary and the constant $C_{k,\delta,\beta}$ is independent of $V$. Let us now introduce a function $w_\phi(t,x) := w(t,x)\phi(t,x)$ which obviously satisfies the equation

$$\partial_t w_\phi = a\Delta_x w_\phi + bw_\phi + h_\phi, \quad w_\phi\big|_{t=0} = 0, \quad w_\phi\big|_{\partial\Omega} = 0, \tag{1.25}$$

where

$$h_\phi := h\phi + w\partial_t\phi - 2\nabla_x\phi\nabla_x w - w\Delta_x\phi. \tag{1.26}$$

Applying now the $L^q$-regularity estimate (see Proposition 1.2) to equation (1.25) and using (1.24), we infer

(1.27)
$$\begin{aligned}\|w_\phi\|^q_{W^{(1,2),q}(\Omega_T)} &\le C\|h_\phi\|^q_{L^q(\Omega_T}\\ &\le C_1(\|h\|^q_{L^q([0,T]\times V_\delta)} + \int_{\Omega_T}[\phi(t,x)]^{q(1-\beta)}(|w(t,x)|^q + |\nabla_x w(t,x)|^q)\,dx\,dt).\end{aligned}$$

Let us assume for the moment that we have proven the interpolation inequality

$$\int_{\Omega_T}\phi^{q(1-\beta)}(|w|^q+|\nabla_x w|^q)\,dx\,dt \le \varepsilon\|w_\phi\|^q_{W^{(1,2),q}(\Omega_T)} + C_\varepsilon\|w\|^q_{L^1([0,T]\times V_\delta)} \tag{1.28}$$

which holds for every $\varepsilon > 0$. Then, inserting it into the right-hand side of (1.27) and fixing $\varepsilon$ to be small enough, we have

$$\|w_\phi\|^q_{W^{(1,2),q}(\Omega_T)} \le C(\|h\|^q_{L^q([0,T]\times V_\delta)} + \|w\|^q_{L^1([0,T]\times V_\delta)}) \tag{1.29}$$

which together with (1.24)(1) implies (1.23) and finishes the proof of the proposition.

Thus, we only need to verify inequality (1.28). Indeed, due to the Hölder inequality,

$$\begin{aligned}\int_{\Omega_T}\phi^{q(1-\beta)}|w|^q\,dx\,dt &= \int_{\Omega_T}(\chi_{V_\delta}(x)|w|)^{\beta q}|w_\phi|^{(1-\beta)q}\,dx\,dt\\ &\le C\|w\|^{\beta q}_{L^1([0,T]\times V_\delta)}\|w_\phi\|^{(1-\beta)q}_{L^s(\Omega_T)}\\ &\le \varepsilon\|w_\phi\|^q_{L^s(\Omega_T)} + C_\varepsilon\|w\|^q_{L^1([0,T]\times V_\delta)}\end{aligned} \tag{1.30}$$

where $\chi_{V_\delta}(x)$ is a characteristic function of the set $V_\delta$ and $s = s(\beta) := \frac{q(1-\beta)}{1-q\beta}$. Now fixing $\beta$ so small that the Sobolev embedding $W^{(1,2),q}(\Omega_T) \subset L^s(\Omega_T)$ holds, we verify inequality (1.28) for the term $\phi^{q(1-\beta)}|w|^q$.

Thus, it now remains to verify (1.28) for the term containing $\nabla_x w$. To this end, we transform this term as follows:

$$\begin{aligned}\int_{\Omega_T}\phi^{q(1-\beta)}|\nabla_x w|^q\,dx\,dt &\le C\int_{\Omega_T}\phi^{1-q\beta}\nabla_x w.\nabla_x w_\phi|\nabla_x w_\phi|^{q-2}\,dx\,dt\\ &\quad + C\int_{\Omega_T}(\phi^{1-\beta}|\nabla_x w|)(\phi^{1-2\beta}|w|)^{q-1}\,dx\,dt.\end{aligned} \tag{1.31}$$

The last integral on the right-hand side can be, in turn, estimated via the Hölder inequality

$$I_1 \le \varepsilon\int_{\Omega_T}\phi^{q(1-\beta)}|\nabla_x w|^q\,dx\,dt + C_\varepsilon\int_{\Omega_T}\phi^{q(1-2\beta)}|w|^q\,dx\,dt. \tag{1.32}$$

The last term on the right-hand of (1.32) can be estimated exactly as (1.30) and the first one coincides with the left-hand side of (1.31), but with arbitrarily small coefficient. This implies

$$\begin{aligned}\int_{\Omega_T}\phi^{q(1-\beta)}|\nabla_x w|^q\,dx &\le C\int_{\Omega_T}\phi^{1-q\beta}\nabla_x w.\nabla_x w_\phi|\nabla_x w_\phi|^{q-2}\,dx\,dt\\ &\quad + \varepsilon\|w_\phi\|^q_{W^{(1,2),q}(\Omega_T)} + C_\varepsilon\|w\|^q_{L^1([0,T]\times V_\delta)}.\end{aligned} \tag{1.33}$$

So, one only needs to estimate the first term on the right-hand side of (1.33). Integrating by parts in that term and again using (1.24)(3), we infer

$$\begin{aligned} I_2 \le C\int_{\Omega_T} \phi^{1-q\beta}|w|\cdot|\Delta_x w_\phi|\cdot|\nabla_x w_\phi|^{q-2}\,dx\,dt \\ + C\int_{\Omega_T} \phi^{1-(q+1)\beta}|w|\cdot|\nabla_x w_\phi|^{q-1}\,dx\,dt \end{aligned} \tag{1.34}$$

(here we have implicitly used that $w\big|_{\partial\Omega} = 0$ and that $q > 2$). Applying now once more the Hölder inequality to both integrals on the right-hand side of (1.34), we finally arrive at

$$\begin{aligned} I_2 &\le \varepsilon\int_{\Omega_T} |\Delta_x(w_\phi)|^q + |\nabla_x w_\phi|^q\,dx\,dt + C_\varepsilon\int_{\Omega_T} \phi^{q(1-(q+1)\beta)}|w|^q\,dx\,dt \\ &\le C\varepsilon\|w_\phi\|^q_{W^{(1,2),q}(\Omega_T)} + C_\varepsilon\int_{\Omega_T} \phi^{q(1-(q+1)\beta)}|w|^q\,dx\,dt. \end{aligned} \tag{1.35}$$

Now estimating the last term on the right-hand side of (1.35) analogously to (1.30), we deduce the analogue of estimate (1.28) for the term $I_2$. Then inserting this estimate into (1.31)–(1.33) and using (1.30), we finish the proof of estimate (1.28). Thus, Proposition 1.3 is proven. □

COROLLARY 1.1. *Let the solution $w(t,x)$ of equation* (1.19) *be defined only for $(t,x) \in [t_0/2,T]\times V_\delta$ and let the coefficients $a$, $b$ and the external force $h$ also be defined only in $[t_0/2,T]\times V_\delta$ and let them satisfy the assumptions of Proposition* 1.3 *in this domain. Then, the solution $w$ satisfies the interior regularity estimate* (1.23) *with the constant $C$ independent of the concrete choice of $V$, $a$, $b$, $h$, and $w$.*

PROOF. Indeed, the function $w_\phi(t,x) := w(t,x)\phi(t,x)$ introduced in the proof of Proposition 1.3 equals zero identically for $(t,x)$ outside of $[3t_0/4,T]\times V_{\delta/2}$. Therefore, we can construct extensions $\tilde a$ and $\tilde b$ of the coefficients $a$ and $b$ from the initial domain of definition $[t_0/2,T]\times V_\delta$ to the whole domain $[0,T]\times\Omega$ in such a way that

$$\begin{cases} (1)\ \tilde a(t,x) = a(t,x),\quad \tilde b(t,x) = b(t,x),\quad (t,x)\in[t_0/2,T]\times V_\delta, \\ (2)\ \|\tilde a\|_{C^\alpha(\Omega_T)} \le C\|a\|_{C^\alpha([t_0/2,T]\times V_\delta)},\quad \|\tilde b\|_{L^r(\Omega_T)} \le C\|b\|_{L^r([t_0/2,T]\times V_\delta)}, \\ (3)\ C_1 \le \tilde a \le C_2, \end{cases} \tag{1.36}$$

where the constant $C$ is independent of $a$, $b$, and $V$ and the constants $C_i$ are the same as in (1.20). Such an extension can be constructed, e.g., via

$$\tilde a(t,x) := C_1(1-\psi(t,x)) + \psi(t,x)a(t,x),\quad \tilde b(t,x) := \psi(t,x)b(t,x)$$

where the cut-off function $\psi$ equals one for $(t,x)\in[3t_0/4,T]\times V_{\delta/2}$ and zero for $(t,x)$ outside of $[t_0/2,T]\times V_\delta$ (this cut-off function exists due to Proposition 1.1).

Thus, due to (1.36)(1), the function $w_\phi$ satisfies the equation

$$\partial_t w_\phi = \tilde a\Delta_x w_\phi + \tilde b w_\phi + h_\phi \tag{1.37}$$

in the whole domain $[0,T]\times\Omega$ and, due to (1.36)(2) and (1.36)(3), the $L^q$-regularity estimate is applicable to (1.37) in $\Omega_T$ and gives (1.27). The rest of the proof of Corollary 1.1 repeats word for word the proof of Proposition 1.4. Corollary 1.1 is proven. □

We conclude by verifying the uniform compactness of the embedding $C^\alpha(V_\delta) \subset C(V)$ which is crucial for our proof of the finite dimensionality given in Section 4.2.

PROPOSITION 1.4. *Let $V \subset \Omega$ be an open bounded set and $\delta > 0$ some positive number. Let us also consider a unit ball $\mathbb{B}_\alpha := B(0,1,C^\alpha([t_0,T] \times V_\delta))$ and its restriction $\Pi_V \mathbb{B}_\alpha$ to the domain $[t_0,T] \times V$ for some $t_0 < T$ and positive $\alpha$. Then, the embedding $\Pi_V \mathbb{B}_\alpha \subset C([t_0,T]\times V)$ is uniformly compact with respect to $V \subset \Omega$ in the following sense: there exists a monotone decreasing function $\varepsilon \to \mathbb{M}(\varepsilon)$ (which depends on $\alpha$, $t_0$, $T$, and $\delta$ but is independent of $V \subset \Omega$) such that*

$$\mathbb{H}_\varepsilon(\Pi_V \mathbb{B}_\alpha, C([t_0,T] \times V)) \leq \mathbb{M}(\varepsilon) \tag{1.38}$$

*holds for every $\varepsilon > 0$.*

PROOF. Let us fix a cut-off function $\phi(x)$ such that $\phi(x) = 1$ for $x \in V$ and $\phi(x) = 0$ for $x \notin V_{\delta/2}$ (see Proposition 1.1). Then, since the norms of derivatives of $\phi$ are uniformly bounded (with respect to $V$), we have the following embedding:

$$\phi \mathbb{B}_\alpha \subset B(0,r,C^\alpha([t_0,T] \times \Omega)) \tag{1.39}$$

where the radius $r$ depends on $\alpha$ and $\delta$ but is independent of $V$.

Let us now fix an arbitrary $\varepsilon > 0$ and find a finite $\varepsilon$-net $W_\varepsilon$ of the metric of $C([t_0,T] \times \Omega)$ in the ball $B(0,r,C^\alpha([t_0,T] \times \Omega))$ (such a net exists since the embedding $C^\alpha \subset C$ is compact). Then, embedding (1.39) guarantees that the finite set $\Pi_V W_\varepsilon$ will be the required $\varepsilon$-net in the set $\Pi_V \mathbb{B}_\alpha$. As usual, increasing the radii of the balls by a factor of two, we can construct the $2\varepsilon$-net with the centers belonging to $\Pi_V \mathbb{B}_\alpha$. Proposition 1.4 is proven. □

## 1.4. The Nemytskii operator and its properties

The investigation of nonlinear equations in the following chapters relies on properties of mappings of the form $u \mapsto f(u)$ in the spaces $C^\alpha(\bar{\Omega})$, $L^p(\Omega)$, and $H^l(\Omega)$.

DEFINITION 1.3. Let $\Omega \subset \mathbb{R}^n$ be a domain. We say that a function

$$\Omega \times \mathbb{R}^m \ni (x,u) \longmapsto f(x,u) \in \mathbb{R}$$

satisfies the Carathéodory conditions if

$$u \longmapsto f(x,u) \textit{ is continuous for almost every } x \in \Omega$$

and

$$x \longmapsto f(x,u) \textit{ is measurable for every } u \in \Omega.$$

Given any $f$ satisfying the Carathéodory conditions and a function $u : \Omega \to \mathbb{R}^m$, we can define another function by composition

$$F(u)(x) := f(x,u(x)). \tag{1.40}$$

The composed operator $F$ is called a Nemytskii operator. In this section we state some important results on the composition of $C^\alpha(\bar{\Omega})$, $L^p(\Omega)$, $H^l(\Omega)$ with nonlinear functions (some of them without proof [62], [97]).

PROPOSITION 1.5. *Let $\Omega \subset \mathbb{R}^n$ be a bounded domain and let*

$$\Omega \times \mathbb{R}^m \ni (x, u) \longmapsto f(x, u) \in \mathbb{R}$$

*satisfy the Carathéodory conditions. In addition, let*

$$|f(x, u)| \leq f_0(x) + c(1 + |u|)^r, \tag{1.41}$$

*where $f_0 \in L^{p_0}(\Omega)$, $p_0 \geq 1$, and $rp_0 \leq p_1$. Then the Nemytskii operator $F$ defined by (1.40) is bounded from $L^{p_1}(\Omega)$ into $L^{p_0}(\Omega)$, and*

$$\|F(u)\|_{0,p_0} \leq C_1 \cdot (1 + \|u\|_{p_1}^r). \tag{1.42}$$

PROOF. By (1.41) and (1.2)

$$\begin{aligned} \|F(u)\|_{0,p_0} &\leq \|f_0(x)\|_{0,p_0} + C\|1\|_{0,p_0} + C\||u|^r\|_{0,p_0} \\ &\leq C' + C\left(\int_\Omega |u|^{rp_0} dx\right)^{\frac{1}{p_0}} \\ &= C' + \|u\|_{0,p_0 r}^r. \end{aligned} \tag{1.43}$$

Since $\Omega$ is bounded, then by the Hölder inequality

$$\|v\|_{0,q} \leq C(\Omega)\|v\|_{0,p} \text{ when } 1 \leq q \leq p,\ v \in L^p(\Omega), \tag{1.44}$$

where $C(\Omega) = (\text{mes}(\Omega))^{\frac{1}{q} - \frac{1}{p}}$. Inequalities (1.43) and (1.44) with $q = rp_0$ and $p = p_1$ imply (1.42). □

It is well known that the notions of continuity and boundedness of a nonlinear operator are independent of one another ([62]). It turns out that the following is valid.

THEOREM 1.3. *Let $\Omega \subset \mathbb{R}^n$ be a bounded domain and let*

$$\Omega \times \mathbb{R}^m \ni (x, u) \longmapsto f(x, u) \in \mathbb{R}$$

*satisfy the Carathéodory conditions. In addition, let $p \in (1, \infty)$ and $g \in L^q(\Omega)$ (where $\frac{1}{p} + \frac{1}{q} = 1$) be given, and let $f$ satisfy*

$$|f(x, u)| \leq C|u|^{p-1} + g(x).$$

*Then the Nemytskii operator $F$ defined by (1.40) is a bounded and continuous map from $L^p(\Omega)$ to $L^q(\Omega)$.*

For a more detailed treatment, the reader should consult [62], [97].

THEOREM 1.4. *Let $\Omega$ be a bounded domain in $\mathbb{R}^n$ with smooth boundary and let*

$$\Omega \times \mathbb{R} \ni (x, u) \mapsto f(x, u) \in \mathbb{R}$$

*satisfy the Carathéodory conditions. Then $f$ induces*

1. *a continuous mapping from $H^s(\Omega)$ into $H^s(\Omega)$ if $f \in C^s$,*
2. *a continuously differentiable mapping from $H^s(\Omega)$ into $H^s(\Omega)$ if $f \in C^{s+1}$, where in both cases $s > n/2$, $s \in \mathbb{N}$.*

PROOF. First we consider the simplest case, that is, $f = f(u)$ is independent of $x$. By the Sobolev embedding theorem, we have $H^s(\Omega) \subset C(\bar{\Omega})$. Hence we have $f(u) \in C(\bar{\Omega})$ for every $u \in H^s(\Omega)$. Moreover, if $u$ is in $C^{(s)}(\bar{\Omega})$, we can obtain the derivatives of $f(u)$ by the chain rule, and in the general case, we can use approximation by smooth functions. Note that all derivatives of $f(u)$ have the form of a product involving a derivative of $f$ and derivatives of $u$. The first factor is in $C(\bar{\Omega})$, while any $l$-th derivative of $u$ lies in $H^{s-l}(\Omega)$, which embeds into $L^{2n/(n-2(s-l))}(\Omega)$ if $s - l < \frac{n}{2}$. We can use this fact and the Hölder inequality to show that all derivatives of $f(u)$ up to order $s$ are in $L^2(\Omega)$; moreover, it is clear from this argument that $f$ is actually continuous from $H^s(\Omega)$ into $H^s(\Omega)$. A proof of the differentiability in this special case is that $f = f(u)$ is based on the relation

$$f(u) - f(v) = \int_0^1 f'_u(v + \theta(u - v))(u - v)d\theta$$

and the same arguments as before.

Let us now consider the general case, that is, $f = f(x, u)$. Let $|\alpha| \leq s$. We must show that

$$u \longmapsto D^\alpha F(u) \tag{1.45}$$

defines a continuous map of $H^s(\Omega)$ into $L^2(\Omega)$. It is not difficult to see that (1.45) is the finite sum of operators of the form

$$u(x) \longmapsto g(x, u(x)) \cdot D^\gamma u(x), \tag{1.46}$$

where $|\gamma| = \gamma_1 + \cdots + \gamma_n \leq s$, while $g$ is a partial derivative of $f$ of order at most $s$. It is obvious that $D^\gamma$ is continuous from $H^s(\Omega)$ into $L^2(\Omega)$ for $|\gamma| \leq s$. On the other hand, the continuous embedding of $H^s(\Omega)$ into $C(\bar{\Omega})$ implies that

$$u(x) \longmapsto g(x, u(x))$$

is continuous from $H^s(\Omega)$ into $C(\bar{\Omega})$. Thus

$$u(x) \longmapsto g(x, u(x)) \cdot D^\gamma u(x)$$

defines a continuous map of $H^s(\Omega)$ into$L^2(\Omega)$ and hence so does $u \longmapsto D^\alpha F(u)$. □

For $p \in \mathbb{N}$, let $\tilde{p}$ be the number of multiindices $\alpha$ with $|\alpha| \leq p$.

COROLLARY 1.2. *An analogous result is valid for a continuity of the operator*

$$F(u)(x) = f(x, u(x), \ldots, D^p u(x)) : H^{s+p}(\Omega) \to H^s(\Omega),$$

*where* $p, s \in \mathbb{N}$ *with* $s > \frac{n}{2}$ *and* $f : \Omega \times \mathbb{R}^{\tilde{p}} \to \mathbb{R}$ *is* $C^s$.

COROLLARY 1.3. *Let* $p, s \in \mathbb{N}$ *with* $s > \frac{n}{2}$ *and let*

$$f : \Omega \times \mathbb{R}^{\tilde{p}} \to \mathbb{R} \text{ be } C^{s+1}.$$

*Then the operator* $F : H^{s+p}(\Omega) \to H^s(\Omega)$ *defined by*

$$F(u)(x) = f(x, u(x), \ldots, D^p u(x))$$

*is Fréchet differentiable from* $H^{s+p}(\Omega)$ *into* $H^s(\Omega)$.

We have the following continuity and $C^1$-differentiability results for a nonlinear differential operator of the form $Au(x) = f(x, u(x), \ldots, D^{2p}u(x))$ in the Hölder spaces. They are based on Theorems 1.5 and 1.6.

Let $p \in \mathbb{N}$ and $\tilde{p}$ denote as before the number of multiindices with $|\alpha| \leq p$. Let $\Omega$ be a bounded domain in $\mathbb{R}^n$.

THEOREM 1.5. *Let the function $f(x,y) = f(x,y_1,\dots,y_{\tilde p})$ be defined on $\bar\Omega\times\mathbb{R}^{\tilde p}$ which satisfies the following conditions:*

(1) $f(x,0)=0$.

(2) *For any* $R>0$, $\sup\limits_{|y|\le R}\left|\frac{\partial^2 f}{\partial y_i\partial y_j}\right| \le C(R)$, $\sup\limits_{|y|\le R}\|f\|_{C^{1,\alpha}(\bar\Omega)} \le C(R)$, *where* $C(R)$ *is constant depending on* $R$.

*Let* $u_1(x),\dots,u_{\tilde p}(x)\in C^\alpha(\bar\Omega)$, $0<\alpha<1$, $\|u_i\|_{C^\alpha(\bar\Omega)}\le R$, $i=1,\dots,\tilde p$. *Then*

$$\|f(x,u_1(x),\dots,u_{\tilde p}(x))\|_{C^\alpha(\bar\Omega)} \le C_1(R)\cdot\sum_{i=1}^{\tilde p}\|u_i\|_{C^\alpha(\bar\Omega)}. \tag{1.47}$$

PROOF. Obviously,

$$\begin{aligned} f(x,y,\dots,y_{\tilde p}) = \int_0^1 \frac{d}{dt} f(x,ty_1,\dots,ty_{\tilde p})dt &= \sum_{j=1}^{\tilde p} y_j\int_0^1 \frac{\partial f(x,ty_1,\dots,ty_{\tilde p})}{\partial y_j}dt \\ &= \sum_{j=1}^{\tilde p}\varphi_j(x,y_1,\dots,y_{\tilde p})\cdot y_j, \end{aligned}$$

where

$$\varphi_j(x,y_1,\dots,y_{\tilde p}) = \int_0^1\frac{\partial f(x,ty_1,\dots,ty_{\tilde p})}{\partial y_j}dt.$$

Hence

$$f(x,u_1(x),\dots,u_{\tilde p}(x)) = \sum_{j=1}^{\tilde p}\varphi_j(x,u_1(x),\dots,u_{\tilde p}(x))\cdot u_j(x).$$

Since $C^\alpha(\bar\Omega)$, $0<\alpha<1$, is a Banach algebra, we have

$$\|f(x,u_1(x),\dots,u_{\tilde p}(x))\|_{C^\alpha} \le \sum_{j=1}^{\tilde p}\|\varphi_j(x,u_1(x),\dots,u_{\tilde p}(x))\|_{C^\alpha}\cdot\|u_j\|_{C^\alpha}.$$

Hence we have to prove that

$$\sup_{|y|\le R}\|\varphi_j(x,u_1(x),\dots,u_{\tilde p}(x))\|_{C^\alpha}\le C_1(R).$$

Indeed

$$\begin{aligned} &|\varphi_j(x+\xi,u_1(x+\xi),\dots,u_{\tilde p}(x+\xi)) - \varphi_j(x,u_1(x),\dots,u_{\tilde p}(x))| \\ &\quad\le |\varphi_j(x+\xi,u_1(x+\xi),\dots,u_{\tilde p}(x+\xi)) - \varphi_j(x,u_1(x+\xi),\dots,u_{\tilde p}(x+\xi))| \\ &\qquad + |\varphi_j(x,u_1(x+\xi),\dots,u_{\tilde p}(x+\xi)) - \varphi_j(x,u_1(x),\dots,u_{\tilde p}(x))|. \end{aligned} \tag{1.48}$$

The first term on the right-hand side of (1.48) is bounded by $C(R)\cdot|\xi|^\alpha$. The second term is bounded by

$$\begin{aligned} &\sup_{|y|\le R}|\frac{\partial\varphi_j}{\partial y_k}|\cdot|\varphi_j(x,u_1(x+\xi),\dots,u_{\tilde p}(x+\xi)) - \varphi_j(x,u_1(x),\dots,u_{\tilde p}(x))| \\ &\quad\le C(R)\cdot R\cdot|\xi|^\alpha. \end{aligned} \tag{1.49}$$

The estimates (1.48) and (1.49) yield (1.47). □

THEOREM 1.6. *Let the function $f(x,y) = f(x, y_1, \dots, y_{\tilde{p}})$ be defined on $\bar{\Omega} \times R^{\tilde{p}}$ satisfy the following conditions:*

(1) $f(x,0) = 0$, $\operatorname{grad}_y f(x,0) = 0$.

(2) *For any $R > 0$, $\sup_{|y| \le R} \|f(x,y)\|_{C^{2,\alpha}(\bar{\Omega})} \le C(R)$ and $\sup_{|y|\le R} |\frac{\partial^3 f}{\partial y_i \partial y_j \partial y_k}| \le C(R)$,*

*where $C(R)$ is a constant depending on $R$. As before let $u_1(x), \dots, u_{\tilde{p}}(x) \in C^\alpha(\bar{\Omega})$ with $\|u_i\|_{C^\alpha(\bar{\Omega})} \le R, i = 1, \dots, \tilde{p}$. Then the following estimate holds:*

$$\|f(x, u_1(x), \dots, u_{\tilde{p}}(x))\|_{C^\alpha(\bar{\Omega})} \le C_2(R) \cdot \sum_{i=1}^{\tilde{p}} \|u_i\|^2_{C^\alpha}. \tag{1.50}$$

PROOF. Obviously we have

$$f(x, y_1, \dots, u_{\tilde{p}}) = \sum_{i,j=1}^{\tilde{p}} g_{ij}(x, y_1, \cdots, y_{\tilde{p}}) \cdot y_i \cdot y_j,$$

so we can write

$$f(x, u_1(x), \dots, u_{\tilde{p}}(x)) = \sum_{i,j=1}^{\tilde{p}} g_{ij}(x, u_1(x), \dots, u_{\tilde{p}}(x)) \cdot u_i(x) \cdot u_j(x)$$

and we have

$$\begin{aligned} &\|f(x, u_1(x), \dots, u_{\tilde{p}}(x))\|_{C^\alpha(\bar{\Omega})} \\ &\le \sum_{i,j=1}^{\tilde{p}} \|g_{ij}(x, u_1(x), \dots, u_{\tilde{p}}(x))\|_{C^\alpha(\bar{\Omega})} \cdot \|u_i\|_{C^\alpha} \cdot \|u_j\|_{C^\alpha}. \end{aligned} \tag{1.51}$$

Due to Theorem 1.5 we obtain

$$\|g_{ij}(x, u_1(x), \dots, u_{\tilde{p}}(x))\|_{C^\alpha(\bar{\Omega})} \le C_0(R). \tag{1.52}$$

Hence the estimates (1.51) and (1.52) yield (1.50):

$$\|f(x, u_1(x), \dots, u_{\tilde{p}}(x))\|_{C^\alpha(\bar{\Omega})} \le C_2(R) \cdot \sum_{i=1}^{\tilde{p}} \|u_i\|^2_{C^\alpha}.$$ □

We apply Theorems 1.5 and 1.6 to the operator

$$Au(x) = f(x, u(x), \dots, D^{2p}u(x)),$$

where the function $f(x, y_1, \dots, y_{\tilde{p}})$ satisfies the conditions of Theorems 1.5 and 1.6, respectively. Hence we have

$$\|Au\|_{C^{2p,\alpha}} \le C(R) \cdot \|u\|_{C^\alpha}.$$

Moreover, it follows from Theorem 1.6 that $A \in C^1$, $A'(0) = 0$, and

$$\|A'(u+h) - A'(u)\|_{L(C^{2p,\alpha}, C^\alpha)} \le C \cdot \|h\|_{C^{2p,\alpha}(\bar{\Omega})}.$$

REMARK 1.1. As shown in the proofs of Theorems 1.5 and 1.6, continuity and differentiability of the operator $Au(x) = f(x, u(x), \dots, D^{2p}u(x))$ between $C^{2p,\alpha}(\bar{\Omega})$ and $C^\alpha(\bar{\Omega})$ remains valid under slightly weaker conditions on a given function $f(x, y_1, \dots, y_{\tilde{p}})$. We leave these as exercises for the reader.

In the investigation of nonlinear boundary value problems related to pseudo-differential operators and in particular the nonlinear Riemann-Hilbert problems as well as the regularity of solutions of degenerate reaction-diffusion systems studied in Chapters 4–11, we need properties of the Nemytskii operators in the spaces $H^s(S^1)$ or $C^{p,\alpha}(S^1)$, where $S^1$ is the unit circle. We recall some of the properties which will be used often in the sequel. The norm in $C^\alpha(M)$ is given by

$$\|f\|_{C^\alpha(M)} = \|f\|_C + \sup_{x\neq y} \frac{|f(x)-f(y)|}{|x-y|^\alpha}, \quad M = S^1.$$

As before, by $C^{k,\alpha}(M)$ we denote the space of Hölder continuous functions, which have derivatives up to order $k$, with $D^k f \in C^\alpha(M)$. Let $F$ be a superposition operator defined by

$$F(u)(x) = f(x, u(x)), \quad x \in M.$$

The following theorems are not hard to prove (although the proofs are not obvious).

THEOREM 1.7. *Let $k \in \mathbb{R}+$. Then the superposition operator $F : E_1 \to E_2$ defined by $F(u)(x) = f(x,u(x))$ acts as a bounded operator in each of the following cases (see also [32]):*

(1) $f \in C(S^1 \times \mathbb{R}, \mathbb{R})$, $\quad E_1 = C(S^1)$, $\quad E_2 = C(S^1)$,
(2) $f \in C^1(S^1 \times \mathbb{R}, \mathbb{R})$, $\quad E_1 = C^\alpha(S^1)$, $\quad E_2 = C^\alpha(S^1)$, $0 < \alpha < 1$.

THEOREM 1.8. *Let $k \in \mathbb{R}+$, $0 < \alpha < 1$. Then the superposition operator $F : E_1 \to E_2$ defined by $F(u)(x) = f(x,u(x))$ is $m$ times continuously differentiable in one of the following cases:*

(1) $D^{0,j} f \in C^k(S^1 \times \mathbb{R}, \mathbb{R})$, $\quad E_1 = C^k(S^1)$, $\quad E_2 = C^k(S^1)$,
(2) $D^{0,j} f \in C^{k+1}(S^1 \times \mathbb{R}, \mathbb{R})$, $\quad E_1 = C^{k,\alpha}(S^1)$, $\quad E_2 = C^{k,\alpha}(S^1)$.

The $j$-th derivative of $F$ is given by

$$D^{0,j} F(x, u(x)) h_1(x) \cdots h_j(x) = D^j F(f)(h_1, \ldots, h_j)(x).$$

Analogous results are valid in Sobolev spaces:

THEOREM 1.9. *Let $X = Y = H^s(S^1)$ $(s \geq 1)$ be the Sobolev space of real functions $x(\tau)$ on the circumference of a circle, where $0 \leq \tau < 2\pi$; $f(\tau, x)$ is a smooth real function, $x \in \mathbb{R}$, $0 \leq \tau < 2\pi$. Then the operator $F : H^s(S^1) \to H^s(S^1)$ defined by $Fx(\tau) = f(\tau, x(\tau))$ is continuous.*

PROOF. It is not difficult to see that

$$\left(\frac{d}{d\tau}\right)^k f(\tau, x(\tau)) = \sup_{\substack{p+q\leq k \\ r_1+\cdots+r_q = k-p \\ r_j \geq 0}} C_{p,q,r_1,\ldots,r_q} \cdot \frac{\partial^{p+q} f(\tau, x(\tau))}{\partial \tau^p \cdot \partial x^q} \cdot x^{(r_1)}(\tau) \cdots x^{(r_q)}(\tau)$$

where $C_{p,q,r_1,\ldots,r_q}$ are some constants. If $x(\tau) \in H^s$, then it follows that the derivatives $\left\{\frac{d^l x(\tau)}{d\tau^l} | 0 \leq l \leq s-1\right\}$ are continuous. Therefore in $\frac{d^s}{d\tau^s} f(\tau, x(\tau))$ all terms without ones are continuous. The last term is equal to $\frac{d^s x(\tau)}{d\tau^s} \times Q(\tau)$ where $Q(\tau)$ is a continuous function, hence also square integrable. As a consequence of these arguments we obtain continuity. □

REMARK 1.2. An analogous result holds for vector functions, and also in the multidimensional case, for functions on an arbitrary smooth compact manifold with boundary.

The following lemma on the smoothness relations between $u$ and $f(u)$ will play a decisive role in Chapters 4–7 and in the modeling of biofilms; see [31].

LEMMA 1.2. *Let the function $f \in C^2(\mathbb{R},\mathbb{R})$ satisfy*

$$C_1|u|^{p-1} \le f'(u) \le C_1|u|^{p-1},$$

$p>1$, *with $C_1$ and $C_2$ some positive constants. Then, for every $s \in (0,1)$ and $1 < q \le \infty$, we have*

$$\|u\|_{W^{s/p,pq}(\Omega)} \le C_p \|f(u)\|^{1/p}_{W^{s,q}(\Omega)}$$

*where the constant $C_p$ is independent of $u$.*

PROOF. Indeed, let $f^{-1}$ be the inverse function to $f$. Then, due to conditions on $f$, the function $G(v) := \operatorname{sgn}(v)|f^{-1}(v)|^p$ is nondegenerate and satisfies

$$C_2 \le G'(v) \le C_1,$$

for some positive constants $C_1$ and $C_2$. Therefore, we have

$$|f^{-1}(v_1) - f^{-1}(v_2)|^p \le C_p|G(v_1) - G(v_2)| \le C'_p|v_1 - v_2|,$$

for all $v_1, v_2 \in \mathbb{R}$. Finally, according to the definition of the fractional Sobolev spaces (see, e.g., [90], [94]),

$$\begin{aligned}\|f^{-1}(v)\|^{pq}_{W^{s/p,qp}(\Omega)} &:= \|f^{-1}(v)\|^{pq}_{L^{pq}(\Omega)} + \int_\Omega\int_\Omega \frac{|f^{-1}(v(x)) - f^{-1}(v(y))|^{pq}}{|x-y|^{n+sq}}\,dx\,dy\\ &\le C\|v\|^q_{L^q(\Omega)} + C'_p\int_\Omega\int_\Omega \frac{|v(x)-v(y)|^q}{|x-y|^{n+sq}}\,dx\,dy\\ &= C''_p\|v\|^q_{W^{s,q}(\Omega)},\end{aligned}$$

where we have implicitly used that $f^{-1}(v) \sim \operatorname{sgn}(v)|v|^{1/p}$. Lemma 1.2 is proved. □

CHAPTER 2

# Global attractors for autonomous evolution equations

## 2.1. Existence theorem for the global attractor

As already mentioned in the introduction, one of the main concepts of the modern theory of DS in infinite dimensions is that of the *global attractor*. We give below its definition for an abstract semigroup $S(t)$ acting on a metric space $\Phi$, although, without loss of generality, the reader may think that $(S(t), \Phi)$ is just a DS associated with one of the PDEs described in the introduction.

To this end, we first recall that a subset $K$ of the phase space $\Phi$ is an attracting set of the semigroup $S(t)$ if it attracts the images of all the *bounded* subsets of $\Phi$, i.e., for every bounded set $B$ and every $\varepsilon > 0$, there exists a time $T$ (depending in general on $B$ and $\varepsilon$) such that the image $S(t)B$ belongs to the $\varepsilon$-neighborhood of $K$ if $t \geq T$. This property can be rewritten in the equivalent form

$$\lim_{t\to\infty} \operatorname{dist}_H(S(t)B, K) = 0, \tag{2.1}$$

where $\operatorname{dist}_H(X, Y) := \sup_{x\in\mathbb{X}} \inf_{y\in\mathbb{Y}} d(x, y)$ is the nonsymmetric Hausdorff distance between subsets of $\Phi$.

We now give the definition of a global attractor, following Babin-Vishik [9].

DEFINITION 2.1. A set $\mathcal{A} \subset \Phi$ is a global attractor for the semigroup $S(t)$ if
(1) $\mathcal{A}$ is *compact* in $\Phi$;
(2) $\mathcal{A}$ is *strictly invariant*: $S(t)\mathcal{A} = \mathcal{A}$, for all $t \geq 0$;
(3) $\mathcal{A}$ is an *attracting* set for the semigroup $S(t)$.

Thus, the second and third properties guarantee that a global attractor, if it exists, is unique and that the DS reduced to the attractor contains all the nontrivial dynamics of the initial system. Furthermore, the first property indicates that the reduced phase space $\mathcal{A}$ is indeed "thinner" than the initial phase space $\Phi$ (we recall that, in infinite dimensions, a compact set cannot contain, e.g., balls and should thus be nowhere dense).

In most applications, one can use the following attractor existence theorem.

THEOREM 2.1. *Let a DS $(S(t), \Phi)$ possess a compact attracting set and let the operators $S(t) : \Phi \to \Phi$ be continuous for every fixed $t$. Then, this system possesses the global attractor $\mathcal{A}$ which is generated by all the trajectories of $S(t)$ which are defined for all $t \in \mathbb{R}$ and are globally bounded.*

The strategy for applying this theorem to concrete equations of mathematical physics is the following. In the first step, one verifies the so-called *dissipative* estimate which usually has the form

$$\|S(t)u_0\|_\Phi \leq Q(\|u_0\|_\Phi)e^{-\alpha t} + C_*, \quad u_0 \in \Phi, \tag{2.2}$$

where $\|\cdot\|_\Phi$ is a norm in the function space $\Phi$ and the positive constants $\alpha$ and $C_*$ and the monotonic function $Q$ are independent of $t$ and $u_0 \in \Phi$ (usually, this estimate follows from energy estimates and is sometimes even used in order to "define" a dissipative system). This estimate obviously gives the existence of an attracting set for $S(t)$ (e.g., the ball of radius $2C_*$ in $\Phi$), which is, however, *noncompact* in $\Phi$. In order to overcome this problem, one usually derives, in the second step, a *smoothing* property for the solutions, which can be formulated as follows:

$$\|S(1)u_0\|_{\Phi_1} \leq Q_1(\|u_0\|_\Phi), \quad u_0 \in \Phi, \tag{2.3}$$

where $\Phi_1$ is another function space which is *compactly* embedded into $\Phi$. In applications, $\Phi$ is usually the space $L^2(\Omega)$ of square integrable functions, $\Phi_1$ is the Sobolev space $H^1(\Omega)$ of the functions $u$ such that $u$ and $\nabla_x u$ belong to $L^2(\Omega)$, and estimate (2.2) is a classical smoothing property for solutions of parabolic equations (for parabolic equations in unbounded domains and for hyperbolic equations, a slightly more complicated *asymptotic* smoothing property should be used instead of (2.2); see [32]).

Since the continuity of the operators $S(t)$ usually poses no difficulty (if the uniqueness is proven), then the above scheme indeed gives the existence of the global attractor for most of the PDE of mathematical physics in bounded domains.

REMARK 2.1. As was shown in [9] the assumption that $S(t) : \Phi \to \Phi$ be continuous for every fixed $t$ can be replaced by the closedness of the graph $\{(u_0, S(t)u_0), u_0 \in \Phi\}$.

REMARK 2.2. Although the global attractor usually has a very complicated geometric structure, there exists one exceptional class of DS for which the global attractor has a relatively simple structure which is completely understood, namely the DS having a global Lyapunov function. We recall that a continuous function $\mathcal{L} : \Phi \to \mathbb{R}$ is a global Lyapunov function if

(1) $\mathcal{L}$ is nonincreasing along the trajectories, i.e., $\mathcal{L}(S(t)u_0) \leq \mathcal{L}(u_0)$, for all $t \geq 0$;

(2) $\mathcal{L}$ is *strictly* decreasing along all nonequilibrium solutions, i.e., $\mathcal{L}(S(t)u_0) = \mathcal{L}(u_0)$ for some $t > 0$ and $u_0$ implies that $u_0$ is an equilibrium of $S(t)$.

It is well known that if a DS possesses a global Lyapunov function, then, at least under the generic assumption that the set $\mathcal{R}$ of equilibria is finite, every trajectory $u(t)$ *stabilizes* to one of these equilibria as $t \to +\infty$. Moreover, every complete bounded trajectory $u(t)$, $t \in \mathbb{R}$, belonging to the attractor is a heteroclinic orbit joining two equilibria. Thus, the global attractor $\mathcal{A}$ can be described as follows [9], [32], [53], [66], [93]:

$$\mathcal{A} = \bigcup_{u_0 \in \mathcal{R}} \mathcal{M}^+(u_0),$$

where $\mathcal{M}^+(u_0)$ is the so-called unstable set of the equilibrium $u_0$ (which is generated by all heteroclinic orbits of the DS which start from the given equilibrium $u_0 \in \mathcal{A}$). It is also known that if the equilibrium $u_0$ is *hyperbolic* (generic assumption [9]), then the set $\mathcal{M}^+(u_0)$ is a $\kappa$-dimensional submanifold of $\Phi$, where $\kappa$ is the instability index of $u_0$. Thus, under the generic hyperbolicity assumption on the equilibria, the attractor $\mathcal{A}$ of a DS having a global Lyapunov function is a finite union of smooth finite-dimensional submanifolds of the phase space $\Phi$. These attractors are called *regular* (following Babin-Vishik (see [9])).

It is also worth emphasizing that, in contrast to general global attractors, regular attractors are robust under perturbations. Moreover, in some cases, it is also possible to verify the so-called *transversality* conditions (for the intersection of stable and unstable manifolds of the equilibria) and, thus, verify that the DS considered is a Morse-Smale system. In particular, this means that the dynamics restricted to the regular attractor $\mathcal{A}$ are also preserved (up to homeomorphisms) under perturbations.

In the sequel we will apply Theorem 2.1 or Remark 2.1 (whenever necessary) to a class of PDEs arising in mathematical physics. We especially emphasize that one of the challenging questions in the theory of attractors is, in which sense are the dynamics on the global attractor finite dimensional? As already mentioned, the global attractor is usually not a manifold but has a rather complicated geometric structure. So, it is natural to use the definitions of dimensions adopted for the study of fractal sets here. We restrict ourselves to the so-called fractal (or box-counting, entropy) dimension, although other dimensions (e.g., Hausdorff, Lyapunov, etc.) are also used in attractor theory. Here the so-called Mañé theorem (which can be considered as a generalization of the classical Whitney embedding theorem for fractal sets) plays an important role in finite-dimensional reduction theory [93].

THEOREM 2.2. *Let $\Phi$ be a Banach space and let $\mathcal{A}$ be a compact set such that $d_f(\mathcal{A}) < N$ for some $N \in \mathbb{N}$. Then, for "almost all" $(2N+1)$-dimensional planes $L$ in $\Phi$, the corresponding projector $\Pi_L : \Phi \to L$ restricted to the set $\mathcal{A}$ is a Hölder continuous homeomorphism.*

Thus, if the finite fractal dimensionality of the attractor is established, then, fixing a hyperplane $L$ satisfying the assumptions of the Mañé theorem and projecting the attractor $\mathcal{A}$ and the DS $S(t)$ restricted to $\mathcal{A}$ onto this hyperplane ($\bar{\mathcal{A}} := \Pi_L \mathcal{A}$ and $\bar{S}(t) := \Pi_L \circ S(t) \circ \Pi_L^{-1}$), we indeed obtain a reduced DS $(\bar{S}(t), \bar{\mathcal{A}})$ which is defined on a finite-dimensional set $\bar{\mathcal{A}} \subset L \sim \mathbb{R}^{2N+1}$. Moreover, this DS will be Hölder continuous with respect to the initial data.

REMARK 2.3. Note that good estimates on the dimension of the attractors in terms of the physical parameters are crucial for the finite-dimensional reduction described above and (consequently) there exists a highly developed machinery for obtaining such estimates. The best known upper estimates are usually obtained by the so-called volume contraction method which is based on the study of the evolution of infinitesimal $k$-dimensional volumes in the neighborhood of the attractor (and, if the DS considered contracts the $k$-dimensional volumes, then the fractal dimension of the attractor is less than $k$; see, e.g., [9], [93]).

REMARK 2.4. Lower bounds on the dimension are usually based on the observation that the global attractor always contains the unstable manifolds of the (hyperbolic) equilibria. Thus, the instability index of a properly constructed equilibrium gives a lower bound on the dimension of the attractor; see, e.g., [9], [93].

Theorem 2.3 below plays a decisive role in the study of the dimension of attractors, which in turn does not require differentiability of the associated semigroup in contrast to [9], [32], [93]. We especially emphasize that for a quite large class of degenerate parabolic systems arising in the modelling of life science problems (see [32]) the associated semigroup is not differentiable. We denote it by $S := S(1)$.

THEOREM 2.3. *Let $H_1$ and $H$ be Banach spaces, let $H_1$ be compactly embedded in $H$, and let $K \subset\subset H$. Assume that there exists a map $S : K \to K$, such that $S(K) = K$ and the following "smoothing" property is valid:*

$$\|S(k_1) - S(k_2)\|_{H_1} \le C\|k_1 - k_2\|_H \tag{2.4}$$

*for every $k_1, k_2 \in K$. Then the fractal dimension of $K$ in $H$ is finite and can be estimated in the following way:*

$$d_F(K, H) \le \mathbb{H}_{\frac{1}{4}C}(B(1, 0, H_1), H) \tag{2.5}$$

*where $C$ is the same as in (2.4) and $B(1, 0, H_1)$ denotes the unit ball in the space $H_1$.*

PROOF. Let $\{B(\varepsilon, k_i, H)\}_{i=1}^{N_\varepsilon}$, $k_i \in K$, be some $\varepsilon$-*covering* of the set $K$ (here and below we denote by $B(\varepsilon, k, V)$ the $\varepsilon$-*ball* in the space $V$, centered in $k$). Then according to (2.4), the system $\{B(C\varepsilon, S(k_i), H_i)\}_{i=1}^{N_\varepsilon}$ of $C\varepsilon$-balls in $H_1$ covers the set $S(K)$ and consequently (since $S(K) = K$) the same system covers the set $K$.

Now cover every $H_1$-ball with radius $C\varepsilon$ by a finite number of $\frac{\varepsilon}{4}$-balls in $H$. By definition, the minimal number of such balls equals

$$N_{\varepsilon/4}(B(C\varepsilon, S(k_i), H_1), H) = N_{\varepsilon/4}(B(C\varepsilon, 0, H_1), H) = N_{\frac{1}{4}C}(B(1, 0, H_1), H) \equiv \mathcal{N}.$$

Note that the centers of the $\varepsilon/4$-covering thus obtained do not necessarily belong to $K$ but we evidently can construct the $\varepsilon/2$-covering with centers in $K$ and with the same number of balls.

Thus, having the initial $\varepsilon$-covering of $K$ in $H$ with the number of balls $N_\varepsilon$, we have constructed the $\varepsilon/2$-covering with the number of balls $N_{\varepsilon/2} = \mathcal{N} N_\varepsilon$.

Consequently, the $\varepsilon$-entropy of the set $K$ possesses the following estimate. In fact the assertion of the theorem is a corollary of this recurrent estimate. Indeed, since $K \subset\subset H$, then there exists $\varepsilon_0$ such that $K \subset B(\varepsilon_0, k_0, H)$ and consequently

$$\mathbb{H}_{\varepsilon_0}(K, N) = 0. \tag{2.6}$$

Iterating the estimate (2.6) $n$ times, we obtain that

$$\mathbb{H}_{\varepsilon_0/2^n}(k, H) \le n \log_2 \mathcal{N}. \tag{2.7}$$

Now fix an arbitrary $\varepsilon > 0$ and choose $n = n(\varepsilon)$ in such a way that

$$\frac{\varepsilon_0}{2^n} \le \varepsilon \le \frac{\varepsilon_0}{2^n - 1}.$$

Then

$$\mathbb{H}_\varepsilon(K) \le \mathbb{H}_{\varepsilon/2^n}(K) \le n \log_2 \mathcal{N} \le \log_2\left(\frac{2\varepsilon_0}{\varepsilon}\right) \log_2 \mathcal{N}. \tag{2.8}$$

Thus (2.5) is an immediate consequence of (2.8). Theorem 2.3 is proved. □

## 2.2. Estimation of time derivatives for nonautonomous perturbations of regular attractors

The aim of this section is to verify the auxiliary estimate for nonautonomous perturbations of regular attractors. To be more precise, consider an ODE in $\mathbb{R}^n$:

$$u'(t) = F(u(t)), \quad u(0) = u_0, \tag{2.9}$$

for some $F \in C^2(\mathbb{R}^n, \mathbb{R}^n)$. We assume that, for every $u_0 \in \mathbb{R}^n$, this equation is globally (for $t \geq 0$) solvable and the associated semigroup $S(t)u_0 := u(t)$ is dissipative, i.e.,

$$\|S(t)u_0\| \leq Q(\|u_0\|)e^{-\beta t} + C_* \tag{2.10}$$

for some positive $\beta$ and $C_*$ and monotone $Q$. Therefore, equation (2.9) possesses a global attractor $\mathcal{A}$ in $\mathbb{R}^n$. Our main assumption is that this attractor is *regular* in the sense of Remark 2.2, i.e., all of the equilibria $u_0 \in \mathcal{R}$ are hyperbolic, every trajectory, belonging to the attractor $\mathcal{A}$ is a heteroclinic orbit connecting two *different* equilibria, and the attractor $\mathcal{A}$ is a finite union of the unstable manifolds $\mathcal{M}^+_{u_0}$ associated with the equilibria $u_0 \in \mathcal{R}$:

$$\mathcal{A} = \bigcup_{u_0 \in \mathcal{R}} \mathcal{M}^+_{u_0}. \tag{2.11}$$

Finally, we assume that the so-called no-cycle condition is satisfied, i.e., the attractor $\mathcal{A}$ does not contain any heteroclinic cycles. As we know, that is always true in the case when (2.9) possesses a global Lyapunov function.

Consider now the following small nonautonomous perturbation of equation (2.9):

$$u' = F(u) + h(t), \quad u(0) = u_0, \quad t \geq 0, \tag{2.12}$$

where the nonautonomous external force is uniformly small:

$$\|h\|_{W^{1,\infty}(\mathbb{R}_+, \mathbb{R}^n)} \leq \varepsilon \ll 1. \tag{2.13}$$

The main result of this section is the following estimate.

PROPOSITION 2.1. *Let the above assumptions hold and let the external force $h \in W^{1,\infty}(\mathbb{R}_+)$ satisfy estimate (2.13) for sufficiently small $\varepsilon > 0$. Then, any solution $u(t)$ of the perturbed problem (2.12) satisfies the following estimate:*

$$\int_0^T \|u'(t)\|\, dt \leq C_1 + C_2 \int_0^T \|h'(t)\|\, dt, \tag{2.14}$$

*where the positive constants $C_1$ and $C_2$ depend only on the norm of $u(0)$ and are independent of $T$ and the concrete choice of $u(0)$ and $h(t)$.*

PROOF. Indeed, using the standard regular attractor perturbation arguments, one can check that for every bounded set $B$ of $\mathbb{R}^n$ and every $\delta > 0$, there exist $T = T(B, \delta)$ and $\varepsilon_0 = \varepsilon_0(B, \delta)$ such that, for every $\varepsilon \leq \varepsilon_0$ and every trajectory $u(t)$ starting from $B$, we can find a sequence $u_1, \dots, u_N$ of *different* equilibria $u_i \in \mathcal{R}$ (of problem (2.9)!) and a sequence of times

$$0 = T_0^+ \leq T_1^- \leq T_1^+ < T_2^- \leq T_2^+ < \cdots < T_N^- < T_N^+ = \infty$$

such that

$$u(t) \in \mathcal{O}_\delta(u_i), \quad t \in (T_i^-, T_i^+), \quad T_i^- - T_{i-1}^+ \leq T, \ i = 1, \dots, N. \tag{2.15}$$

In other words, the sequence of equilibria $u_i$ and the values of $T_l^\pm$ depend on the concrete choice of $u(0)$ and $h$, but the number $N$ of equilibria is bounded by the whole number $\#\mathcal{R} < \infty$ of possible equilibria (since the equilibria must be different) and the differences $T_i^+ - T_{i-1}^-$ are also uniformly bounded by $T$; see [9], [42] for the details.

Thus, any trajectory starting from $B$ spends at most time $T_{\text{out}} := \#\mathcal{R} \cdot T$ outside of the $\delta$-neighborhood $\mathcal{O}_\delta(\mathbb{R})$ of the equilibria set $\mathcal{R}$ and this time depends only on $B$ and $\delta$. For this reason, the part of the trajectory lying outside of $\mathcal{O}_\delta(\mathbb{R})$ gives only a *finite and uniformly bounded* impact on the integral (2.14) (which can be included in the constant $C_1$). So, we only need to estimate the left-hand side of (2.14) for the case where $u(t)$ belongs to a small neighborhood of a single fixed equilibrium $u_0 \in \mathcal{R}$ only.

To this end, we will use the hyperbolicity assumption on $u_0$. Indeed, the implicit function theorem implies the existence of $\varepsilon_0 > 0$ and $\delta > 0$ such that, for every $\varepsilon \le \varepsilon_0$, there exists a unique solution $U_{u_0,h}(t)$ of (2.12) belonging to the $\delta$-neighborhood of $u_0$ for all $t$. Moreover, this solution, in fact, belongs to the $C\varepsilon$-neighborhood of $u_0$ and the following estimate holds:

$$|U'_{u_0,h}(t)| \le C \int_{\mathbb{R}} e^{-\kappa|t-s|}|h'(s)|\,ds, \tag{2.16}$$

where the constant $C$ and the hyperbolicity exponent $\kappa$ are independent of the concrete choice of $u_0 \in \mathcal{R}$ and the external force $h$ satisfying (2.13); see [50].

Furthermore, since $u_0$ is hyperbolic, the trajectory $U_{u_0,h}(t)$ will also be hyperbolic and we will have an exponential dichotomy in a small $\delta$-neighborhood of $U_{u_0,h}$. In particular, every trajectory $u(t)$ belonging to $\mathcal{O}_\delta(u_0)$ for $t \in [0,S]$, $S \gg 1$, will tend exponentially to $U_{u_0,h}(t)$ inside of the interval

$$|u(t) - U_{u_0,h}(t)| + |u'(t) - U'_{u_0,h}(t)| \le C(e^{-\kappa|t|} + e^{-\kappa|S-t|}) \tag{2.17}$$

and $C$ and $\kappa$ are independent of the concrete choice of $u$ and $h$; see [50] for the details. Therefore,

$$\int_0^t |u'(s)|\,ds \le C + \int_0^t |U'_{u_0,h}(s)|\,ds \tag{2.18}$$

for $t \in [0,S]$ and $u(t) \in \mathcal{O}_\delta(u_0)$.

Thus, we have proved that

$$\int_0^T |u'(s)|\,ds \le C_1 + C_2 \sum_{u_0 \in \mathcal{R}} \int_0^T |U'_{u_0,h}(s)|\,ds, \tag{2.19}$$

where the constants $C_i$ depend only on the radius of $B$.

In order to deduce (2.14) from (2.19), we will use estimate (2.16). Indeed, integrating it over $t \in [0,T]$ and using that $h$ can be extended for $t \le 0$ by zero, we have

$$\begin{aligned}\int_0^T |U'_{u_0,h}(t)|\,dt \le& C\int_0^T |h'(t)|\,dt + C\int_T^\infty e^{-\kappa|t-T|}|h'(s)|\,ds\\ \le& C_1 + C\int_0^T |h'(t)|\,dt.\end{aligned} \tag{2.20}$$

Inserting this estimate into the right-hand side of (2.19), we obtain (2.14) and finish the proof of the proposition. □

REMARK 2.5. As we can see from the proof, estimate (2.14) has a general nature which is not restricted by the class of ordinary differential equations. However, since we intend to use it only for the ODE case (see Chapter 11), in order to avoid the technicalities related to the formulation for a "general" PDE, we restrict our consideration to the case of an ODE (see also [44]).

CHAPTER 3

# Exponential attractors

This chapter is devoted to the systematic study of exponential attractors both for autonomous and nonautonomous dynamical systems. We deal with the existence theorems as well as the perturbation theory of the exponential attractors. Moreover, we present some recent development on pull-back exponential attractors.

## 3.1. Exponential attractors for autonomous systems

In this section we discuss the behavior as time goes to infinity of partial differential equations of the form

$$\frac{\partial u}{\partial t} = F(t, u), \tag{3.1}$$

in a Banach space $H$. When the system is autonomous, which will be the topic of Sections 3.1 and 3.2, i.e., when the time does not appear explicitly in (3.1) ($F(t, u) \equiv F(u)$), then we showed that the long-time behavior of the system can be described in terms of the global attractor $\mathcal{A}$ (see Definition 2.1). We already noticed that it follows from conditions (2) and (3) of Definition 2.1 that the global attractor, if it exists, is unique. Furthermore, it follows from (1) that it is essentially thinner than the original phase space $H$; indeed, in general, $H$ is an infinite-dimensional function space and, in infinite dimensions, a compact set cannot contain a ball and be nowhere dense. It is not difficult to prove that the global attractor is the smallest (by inclusion) closed set enjoying the attraction property (3); it is also the largest bounded invariant set. As we have seen in Chapter 2 (see also the references therein) for the most nondegenerate equations of mathematical physics in a bounded domain the global attractor has a finite fractal dimension, so that, even though the initial phase space is infinite dimensional, the dynamics, reduced to the global attractor, are, in some proper sense, finite dimensional and can be described by a finite number of parameters. It thus follows that the global attractor appears as a suitable object in view of the study of the long-time behavior of the system (see, e.g., [9], [19], [32], [53], [66], [72], [86], [93]). However, as we indicated in the introduction, the global attractor may attract the trajectories slowly, as a consequence of which the global attractor may be sensitive to perturbations; a given system is only an approximation of reality and it is thus essential that the objects that we study be robust under small perturbations. Actually, in general, the global attractor is upper semicontinuous with respect to perturbations, i.e.,

$$\operatorname{dist}_H(\mathcal{A}_\varepsilon, \mathcal{A}_0) \to 0 \quad \text{as } \varepsilon \to 0,$$

where $\mathcal{A}_0$ is the global attractor associated with the nonperturbed system and $\mathcal{A}_\varepsilon$ is that associated with the perturbed system, $\varepsilon > 0$ being the perturbation parameter.

Now, the lower semicontinuity, i.e.,

$$\mathrm{dist}_H(\mathcal{A}_0, \mathcal{A}_\varepsilon) \to 0 \quad \text{as } \varepsilon \to 0,$$

which roughly speaking means that the global attractor cannot implode, is much more difficult to prove (see, e.g., [85]). Furthermore, this property may not hold. This is in particular the case when the perturbed and nonperturbed problems do not have the same equilibria (stationary solutions). This can already be seen in finite dimensions by considering the following ordinary differential equation (see [85]):

$$x' = -x(1-x)^2 - \varepsilon, \quad \varepsilon \geq 0.$$

Then, the global attractor

$$\mathcal{A}_\varepsilon = \begin{cases} [0,1], & \text{for } \varepsilon = 0, \\ \{0\}, & \text{for } \varepsilon > 0, \end{cases}$$

so that a bifurcation phenomenon takes place at $\varepsilon = 0$, and, consequently, $\mathcal{A}_\varepsilon = \{0\}$ is not lower semicontinuous. It thus follows that the global attractor may change drastically under small perturbations. Furthermore, in many situations, the global attractor may not be observable in experiments or in numerical simulations. Indeed, we can consider for instance the following Chafee-Infante equation in one spatial dimension:

$$\frac{\partial u}{\partial t} - \nu \frac{\partial^2 u}{\partial x^2} + u^3 - u = 0, \quad x \in [0,1], \ \nu > 0,$$
$$u(0,t) = u(1,t) = -1, \quad t \geq 0.$$

Then, due to the boundary conditions, $\mathcal{A} = \{-1\}$. Now, this problem possesses many metastable "almost stationary" equilibria which live up to a time $t_\star \sim e^{\nu^{-\frac{1}{2}}}$. Thus, for $\nu$ small, one will not see the global attractor in numerical simulations. Finally, in some situations, the global attractor may fail to capture important transient behavior. This can be observed, e.g., on some models of one-dimensional Burgers equations with a weak dissipation term (see [11]). In that case, the global attractor is trivial, it is reduced to one exponentially attracting point, but the system presents very rich and important transient behavior, which resembles some modified version of the Kolmogorov law. Another example for which one observes important transient behavior which is not contained in the global attractor is a chemotaxis equation (see [3], [4]).

We can mention that another possible object is an inertial manifold (see [93]). An inertial manifold $\mathcal{M}$ is a smooth (at least Lipschitz) finite-dimensional manifold which satisfies the following asymptotic completeness property: $\forall u \in H$, $\exists v \in \mathcal{M}$ such that

$$\|S(t)u - S(t)v\|_H \leq Q(\|u\|_H)e^{-\alpha t}, \quad t \geq 0,$$

where the monotonic function $Q$ and the constant $\alpha > 0$ are independent of $u$ and $v$. Actually, an inertial manifold would be a perfect object in view of the remarks made above. Indeed, since it is a smooth (or Lipschitz) manifold and owing to the asymptotic completeness, the dynamics, reduced to $\mathcal{M}$, are equivalent in a strong way to that of the initial system and can be described by a finite system of ordinary differential equations (called the inertial form). Unfortunately, all the known constructions of inertial manifolds are based on a very restrictive condition, namely, the so-called spectral gap condition (see, e.g., [93]). Consequently, the existence of an inertial manifold is not known in general.

Thus, it follows from the above considerations that it should be useful to have a (possibly) larger object which contains the global attractor, attracts the trajectories at a fast rate, is still finite dimensional, and is more robust under perturbations.

Our purpose in what follows is to study such an object, namely an exponential attractor, as proposed by A. Eden, C. Foias, B. Nicolaenko, and R. Temam in [29]. Let $S(t)$, $t \geq 0$, be the semigroup associated with the problem

$$\frac{\partial u}{\partial t} = F(u), \tag{3.2}$$

$$u\big|_{t=0} = u_0, \tag{3.3}$$

in a Banach space $H$ (in particular, we assume that (3.2)–(3.3) is well-posed), $u_0 \in H$. We have the following definition.

DEFINITION 3.1. A set $\mathcal{M}$ is an exponential attractor for $S(t)$ in $H$ if

(i) it is compact in $H$,
(ii) it is positively invariant, i.e., $S(t)\mathcal{M} \subset \mathcal{M}$, $\forall t \geq 0$,
(iii) it has finite fractal dimension,
(iv) it attracts exponentially fast the bounded sets of initial data in the following sense: there exists a monotonic function $Q$ and a constant $\alpha > 0$ such that

$$\forall B \subset H \text{ bounded, } \operatorname{dist}_H(S(t)B, \mathcal{M}) \leq Q(\|B\|_H)e^{-\alpha t}, \quad t \geq 0.$$

It follows from this definition that an exponential attractor always contains the global attractor [actually, it follows from the definition that if $S(t)$ possesses an exponential attractor $\mathcal{M}$, then it also possesses the global attractor $\mathcal{A} \subset \mathcal{M}$; indeed, $\mathcal{M}$ is a compact attracting set (see, e.g., [9]; the continuity of $S(t)$, $\forall t \geq 0$, generally holds)]. Thus, an exponential attractor is still finite dimensional, like the global attractor, and moreover, one now has an explicit (exponential) control on the rate of attraction of the trajectories.

(i) Actually, proving the existence of an exponential attractor is also one way of proving the finite (fractal) dimensionality of the global attractor. In general, in order to prove the finite dimensionality of the global attractor, one uses the so-called volume contraction method, based on the study of the evolution of infinitesimal $k$-dimensional volumes in a neighborhood of the global attractor: if the semigroup contracts the $k$-dimensional volumes, then the fractal dimension of the global attractor is less than $k$. This method usually gives the best estimates on the dimension in terms of the physical parameters of the problem (see, e.g., [93]); see also [32] for the derivation of sharp estimates on the dimension of exponential attractors for reaction-diffusion systems. However, the volume contraction method requires some differentiability of the associated semigroup. As we shall see below, the construction of exponential attractors requires weaker assumptions, namely, some Lipschitz or Hölder property, which can be useful when the differentiability is not known.
(ii) The choice of the fractal dimension over other dimensions, e.g., the Hausdorff dimension, in Definition 3.1 is related, as mentioned in Section 2.1, to the Mañé theorem which gives some indications on the existence of a reduced finite-dimensional system which is Hölder continuous (but, unfortunately, not Lipschitz continuous) with respect to the initial data; see [29].

Now, the main drawback of exponential attractors is that an exponential attractor, if it exists, is not unique. Therefore, the question of the best choice, if it makes sense, of an exponential attractor is a crucial one. In order to overcome this drawback, one possibility consists in finding a "simple" algorithm $S \mapsto \mathcal{M}(S)$ which maps a given mapping or semigroup $S$ onto an exponential attractor $\mathcal{M}(S)$; by "simple", we have in mind, in particular, the numerical realization of such an algorithm. We will touch upon this question in Sections 3.2 and 3.3.

The first construction of exponential attractors was due to A. Eden, C. Foias, B. Nicolaenko and R. Temam [29]. This construction is based on the so-called squeezing property which, roughly speaking, says that either the higher modes are dominated by the lower ones or that the flow is contracted exponentially. More precisely:

DEFINITION 3.2. Let $H$ be a Hilbert space. We consider a dynamical system $(S(t), H)$. For some fixed $0 < t^* < \infty$, assume that for $S = S(t^*)$, either

$$||Su_1 - Su_2|| \leq \delta ||u_1 - u_2|| \tag{3.4}$$

or

$$||(I - P)(Su_1 - Su_2)|| \leq ||P(Su_1 - Su_2)|| \tag{3.5}$$

holds for any pair $(u_1, u_2)$ in $H \times H$ with some fixed constant $0 \leq \delta < \frac{1}{4}$ and some fixed orthogonal projection $P$ from $H$ onto a finite-dimensional subspace $Z$ in $X$, say dim $Z = N < \infty$. The condition that either (3.4) or (3.5) holds is called a *squeezing property* for $S = S(t^*)$.

Let $X$ be a compact subset of $H$ such that $S(t)X \subseteq X$ for all $t \geq 0$.

THEOREM 3.1. *Let $(S(t), H)$ be a Lipschitz continuous dynamical system that for some fixed $t^*$ enjoys the squeezing property. Then this dynamical system possesses an exponential attractor $\mathcal{M}$ in $X$.*

A proof of this theorem can be found in [29]. Note that the assertion of Theorem 3.1 remains valid if we replace the Hilbert space $H$ by a bounded subset $X$ in $H$ (i.e., $\mathcal{M}$ satisfies all the assumptions of Definition 3.1 with $H$ replaced by $X$. Application of this theorem to PDEs can be found in [29], [32]; also see the references there.

REMARK 3.1. We emphasize that the proof of Theorem 3.1 requires the Hilbert structure and is nonconstructible (indeed, Zorn's lemma is used in order to construct the appropriate exponential attractor) and is only valid in Hilbert spaces (since it makes essential use of orthogonal projectors of finite rank). Furthermore, based on this construction, it is possible to prove the lower semicontinuity of proper exponential attractors under perturbations, but only up to some time shift, so that, essentially, one only proves that

$$\text{dist}_H(\mathcal{A}_0, \mathcal{M}_\varepsilon) \to 0 \quad \text{as } \varepsilon \to 0,$$

where $\mathcal{A}_0$ is the global attractor associated with the nonperturbed system and $\mathcal{M}_\varepsilon$ is an exponential attractor associated with the perturbed one, which is not satisfactory.

Here we propose a second construction (see Theorem 3.2 below) which is valid in Banach spaces as well (see also [28] for another construction of exponential attractors valid in Banach spaces; this construction consists in adapting that of [29] to a Banach space setting and has thus some of the drawbacks mentioned above). The key point in our construction is a smoothing property on the difference of two solutions, which generalizes in some sense (in particular, to a Banach space setting) the techniques proposed by O. A. Ladyzhenskaya in order to prove the finite dimensionality of the global attractor; see, e.g., [67].

THEOREM 3.2. *Let $H$ and $H_1$ be two Banach spaces such that $H_1$ is compactly embedded in $H$. Also let $X$ be a bounded subset of $H$. We consider a nonlinear map $L : X \to X$ such that $L$ can be decomposed into a sum of two maps*

$$L = L_0 + K, \qquad L_0 : X \to H, \quad K : X \to H,$$

*where $L_0$ is a contraction, i.e.,*

$$\|L_0(x_1) - L_0(x_2)\|_H \leq \alpha \|x_1 - x_2\|_H, \ \forall x_1, x_2 \in X \ \text{ and with } \alpha < 1/2, \tag{3.6}$$

*and $K$ satisfies the condition*

$$\|K(x_1) - K(x_2)\|_{H_1} \leq C\|x_1 - x_2\|_H, \ \forall x_1, x_2 \in X. \tag{3.7}$$

*Then, the map $L : X \to X$ possesses an exponential attractor.*

PROOF. Let us fix $\theta > 0$ such that $2(\alpha + \theta) < 1$. Since $X$ is bounded, there exists a ball $B(R, x_0, H)$ of radius $R$ centered at $x_0 \in X$ in $H$ which contains $X$. We set $E_0 = V_0 = \{x_0\}$. It follows from (3.7) that the $H_1$-ball $B(CR, K(x_0), H_1)$ covers the image $K(X)$. We now cover this ball by a finite number of $\theta R$-balls in $H$ with centers $y_i$ (it is possible to do so because the embedding $H_1 \subset H$ is compact). Moreover, the minimal number of balls in this covering can be estimated as follows:

$$\begin{aligned} N_{\theta R}(B(CR, K(x_0), H_1), H) &= N_{\theta R}(B(CR, 0, H_1), H) \\ &= N_{\theta/C}(B(1, 0, H_1), H) \\ &\equiv N(\theta). \end{aligned} \tag{3.8}$$

(It is essential for us that this number be independent of $R$.) Thus, we have constructed a $\theta R$-covering of $K(X)$. It follows from assumption (3.6) that the family of balls with centers $y_i + L_0(x_0)$ and with radius $(\alpha+\theta)R$ covers $L(X)$. Now, the centers of the balls in this covering may go out of $L(X)$ and even out of $X$. To avoid this problem, we increase the radius twice and construct the new $2(\alpha+\theta)$-covering $\{B(2(\alpha+\theta)R, x_1^i, H)\}$, $i = 1, \dots, N(\theta)$, of $L(X)$ so that $x_1^i \in L(X)$. We then set $V_1 = \{x_1^i, i = 1, \dots, N(\theta)\}$.

Applying the above procedure to every ball in this new covering, we obtain the $(2(\alpha+\theta))^2 R$-covering of $L^2(X)$ with $N(\theta)^2$ balls. We denote by $V_2$ the set of their centers. Repeating this procedure, we finally construct a sequence of sets $V_k \subset L^k(X)$ such that

$$\operatorname{dist}_H(L^k(X), V_k) \leq R(2(\alpha+\theta))^k \quad \text{and} \quad \#V_k \leq N(\theta)^k. \tag{3.9}$$

To obtain the invariance, we now introduce the sequence of sets $E_k = L(E_{k-1}) \cup V_k$ and we set

$$E_\infty = \bigcup_{k=1}^{\infty} E_k \ ; \ \ \mathcal{M} = [E_\infty]_H \, , \tag{3.10}$$

where $[\cdot]_H$ denotes the closure in $H$. Let us verify that $\mathcal{M}$ is an exponential attractor for $L$ on $X$. Indeed, the invariance follows immediately from our construction. Since $V_k \subset \mathcal{M}$ and $2(\alpha+\theta) < 1$, the exponential attraction property is a consequence of (3.9). Thus, it remains to estimate the dimension of $\mathcal{M}$ or, equivalently, that of $E_\infty$.

We note that $L(X) \subset X$ and that

$$\bigcup_{k\geq n} E_k \subset L^n(X) \subset \bigcup_{v\in V_n} B(v, R(2(\theta+\alpha))^n, H).$$

We fix $\varepsilon > 0$ and we choose the smallest integer $n$ such that $R(2(\alpha+\theta))^n \leq \varepsilon$. Then

$$N_\varepsilon(E_\infty, H) \leq N_\varepsilon(\bigcup_{k\leq n} E_k) + N_\varepsilon(\bigcup_{k>n} E_k) \leq \sum_{k\leq n} \#E_k + \#V_{n+1} \leq C_2 N(\theta)^n.$$

Here, we have used the fact that $\#E_k \leq C_1 N(\theta)^n$, which can be easily deduced from the recurrent formula $\#E_n \leq \#E_{n-1} + N(\theta)^n$. Thus, $\dim_F(X,H) \leq \frac{\log_2 N(\theta)}{\log_2 \frac{1}{2(\theta+\alpha)}}$, and Theorem 3.2 is proved. □

REMARK 3.2. Analogous sufficient conditions are given in Chapter 2 for the existence of the global attractor. For a continuous semigroup $S(t)$ acting on $X$, assuming that the map $(t,x) \mapsto S(t)x$ is Lipschitz (or even Hölder continuous) on $[0,t^*] \times X$, following [29], we can prove that the set $\mathcal{M} := \bigcup_{t\in[0,t^*]} S(t)\mathcal{M}^*$ is an exponential attractor for $S(t)$ on $X$. Here $\mathcal{M}^*$ is an exponential attractor for the discrete dynamical system generated by the mapping $S^*{:=}S(t^*)$ which satisfies the smoothing property (3.6) and (3.7). We emphasize that, in the application of Theorem 3.2 and its generalizations (see Sections 3.4 and 3.6 of this monograph and [32] and the references therein) to the equations of mathematical physics, usually the set $X$ is a positive, invariant, bounded, absorbing set. Therefore, if we consider a bounded, positively invariant subset $X$ in $H$, instead of the whole space $H$, we do not impose any restriction.

REMARK 3.3. It is worth mentioning that, in a Hilbert setting, i.e., when $H$ and $H_1$ are Hilbert spaces, (3.7) implies the squeezing property (for simplicity of presentation we took $L_0 = 0$). Indeed, let $H$ and $H_1$ be two Hilbert spaces such that the embedding $H_1 \subset H$ is compact and let us assume that the map $L : X \to X$, where $X$ is a bounded subset of $H$, enjoys the smoothing property

$$\|Lu_1 - Lu_2\|_{H_1} \leq C\|u_1 - u_2\|_H, \ \forall u_1, \ u_2 \in X.$$

Then let $M : H \to H_1$ be a linear mapping such that $\|Mu\|_{H_1} = \|u\|_H$, $\forall u \in H$. The mapping $M$ is obviously compact. Therefore, we can consider the projectors $P_n$ based on the spectrum of $M$; i.e., $P_n : H \to H_n = \text{Span}\{e_1,\dots,e_n\}$ is an orthogonal projector, $Me_n = \lambda_n e_n$, $\lambda_n \to +\infty$ as $n \to +\infty$. Obviously, we have $P_n u \to u$ for each $u \in H$ as $n \to +\infty$. Let us now check the squeezing property. We have, for $u_1$ and $u_2$ in $X$,

$$\begin{aligned} \|(I-P_n)(Lu_1 - Lu_2)\|_H &= \|M(I-P_n)(Lu_1 - Lu_2)\|_{H_1} \\ &\leq C\|M(I-P_n)\|_{\mathcal{L}(H_1,H_1)}\|u_1 - u_2\|_H. \end{aligned} \tag{3.11}$$

Now let $\delta$ belong to $(0, \frac{1}{4})$. If $\|Lu_1 - Lu_2\|_H \leq \delta\|u_1 - u_2\|_H$, then the squeezing property is satisfied. So, let us assume that

$$\|Lu_1 - Lu_2\|_H > \delta\|u_1 - u_2\|_H, \text{ i.e., } \|u_1 - u_2\|_H < \frac{1}{\delta}\|Lu_1 - Lu_2\|_H. \tag{3.12}$$

We have to show that there exists $n_0(= n_0(\delta)) \in N$ such that

$$\|(I - P_{n_0})(Lu_1 - Lu_2)\|_H \leq \|P_{n_0}(Tu_1 - Tu_2)\|_H.$$

Indeed, it follows from (3.11) and (3.12) that

$$\begin{aligned}\|(I-P_n)(Lu_1-Lu_2)\|_H &\le \frac{L}{\delta}\|M(I-P_n)\|_{\mathcal{L}(H_1,H_1)}\|Lu_1-Lu_2\|_H\\ &\le \varepsilon_n(\|P_n(Lu_1-Lu_2)\|_H+\|(I-P_n)(Lu_1-Lu_2)\|_H),\end{aligned}$$

where $\varepsilon_n=\frac{L}{\delta}\|M(I-P_n)\|_{\mathcal{L}(H_1,H_1)}$, which yields

$$(1-\varepsilon_n)\|(I-P_n)(Lu_1-Lu_2)\|_H\le\varepsilon_n\|P_n(Lu_1-Lu_2)\|_H.$$

So, we will have proved the squeezing property if $\varepsilon_n\to 0$ as $n\to+\infty$ (we note that $\varepsilon_n$ does not depend on the choice of $u_1$ and $u_2$). Indeed, we have

$$\begin{aligned}\|M(I-P_n)\|_{\mathcal{L}(H_1,H_1)} &= \sup_{u\neq 0}\frac{(M(I-P_n)u,u)_{H_1}}{(u,u)_{H_1}}\\ &= \sup_{v\neq 0}\frac{(M(I-P_n)Mv,Mv)_{H_1}}{(Mv,Mv)_{H_1}}\\ &= \sup_{v\neq 0}\frac{((I-P_n)Mv,v)_H}{(v,v)_H}\\ &= \|(I-P_n)M\|_{\mathcal{L}(H,H)}\end{aligned}\tag{3.13}$$

and we have $\varepsilon_n\to 0$ as $n\to+\infty$, hence the squeezing property.

## 3.2. Perturbation of exponential attractors: Autonomous case

In this section, we give a construction of a family of exponential attractors for abstract semigroups which will be Hölder continuous with respect to perturbations (the nonautonomous case will be considered in the next section). To this end, we first define the admissible class of semigroups.

DEFINITION 3.3. Let $H$ and $H_1$ be two Banach spaces such that $H_1$ is compactly embedded into $H$ and let $B$ be a bounded subset of $H_1$. For given positive constants $\delta$ and $K$, we define a class $\mathbb{S}_{\delta,K}(B)$ of nonlinear operators $S:H_1\to H_1$ as follows. An operator $S$ belongs to $\mathbb{S}_{\delta,K}(B)$ if:

(1) The operator $S$ maps a $\delta$-neighborhood $\mathcal{O}_\delta(B)$ of the set $B$ into $B$:

$$S:\mathcal{O}_\delta(B)\to B,\tag{3.14}$$

where the neighborhood is taken in the topology of $H_1$.

(2) For all points $h_1,h_2\in\mathcal{O}_\delta(B)$, we have

$$\|Sh_1-Sh_2\|_{H_1}\le K\|h_1-h_2\|_H.\tag{3.15}$$

Here, the metric in the space $\mathbb{S}_{\delta,K}(B)$ is defined by

$$\|S_1-S_2\|_{\mathbb{S}}:=\sup_{h\in\mathcal{O}_\delta(B)}\|S_1h-S_2h\|_{H_1}.\tag{3.16}$$

For every $S\in\mathbb{S}_{\delta,K}(B)$, the associated semigroup $\{S(n),\ n\in\mathbb{N}\}$ is defined as the iterations of the map $S$.

The main result of this section is the following theorem which gives a Hölder continuous family of exponential attractors associated with the maps $S\in\mathbb{S}_{\delta,K}(B)$.

THEOREM 3.3. *For every $S \in \mathbb{S}_{\delta,K}(B)$, there exists an exponential attractor $\mathcal{M}_S$ which satisfies the following conditions:*

(1) *The set $\mathcal{M}_S$ is a compact finite-dimensional subset of $H_1$; i.e.,*

$$\dim_F(\mathcal{M}_S, H_1) \le C_1, \tag{3.17}$$

*where $\dim_F$ is the fractal dimension in $H_1$.*

(2) *The set $\mathcal{M}_S$ is semiinvariant with respect to $S$, i.e., $S\mathcal{M}_S \subset \mathcal{M}_S$.*

(3) *This set enjoys the following exponential attraction property:*

$$\operatorname{dist}_{H_1}(S(n)B, \mathcal{M}_S) \le C_2 e^{-\alpha n}, \quad n \in \mathbb{N}, \text{ for some } \alpha > 0. \tag{3.18}$$

(4) *The map $S \mapsto \mathcal{M}_S$ is Hölder continuous in the following sense:*

$$\operatorname{dist}^{symm}_{H_1}(\mathcal{M}_{S_1}, \mathcal{M}_{S_2}) \le C_3 \|S_1 - S_2\|^{\kappa}_{\mathbb{S}}, \tag{3.19}$$

*where the positive constants $C_i$, $i = 1, 2, 3$, $\alpha$, and $\kappa$ only depend on $B$, $H$, $H_1$, $\delta$, and $K$ but are independent of the concrete choice of $S \in \mathbb{S}_{\delta,K}(B)$.*

PROOF. We first fix a finite covering of the set $B$ by $\delta/K$-balls in the space $H$ (such a covering exists since the embedding $H_1 \subset H$ is compact). Let $V_0 := \{h_1, \dots, h_{N_0}\} \subset B$ be the centers of these balls.

We also fix some $S \in \mathbb{S}_{\delta,K}(B)$ and set $V_1 = V_1(S) := SV_0$. Then, according to estimate (3.15), the system of $\delta$-balls in the space $H_1$ centered at the points of $V_1$ covers the image $SB$. Moreover, according to (3.14), every ball $B(\delta, h, H_1)$ centered at $h \in V_1$ belongs to $\mathcal{O}_\delta(B)$ (here and below, we denote by $B(r, h, V)$ the $r$-ball in $V$ centered at $h$).

We now construct, for every $n \in \mathbb{N}$, a special covering of $S(n)B$ by $\delta 2^{-n+1}$-balls centered at $V_n \subset B$ by using an inductive procedure. For $n = 1$, the required set $V_1$ has already been constructed. We assume that, for $n = k$, we already have the required system of $H_1$-balls of radius $\delta 2^{-k+1}$ centered at $V_k \subset B$ which covers $S(k)B$. In order to construct the next set $V_{k+1}$, we first need to fix a model covering of the unit ball $B(1, 0, H_1)$ in the space $H_1$ by $1/(2K)$-balls in the space $H$ (such a covering exists due to the compactness of the embedding $H_1 \subset H$). Let $\mathbb{U} := \{u_1, \dots, u_N\} \subset B(1, 0, H_1)$ be the centers of these balls. Then, we cover every ball $B(\delta 2^{-k+1}, h, H_1)$ with $h \in V_k$ by $N$ balls of radius $\delta 2^{-k+1}/(2K) = \delta 2^{-k}/K$ in the space $H$, defining the centers $V'_k(h)$ of these balls by the following formula:

$$V'_k(h) := h + \delta 2^{-k+1}\mathbb{U}, \quad h \in V_k \tag{3.20}$$

(it is essential for our construction always to use *the same* model covering $\mathbb{U}$ which is also independent of the choice of $S \in \mathbb{S}_{\delta,K}(B)$). Thus, the balls in the space $H$ with radius $\delta 2^{-k}/K$ centered at all the points of $V'_k := \bigcup_{h \in V_k} V'_k(h)$ cover $\bigcup_{h \in V_k} B(\delta 2^{-k+1}, h, H_1)$ and, consequently, they also cover $S(k)B$ by the induction assumption. We finally set $V_{k+1}(h) := SV'_k(h)$ and $V_{k+1} := \bigcup_{h \in V_k} V_{k+1}(h)$. Then, according to (3.15), the $\delta 2^{-k}$-balls in the space $H_1$ centered at $V_{k+1}$ cover $S(k+1)B$. Thus, by induction, the sets $V_n$ are constructed for all $n \in \mathbb{N}$.

Furthermore, according to the above construction, we have

$$\begin{cases} (1) \ \#V_k = N_0 \cdot N^{k-1}, \ k \in \mathbb{N}, \\ (2) \ \operatorname{dist}_{H_1}(S(k)B, V_k) \le \delta 2^{-k+1}, \\ (3) \ \operatorname{dist}^{symm}_{H_1}(V_{k+1}, SV_k) \le \delta 2^{-k}. \end{cases} \tag{3.21}$$

Indeed, the first and the second assertions of (3.21) are straightforward and the third one follows from the fact that, by construction, $\operatorname{dist}_{H_1}(V_{k+1}(h), Sh) \le \delta 2^{-k}$ for all $h \in V_k$.

We now define the sets $E_k = E_k(S)$ by

$$E_1 := V_1, \quad E_{k+1} := V_{k+1} \cup SE_k, \quad k \in \mathbb{N}. \tag{3.22}$$

Then, obviously,

$$\begin{cases} (1) \ \#E_k \le kN_0 \cdot N^k, \\ (2) \ SE_k \subset E_{k+1}, \\ (3) \ \operatorname{dist}_{H_1}(S(k)B, V_k) \le \delta 2^{-k+1}. \end{cases} \tag{3.23}$$

We finally define the required exponential attractor $\mathcal{M} = \mathcal{M}_S$ as follows:

$$\mathcal{M} = [\mathcal{M}']_{H_1}, \quad \mathcal{M}' := \bigcup_{k=1}^{\infty} E_k, \tag{3.24}$$

where $[\cdot]_V$ denotes the closure in the space $V$. Thus, it only remains to verify that the set $\mathcal{M}$ defined by (3.24) satisfies all the assertions of Theorem 3.3. To this end, we need to control the distance between $E_k$ and $S(n)B$.

LEMMA 3.1. *Let the above assumptions hold. Then, there exist positive constants $C$ and $\alpha < 1$, depending only on $K$, such that*

$$\operatorname{dist}_{H_1}(E_k, S(n)B) \le C\delta 2^{-\alpha k} \tag{3.25}$$

*for all $k \in \mathbb{N}$ and $n \le \alpha k$.*

PROOF. We first note that, due to the Lipschitz continuity of $S$ provided by estimate (3.15), we have

$$\operatorname{dist}^{symm}_{H_1}(S(m)A, S(m)C) \le K^m \operatorname{dist}^{symm}_{H_1}(A, C) \tag{3.26}$$

for all $A, C \subset B$ and all $m \in \mathbb{N}$ (without loss of generality, we may assume that $\|\cdot\|_H \le \|\cdot\|_{H_1}$). Thus, iterating the third estimate of (3.21) and using (3.26), we obtain

$$\begin{aligned} &\operatorname{dist}^{symm}_{H_1}(S(n-1)V_{k+1-n}, S(n)V_{k-n}) \le \delta 2^{-k+n} K^{n-1}, \\ &\operatorname{dist}^{symm}_{H_1}(S(n-2)V_{k+1-(n-1)}, S(n-1)V_{k-(n-1)}) \le \delta 2^{-k+(n-1)} K^{n-2}, \\ &\vdots \\ &\operatorname{dist}^{symm}_{H_1}(V_{k+1}, SV_k) \le \delta 2^{-k}, \end{aligned} \tag{3.27}$$

and, consequently, using the triangle inequality, we have

$$\operatorname{dist}^{symm}_{H_1}(S(m)V_{k+1-m}, S(n)V_{k-n}) \le \sum_{l=0}^{n-1} \delta 2^{-k+l} K^l \le C\delta 2^{-k}(2K)^n \tag{3.28}$$

for all $m \le n-1$, $n \le k$, and $k \in \mathbb{N}$. We now recall that, due to the construction of $V_k$, we have $V_k \subset B$ and, due to assumption (3.14), $SB \subset B$. Thus, (3.28) implies that

$$\operatorname{dist}_{H_1}(S(m)V_{k+1-m}, S(n)B) \le C\delta 2^{-k}(2K)^n, \quad n \le k, \quad m \le n-1. \tag{3.29}$$

We now restrict the possible values of $n$ as follows:

$$n \le \alpha k, \quad \alpha := \frac{1}{1+\log_2(2K)}. \tag{3.30}$$

Then, (3.29) yields

$$\operatorname{dist}_{H_1}(S(m)V_{k-m}, S(n)B) \le 2C\delta 2^{-\alpha k}, \quad n \le \alpha k, \quad m \le n-1. \tag{3.31}$$

Since, by definition, $E_k := \bigcup_{m=0}^{k-1} S(m)V_{k-m}$ and $S(m)V_{k-m} \subset S(m)B \subset S(n)B$ if $m \ge n$, then, estimate (3.31) implies (3.25), which finishes the proof of Lemma 3.1. □

We are now ready to verify that the attractor $\mathcal{M}$ defined by (3.24) satisfies all the assertions of Theorem 3.3. Indeed, semiinvariance and the exponential attraction property (3.18) are straightforward consequences of (3.23) and (3.24).

Let us verify the finite dimensionality. Let $\varepsilon > 0$ be sufficiently small. We need to construct a covering of $\mathcal{M}$ (or, equivalently, of $\mathcal{M}'$) by a finite number of $\varepsilon$-balls in $H_1$. We fix $n$ such that the $\varepsilon/2$-balls centered at $V_n$ cover $S(n)B$. According to (3.21)(2), we then have $n \sim n(\varepsilon) := \log_2 \frac{2\delta}{\varepsilon}$. Moreover, according to Lemma 3.1, the $\varepsilon$-balls centered at $V_n$ cover every $E_k$ with $k \ge k(\varepsilon) := \alpha^{-1}\log_2 \frac{4C\delta}{\varepsilon}$. Thus, the minimal number $N_\varepsilon(\mathcal{M}, H_1)$ of $\varepsilon$-balls which are necessary to cover $\mathcal{M}$ can be estimated as follows:

$$N_\varepsilon(\mathcal{M}, H_1) \le \#V_{n(\varepsilon)} + \sum_{k \le k(\varepsilon)} \#E_k \le N_0(k(\varepsilon)+1)^2 \cdot N^{k(\varepsilon)+1}, \tag{3.32}$$

where we have used (3.21) and (3.23) in order to estimate $\#E_k$. Consequently,

$$\dim_F(\mathcal{M}, H_1) = \limsup_{\varepsilon \to 0} \frac{\log_2 N_\varepsilon(\mathcal{M}, H_1)}{\log_2 \frac{1}{\varepsilon}} \le \alpha^{-1}\log_2 N < \infty. \tag{3.33}$$

Thus, the compactness and the finite dimensionality of $\mathcal{M}$ are also verified and it only remains to verify the Hölder continuity of the function $S \mapsto \mathcal{M}_S$. To this end, we need one more lemma.

LEMMA 3.2. *Let the above assumptions hold. Then, for every $S_1, S_2 \in \mathbb{S}_{\delta,K}(B)$,*

$$\operatorname{dist}_{H_1}^{symm}(E_k(S_1), E_k(S_2)) \le CK^k \|S_1 - S_2\|_{\mathbb{S}}, \quad k \in \mathbb{N}, \tag{3.34}$$

*where the constant $C$ only depends on $K$.*

PROOF. We first verify (3.34) for the sets $V_k(S_i)$ by induction. We denote the left-hand side of (3.34) by $d_k$. Then, since $V_0$ is the same for all $S \in \mathbb{S}_{\delta,K}(B)$, we have $d_1 \le \|S_1 - S_2\|_{\mathbb{S}}$. We now assume that the required estimate is already fulfilled for $k = n$. Then, for every $h^1 \in V_n(S_1)$, there exists $h^2 \in V_n(S_2)$ such that $\|h^1 - h^2\|_{H_1} \le d_n$, and vice versa. Furthermore, according to our construction of the sets $V_k'$ (and, more precisely, due to the fact that our model covering $\mathbb{U}$ is independent of $S$), we conclude that, for every $\bar h^1 \in V_k'(h^1)$, there exists $\bar h^2 \in V_k'(h^2)$ such that

$$\|\bar h^1 - \bar h^2\|_{H^1} \le d_n. \tag{3.35}$$

Thus, it only remains to estimate $S_1\bar h^1 - S_2\bar h^2$:

$$\begin{aligned}\|S_1\bar h^1 - S_2\bar h^2\|_{H_1} &\le \|S_1\bar h^1 - S_1\bar h^2\|_{H_1} + \|S_2\bar h^2 - S_1\bar h^2\|_{H_1}\\ &\le K\|h^1 - h^2\|_{H_1} + \|S_1 - S_2\|_{\mathbb{S}}.\end{aligned} \tag{3.36}$$

Consequently, (3.35) and (3.36) imply that

$$d_{n+1} \leq K d_n + \|S_1 - S_2\|_{\mathbb{S}}, \quad d_1 \leq \|S_1 - S_2\|_{\mathbb{S}}. \tag{3.37}$$

Solving (3.37), we infer

$$\operatorname{dist}^{symm}_{H_1}(V_k(S_1), V_k(S_2)) \leq CK^k \|S_1 - S_2\|_{\mathbb{S}}, \quad k \in \mathbb{N}. \tag{3.38}$$

The required estimate for the sets $E_k$ is a straightforward consequence of (3.38) and of the obvious estimate

$$\operatorname{dist}^{symm}_{H_1}(S_1 A, S_2 C) \leq K \operatorname{dist}^{symm}_{H_1}(A, C) + \|S_1 - S_2\|_{\mathbb{S}}, \tag{3.39}$$

which is valid for every $S_i \in \mathbb{S}_{\delta,K}(B)$, $i = 1, 2$, and every $A, C \subset B$. This finishes the proof of Lemma 3.2. □

We are now ready to verify the required Hölder continuity of the map $S \mapsto \mathcal{M}_S$. Indeed, let $S_1, S_2 \in \mathbb{S}_{\delta,K}(B)$. Our aim is to prove that

$$\operatorname{dist}_{H_1}(\mathcal{M}_{S_1}, \mathcal{M}_{S_2}) \leq C_1 \|S_1 - S_2\|^{\kappa}_{\mathbb{S}} \tag{3.40}$$

for some positive constants $C_1$ and $\kappa$ (the estimate for the Hausdorff distance between $\mathcal{M}_{S_2}$ and $\mathcal{M}_{S_1}$ follows immediately from (3.40) by exchanging $S_1$ and $S_2$). Moreover, in order to prove (3.40), it is sufficient to verify it for every $E_k(S_1)$, $k \in \mathbb{N}$ (see (3.24)).

We assume that $k \in \mathbb{N}$ and $u_0 \in E_k(S_1)$ are arbitrary. Then, according to Lemma 3.2, we have

$$\operatorname{dist}_{H_1}(u_0, \mathcal{M}_{S_2}) \leq CK^k \|S_1 - S_2\|_{\mathbb{S}}. \tag{3.41}$$

On the other hand, according to Lemma 3.1, for every $n \leq \alpha k$, there exists $u_0' \in S_1(n)B$ such that

$$\|u_0 - u_0'\|_{H_1} \leq C\delta 2^{-\alpha k}. \tag{3.42}$$

Let $\bar{u} \in B$ be such that $u_0' = S_1(n)\bar{u}$ and set $u_0'' := S_2(n)\bar{u}$. Then obviously

$$\|u_0' - u_0''\|_{H_1} \leq CK^n \|S_1 - S_2\|_{\mathbb{S}}. \tag{3.43}$$

Moreover, due to estimate (3.18), we have

$$\operatorname{dist}_{H_1}(u_0'', \mathcal{M}_{S_2}) \leq \delta 2^{-n+1}. \tag{3.44}$$

Combining now (3.42)–(3.44) and using the triangle inequality, we deduce that

$$\operatorname{dist}_{H_1}(E_k(S_1), \mathcal{M}_{S_2}) \leq C_1(2^{-\alpha k} + 2^{-n} + K^n \|S_1 - S_2\|_{\mathbb{S}}), \tag{3.45}$$

where $C_1$ only depends on $K$ and $\delta$. We shall use estimate (3.41) for $k \leq n/\alpha$ and estimate (3.45) otherwise, where $n = n(\|S_1 - S_2\|_{\mathbb{S}})$ will be fixed below. Then, we have

$$\begin{aligned} \operatorname{dist}_{H_1}(E_k, \mathcal{M}_{S_2}) &\leq C_1(2 \cdot 2^{-n} + K^n \|S_1 - S_2\|_{\mathbb{S}}) + CK^{n/\alpha} \|S_1 - S_2\|_{\mathbb{S}} \\ &\leq C_2(2^{-n} + K^{n/\alpha} \|S_1 - S_2\|_{\mathbb{S}}). \end{aligned} \tag{3.46}$$

Now fixing $n$ on the right-hand side of (3.46) in an optimal way, i.e.,

$$n \sim \frac{\alpha}{\alpha + \log_2 K} \log_2 \frac{1}{\|S_1 - S_2\|_{\mathbb{S}}},$$

we obtain estimate (3.40) with $\kappa := \frac{\alpha}{\alpha + \log_2 K}$ and finish the proof of Theorem 3.3. □

We now discuss several relaxations of the assumptions of Theorem 3.3.

REMARK 3.4. We first note that, according to Definition 3.3, the smoothing property (3.15) should hold in a $\delta$-neighborhood $\mathcal{O}_\delta(B)$ of the set $B$. However, very often in applications, one can verify (3.15) only for *smoother* initial data belonging to a compact absorbing (exponentially attracting) set in $H_1$. Thus, this assumption can be a rather essential restriction. In order to overcome this restriction, we note that we have factually used properties (3.14) and (3.15) in the "neighborhood"

$$\mathcal{O}'_\delta(B) := B + \bigcup_{\nu\in[0,\delta]} \nu\mathbb{U}, \tag{3.47}$$

where $\mathbb{U}$ consists of the centers of the model covering of the unit ball of $H_1$ by $1/(2K)$-balls in the space $H_1$. So, Theorem 3.3 remains valid if we replace the neighborhood $\mathcal{O}_\delta(B)$ by $\mathcal{O}'_\delta(B)$. In particular, if $B$ is a bounded set in a stronger space $H_2 \subset H_1$ and the centers $\mathbb{U}$ can be taken in $H_2$, then the $\delta$-neighborhood of $B$ can also be chosen in the metric of $H_2$. Moreover, the corresponding exponential attractors $\mathcal{M}_S$ will belong to $H_2$.

REMARK 3.5. It is also worth noting that we have actually verified an inequality slightly stronger than (3.17). Indeed, let us consider the Kolmogorov $\varepsilon$-entropy of the exponential attractor constructed in Theorem 3.3. Then, according to (3.32), we have

$$\mathbb{H}_\varepsilon(\mathcal{M}_S, H_1) \le C_1 \log_2 \frac{1}{\varepsilon} + C_2, \quad \varepsilon \to 0, \tag{3.48}$$

where the constants $C_1$ and $C_2$ are independent of the concrete choice of $S$. Thus, not only the fractal dimension of the exponential attractors, but also the minimal number of $\varepsilon$-balls which are necessary to cover them for every fixed positive $\varepsilon$, are uniformly bounded.

We now recall that the smoothing property (3.15) is typical for parabolic equations in bounded domains, but it is not usually satisfied for more general (e.g., hyperbolic) problems. In order to overcome this restriction, we need to generalize the class $\mathbb{S}_{\delta,K}(B)$ of admissible maps.

REMARK 3.6. Let $H$ and $H_1$ be two Banach spaces such that $H_1$ is compactly embedded into $H$ and let $B$ be a closed subset of $H_1$. An operator $S : H_1 \to H_1$ belongs to the class $\mathbb{S}_{\delta,\varepsilon,K}(B)$ for some positive constants $\varepsilon$, $K$, and $\delta$ if:

(1) Condition (3.14) is satisfied.

(2) A generalized version of (3.15), namely

$$\|Sh_1 - Sh_2\|_{H_1} \le (1-\varepsilon)\|h_1 - h_2\|_{H_1} + K\|h_1 - h_2\|_H, \tag{3.49}$$

holds for every $h_1, h_2 \in \mathcal{O}_\delta(B)$.

The following theorem is the analogue of Theorem 3.3 for the class $\mathbb{S}_{\delta,\varepsilon,K}(B)$.

THEOREM 3.4. *We assume that the set $B$ can be covered by a finite number of $\delta$-balls in the space $H_1$ with centers $V_0 \subset B$. Then, for every $S \in \mathbb{S}_{\delta,\varepsilon,K}(B)$, there exists an exponential attractor $\mathcal{M}_S \subset H_1$ which satisfies all the assertions of Theorem 3.3 (with constants $C_i$, $i = 1, \dots, 4$, $\alpha$, and $\kappa$ depending only on $H$, $H_1$, $B$, $\delta$, $\varepsilon$, and $K$).*

PROOF. The proof of this theorem is very similar to that of Theorem 3.3. We thus only indicate below how to construct the sets $V_k(S)$, leaving the details to the reader. Indeed, the initial set $V_0 = V_0(S) \subset B$, which gives a $\delta$-covering of

$B$, is already constructed by assumption. We now assume that we have already constructed the set $V_k \subset B$ for some $k = n$ such that the $\delta_n := \delta(1-\varepsilon/2)^n$-balls centered at $V_n$ cover $S(n)B$. We introduce the model covering of the unit ball $B(1,0,H_1)$ by a finite number of $\varepsilon/(2K)$-balls in the space $H$ and let $\mathbb{U} \subset B(1,0,H)$ be the centers in this covering. We cover (as in the proof of Theorem 3.3) every ball $B(\delta_n, h, H_1)$, $h \in V_n(S)$, by a finite number of $\delta_n\varepsilon/(2K)$-balls in the space $H$ centered at

$$V_n'(h) := h + \delta_n \mathbb{U} \tag{3.50}$$

and set $V_{n+1}(h) := SV_n'(h)$, $V_{n+1} := \bigcup_{h\in V_n} V_{n+1}(h)$. Then, formula (3.49) implies that the system of $\delta_{n+1}$-balls (where $\delta_{n+1} = (1-\varepsilon)\delta_n + K\delta_n\varepsilon/(2K) = \delta(1-\varepsilon/2)^{n+1}$) covers $S(n+1)B$. Thus, the system of sets $V_n(S)$ is constructed for every $n \in \mathbb{N}$. The rest of the proof repeats word for word that of Theorem 3.3 (with very minor changes related to the fact that the exponent $2^{-n}$ is now replaced by $(1-\varepsilon/2)^n$). □

To conclude, we give a slightly different form of the asymptotic smoothing property (3.49) which gives exponential attractors in the weaker space $H$ and which is useful for the exponential attractor theory of hyperbolic equations.

DEFINITION 3.4. Let $H$ and $H_1$ be two Banach spaces such that $H_1$ is compactly embedded into $H$ and let $B$ be a bounded subset *of the space* $H$. An operator $S: H \to H$ belongs to the class $\mathbb{S}'_{\delta,\varepsilon,K}(B)$ if:

(1) Assumption (3.14) holds for the $\delta$-neighborhood $\mathcal{O}_\delta(B)$ in the space $H$.

(2) For every $h_1, h_2 \in \mathcal{O}_\delta(B)$, the difference $v := Sh_1 - Sh_2$ can be split into a sum $v = v_1 + v_2$ such that

$$\|v_1\|_H \le (1-\varepsilon)\|h_1 - h_2\|_H, \qquad \|v_2\|_{H_1} \le K\|h_1 - h_2\|_H. \tag{3.51}$$

In contrast to (3.16), the distance between two maps $S_1, S_2 \in \mathbb{S}'_{\delta,\varepsilon,K}(B)$ is now defined by using the space $H$ instead of $H_1$.

The following theorem is the analogue of Theorems 3.3 and 3.4 for the class $\mathbb{S}'_{\delta,\varepsilon,K}(B)$.

THEOREM 3.5. *We assume that the set $B$ can be covered by a finite number of $\delta$-balls in the space $H$ with centers $V_0 \subset \mathcal{O}_\delta(B)$. Then, for every $S \in \mathbb{S}'_{\delta,\varepsilon,K}(B)$, there exists an exponential attractor $\mathcal{M}_S \subset \mathcal{O}_\delta(B)$ which satisfies all the assertions of Theorem* 3.3 *(in which the space $H_1$ is replaced by $H$).*

The proof of this theorem is completely analogous to those of Theorems 3.3 and 3.4 and we leave it to the reader.

REMARK 3.7. The idea of replacing the initial neighborhood $\mathcal{O}_\delta(B)$ by the special "neighborhood" $\mathcal{O}'_\delta(B)$ (see (3.47)) described in Remark 3.4 remains valid for the classes $\mathbb{S}_{\delta,\varepsilon,K}(B)$ and $\mathbb{S}'_{\delta,\varepsilon,K}(B)$ as well.

## 3.3. Perturbation of exponential attractors: Nonautonomous case

In this section, we extend the results of Section 3.2 to nonautonomous dynamical systems. We first recall that, in the nonautonomous case, we should consider, instead of semigroups, the so-called dynamical processes. By definition, a dynamical process $U$ in the phase space $\Phi$ is a two-parametric family of maps $\{U(l,m) : \Phi \to \Phi,\ l,m \in \mathbb{Z},\ l \geq m\}$ such that

$$U(l,k) \circ U(k,m) = U(l,m), \quad U(m,m) = \mathrm{Id}, \quad l,k,m \in \mathbb{Z}, \quad l \geq k \geq m. \tag{3.52}$$

We set $U(n) := U(n+1,n)$. Then, every dynamical process $U$ is uniquely determined by the one-parametric family of maps $\{U(n)\}_{n\in\mathbb{Z}}$ by

$$U(n+k,n) = U(n+k-1) \circ U(n+k-2) \circ \cdots \circ U(n), \quad n \in \mathbb{Z}, \quad k \in \mathbb{N}, \tag{3.53}$$

and, vice versa, i.e., every such family generates the associated dynamical process $U$ defined by (3.53). Thus, one can identify the dynamical process $U$ with the one-parametric family $\{U(n)\}_{n\in\mathbb{Z}}$. We also recall that the case where the maps $U(n)$ are independent of $n$, $U(n) \equiv S$, $n \in \mathbb{Z}$, corresponds to the autonomous case considered in the previous section. Indeed, in that case, obviously

$$U(n+k,n) = S(k),$$

where $S(k)$ is the semigroup generated by the map $S$.

The following theorem seems to be a natural generalization of Theorem 3.3 to the nonautonomous case.

THEOREM 3.6. *Let the spaces $H_1$ and $H$ and the set $B$ be the same as in Definition 3.3. Then, for every dynamical process $U$ in $H_1$ such that $U(n) \in \mathbb{S}_{\delta,K}(B)$ for every $n \in \mathbb{Z}$ (for some $\delta$ and $K$ which are independent of $n$), there exists a nonautonomous exponential attractor $n \mapsto \mathcal{M}_U(n)$, $n \in \mathbb{Z}$, which satisfies the following properties:*

(1) *The attractor $\mathcal{M}_U(n) \subset B$ for every $n \in \mathbb{Z}$ and its fractal dimension is finite; i.e.,*

$$\dim_F(\mathcal{M}_U(n), H_1) \leq C_1, \tag{3.54}$$

*where the constant $C_1$ is independent of $n$.*

(2) *The family $\mathcal{M}_U(n)$ is semiinvariant with respect to $U$; i.e.,*

$$U(k,m)\mathcal{M}_U(m) \subset \mathcal{M}_U(k) \tag{3.55}$$

*for all $k,m \in \mathbb{Z}$, $k \geq m$.*

(3) *This family enjoys a uniform exponential attraction property of the following form:*

$$\mathrm{dist}_{H_1}(U(n+k,n)B, \mathcal{M}_U(n+k)) \leq C_2 e^{-\alpha k}, \tag{3.56}$$

*where the constants $C_2$ and $\alpha$ are independent of $n \in \mathbb{Z}$ and $k \in \mathbb{N}$.*

(4) *The map $U \mapsto \mathcal{M}_U(n)$ is uniformly Hölder continuous in the following sense: For all dynamical processes $U_1$ and $U_2$ such that $U_i(n) \in \mathbb{S}_{\delta,K}(B)$, $n \in \mathbb{Z}$, $i = 1,2$, we have*

$$\mathrm{dist}^{symm}_{H_1}(\mathcal{M}_{U_1}(n), \mathcal{M}_{U_2}(n)) \leq C_3 \sup_{l\in(-\infty,n)} \left\{ e^{-\beta(n-l)} \|U_1(l) - U_2(l)\|^{\kappa}_{\mathbb{S}} \right\}, \tag{3.57}$$

*where the positive constants $C_i$, $i = 1,2,3$, $\alpha$, $\beta$, and $\kappa$ only depend on $B$, $H$, $H_1$, $\delta$, and $K$ but are independent of $n$ and of the concrete choice of the $U_i$.*

REMARK 3.8. Estimate (3.57) confirms that the attractor $\mathcal{M}_U(n)$ satisfies the causal principle, i.e., $\mathcal{M}_U(n)$ is independent of $U(k)$, $k \geq n$. Moreover, (3.57) also shows that the influence of the past decays exponentially with respect to time, in complete agreement with our physical intuition.

PROOF. The proof of this theorem is similar to that of Theorem 3.3 in the autonomous case, except that we now need to consider "time-dependent" analogues of $V_k$, $E_k$, and $\mathcal{M}$. As in Theorem 3.3, we fix a finite covering of the set $B$ by $\delta/K$-balls in the space $H$ and let $V_0 := \{h_1, \ldots, h_{N_0}\} \subset B$ be the centers of these balls. We also fix an arbitrary dynamical process $U$ satisfying the assumptions of the theorem.

We now set $V_1(n) = V_1^U(n) := U(n-1)V_0$, $n \in \mathbb{Z}$. Since all the maps $U(n)$ satisfy (3.15), the system of $\delta$-balls in the space $H_1$ centered at $V_1(n)$ covers the set $U(n-1)B = U(n, n-1)B$ for all $n \in \mathbb{Z}$.

As in Theorem 3.3, our aim is to construct the family of sets $V_k(n) = V_k^U(n)$ by induction with respect to $k$ such that the $\delta 2^{-k+1}$-balls in the space $H_1$ centered at $V_k(n)$ cover $U(n, n-k)B$ (for all $n \in \mathbb{Z}$). For $k = 1$, these sets have already been constructed. We now assume that the required sets are already constructed for some $k = l$ and we let the model covering $\mathbb{U}$ be the same as in the proof of Theorem 3.3. Then, for every $n \in \mathbb{Z}$, we cover every ball $B(\delta 2^{-l+1}, h, H_1)$ with $h \in V_l(n)$ by $N := \#\mathbb{U}$ balls of radius $\delta 2^{-l+1}/(2K)$ in the space $H$ centered at $V_l'(h) := h + \delta 2^{-l+1}\mathbb{U}$ (see (3.20)). Thus, according to the assumption of induction, the system of $\delta 2^{-l+1}/(2K)$-balls in the space $H$ centered at $V_l'(n) := \bigcup_{h \in V_l(n)} V_l'(h)$ covers $U(n, n-l)B$, $n \in \mathbb{Z}$. We finally set $V_{l+1}(n+1) := U(n)V_l'(n)$. Since all the maps $U(n)$ satisfy (3.15), the $\delta 2^{-l}$-balls in the space $H_1$ centered at $V_{l+1}(n+1)$ cover $U(n) \circ U(n, n-l)B = U(n+1, n-l)B$ ($n \in \mathbb{Z}$) and condition (3.14) of Definition 3.3 for $U(n)$ guarantees that $V_{l+1}(n+1) \subset B$. Thus, the required "nonautonomous sets" $V_k(n)$ are constructed for every $n \in \mathbb{Z}$ and $k \in \mathbb{N}$.

Furthermore, the above construction gives the following analogue of (3.21):

$$
\begin{cases} (1) \ \ V_k(n) = N_0 \cdot N^{k-1}, \quad N_0 := \#V_0, \ N := \#\mathbb{U}, \ k \in \mathbb{N}, \ n \in \mathbb{Z}, \\ (2) \ \ \operatorname{dist}_{H_1}(U(n, n-k)B, V_k(n)) \leq \delta 2^{-k+1}, \\ (3) \ \ \operatorname{dist}_{H_1}^{symm}(V_{k+1}(n+1), U(n)V_k(n)) \leq \delta 2^{-k}. \end{cases} \tag{3.58}
$$

As in the autonomous case, we define the sets $E_k(n) = E_k^U(n)$ by

(3.59)
$$
E_1(n) := V_1(n), \quad E_{k+1}(n+1) := V_{k+1}(n+1) \cup U(n)E_k(n), \quad k \in \mathbb{N}, \quad n \in \mathbb{Z}.
$$

Then, analogously to (3.23), we have

$$
\begin{cases} (1) \ \ \#E_k(n) \leq kN_0 \cdot N^k, \\ (2) \ \ U(n)E_k(n) \subset E_{k+1}(n+1), \\ (3) \ \ \operatorname{dist}_{H_1}(U(n, n-k)B, E_k(n)) \leq \delta 2^{-k+1}, \end{cases} \tag{3.60}
$$

and the required attractor $\mathcal{M}(n) = \mathcal{M}_U(n)$ can be defined analogously to (3.24):

$$
\mathcal{M}(n) = \left[\mathcal{M}'(n)\right]_{H_1}, \quad \mathcal{M}'(n) := \bigcup_{k=1}^{\infty} E_k(n), \quad n \in \mathbb{Z}. \tag{3.61}
$$

It only remains to verify that the attractor $\mathcal{M}(n)$ defined by (3.61) satisfies all the assertions of the theorem. Indeed, semiinvariance (3.55) is a straightforward consequence of (3.60)(2) and (3.61), and the uniform exponential attraction property follows from (3.60)(3). In order to verify the finite dimensionality and the Hölder continuity, we need the following natural analogue of Lemma 3.1.

LEMMA 3.3. *Let the assumptions of Theorem* 3.6 *hold. Then, there exist positive constants $C$ and $\alpha<1$ depending only on $K$ such that*

$$\operatorname{dist}_{H_1}(E_k(n), U(n,n-l)B) \le C\delta 2^{-\alpha k} \tag{3.62}$$

*for all $k\in\mathbb{N}$, $l\le \alpha k$, and $n\in\mathbb{Z}$.*

The proof of estimate (3.62) is based on the obvious nonautonomous analogue of estimate (3.26), namely,

$$\operatorname{dist}^{symm}_{H_1}(U(n,n-m)A, U(n,n-m)C) \le K^m \operatorname{dist}^{symm}_{H_1}(A,C) \tag{3.63}$$

(which holds for all $A,C\subset B$, $n\in\mathbb{Z}$, and $m\in\mathbb{N}$) and iterations of estimate (3.58)(3); it can be obtained by repeating word for word the proof of Lemma 3.1 and we thus leave it to the reader.

Having estimate (3.63), it is not difficult to verify (arguing exactly as in the proof of Theorem 3.3) that the sets $\mathcal{M}(n)$ are indeed finite dimensional and satisfy estimate (3.54) with exactly the same constant $C_1$ as in the autonomous case. So, it only remains to verify the Hölder continuity (3.57). To this end, we need the following natural generalization of Lemma 3.2.

LEMMA 3.4. *Let the assumptions of Theorem* 3.6 *hold. Then, for all dynamical processes $U_1$ and $U_2$ and for every $k\in\mathbb{N}$ and $n\in\mathbb{Z}$,*

$$\operatorname{dist}^{symm}_{H_1}(E^{U_1}_k(n), E^{U_2}_k(n)) \le C \sup_{l\in(0,k]} \|U_1(n-l)-U_2(n-l)\|_{\mathbb{S}}, \tag{3.64}$$

*where the constant $C$ is the same as in Lemma* 3.2.

The proof of this estimate also repeats word for word that of Lemma 3.2 in the autonomous case and we also leave it to the reader (we note that only the maps $U(n-1),\dots,U(n-k)$ are involved in the construction of the sets $E^U_k(n)$. That is the reason why we have the supremum with respect to $l\in(0,k]$ on the right-hand side of (3.64)).

We are now ready to verify the Hölder continuity (3.57) and finish the proof of Theorem 3.6. Let $U_1$ and $U_2$ be two dynamical processes satisfying the assumptions of the theorem. Then, as in Theorem 3.3, it is sufficient to verify that, for every $k\in\mathbb{N}$ and every $n\in\mathbb{Z}$,

$$\operatorname{dist}_{H_1}(E^{U_1}_k(n), \mathcal{M}^{U_2}(n)) \le CK^k \sup_{l\in(-\infty,n)} e^{-\beta(n-l)}\|U_1(l)-U_2(l)\|^{\kappa}_{\mathbb{S}}. \tag{3.65}$$

Indeed, let $u_0\in E^{U_1}_k(n)$ be arbitrary. Then, according to Lemma 3.4, we have

$$\operatorname{dist}_{H_1}(u_0, \mathcal{M}_{U_2}(n)) \le CK^k \sup_{l\in(0,k]} \|U_1(n-l)-U_2(n-l)\|_{\mathbb{S}}. \tag{3.66}$$

On the other hand, according to Lemma 3.3, for every $l\le\alpha k$, there exists $u_0'\in U_1(n,n-l)B$ such that

$$\|u_0-u_0'\|_{H_1} \le C\delta 2^{-\alpha k}. \tag{3.67}$$

Let $\bar{u} \in U_1(n, n-l)B$ be such that $u_0' = U(n, n-l)\bar{u}$ and set $u_0'' := U_2(n, n-l)\bar{u}$. Then, obviously,

$$\|u_0' - u_0''\|_{H_1} \leq CK^l \sup_{m\in(0,l]} \|U_1(n-m) - U_2(n-m)\|_{\mathbb{S}}. \tag{3.68}$$

We now recall that $\mathcal{M}^{U_2}(n)$ attracts exponentially the images of $B$ with respect to $U_2$, and consequently (see (3.60)(3)),

$$\operatorname{dist}_{H_1}(u_0'', \mathcal{M}^{U_2}(n)) \leq \delta 2^{-l+1}. \tag{3.69}$$

Now estimating the right-hand side of (3.68) as

$$\begin{aligned} &K^l \sup_{m\in(0,l]} \|U_1(n-m) - U_2(n-m)\|_{\mathbb{S}} \\ &\quad \leq K^l e^{\beta l} \sup_{m\in(0,l]} e^{-\beta m}\|U_1(n-m) - U_2(n-m)\|_{\mathbb{S}} \\ &\quad \leq (Ke^{\beta})^l \sup_{m\in(-\infty,n)} e^{-\beta(n-m)}\|U_1(m) - U_2(m)\|_{\mathbb{S}} \\ &\quad := K_1^l \cdot D_n(U_1, U_2) \end{aligned} \tag{3.70}$$

(which is valid for some positive $\beta$) and using the triangle inequality, we have

$$\operatorname{dist}_{H_1}(E_k^{U_1}(n), \mathcal{M}^{U_2}(n)) \leq C_1(2^{-\alpha k} + 2^{-l} + K_1^l \cdot D_n(U_1, U_2)), \tag{3.71}$$

where $C_1$ only depends on $K$ and $\delta$. As in the proof of Theorem 3.3, we use estimate (3.66) for $k \leq l/\alpha$ and estimate (3.71) otherwise. Then, analogously to (3.46), we deduce that

$$\operatorname{dist}_{H_1}(E_k^{U_1}(n), \mathcal{M}_{U_2}(n)) \leq C_2(2^{-l} + K_1^{l/\alpha} \cdot D_n(U_1, U_2)). \tag{3.72}$$

Now fixing $l = l(D_n(U_1, U_2))$ in an optimal way, i.e.,

$$l \sim \frac{\alpha}{\alpha + \log_2 K_1} \log_2 \frac{1}{D_n(U_1, U_2)},$$

we obtain (3.65), finishing the proof of Theorem 3.6. □

We now study the dependence of the exponential attractors $\mathcal{M}_U(n)$ on $n \in \mathbb{Z}$. To this end, we first introduce the group of temporal translations $\{T_k,\ k \in \mathbb{Z}\}$ acting on the space of all dynamical processes by

$$(T_k U)(m, n) := U(m+k, n+k), \quad k, m, n \in \mathbb{Z}, \quad m \geq n. \tag{3.73}$$

Then, obviously, $(T_k U)(n) = U(n+k)$. The following simple corollary gives the translation invariance of the exponential attractors $\mathcal{M}_U(n)$ constructed in the previous theorem.

COROLLARY 3.1. *Let the assumptions of Theorem 3.6 hold. Then, for every dynamical process $U$ satisfying the assumptions of that theorem, the associated exponential attractor satisfies the following cocycle identity:*

$$\mathcal{M}_U(n+k) = \mathcal{M}_{T_k U}(n) \tag{3.74}$$

*for all $k, n \in \mathbb{Z}$.*

Indeed, identity (3.74) follows immediately from the explicit construction of $\mathcal{M}_U(n)$ given in Theorem 3.6.

We now note that, in the autonomous case, where $U(n) \equiv S$, $n \in \mathbb{Z}$, the corresponding exponential attractor $n \mapsto \mathcal{M}_U(n)$ is independent of $n$ and thus coincides with the "autonomous" exponential attractor $\mathcal{M}_S$ of the associated semigroup $S(n)$ constructed in Theorem 3.3. Indeed, according to (3.67) and (3.74),

$$\operatorname{dist}^{symm}_{H_1}(\mathcal{M}_U(n+k),\mathcal{M}_U(n)) \le C \sup_{m\in(-\infty,n)} \|U(m+k)-U(m)\|^{\kappa}_{\mathbb{S}} = 0.$$

More generally, if the process $U$ is periodic-in-time, i.e., $U(n+T)\equiv U(n)$ for all $n\in\mathbb{Z}$ and some period $T\in\mathbb{N}$, then analogous reasonings show that the associated exponential attractor $n\mapsto\mathcal{M}_U(n)$ is also time-periodic with the same period. The next corollary addresses another important case, namely the situation where the process $U$ is *asymptotically* autonomous.

COROLLARY 3.2. *Let $U$ be a dynamical process satisfying the assumptions of Theorem 3.6. We assume, in addition, that the process $U$ is a heteroclinic orbit joining two autonomous semigroups $S_1(n)$ and $S_2(n)$ ($S_1,S_2\in\mathbb{S}_{\delta,K}(B)$), i.e.,*

$$\lim_{n\to-\infty}\|U(n)-S_1\|_{\mathbb{S}} = \lim_{n\to+\infty}\|U(n)-S_2\|_{\mathbb{S}} = 0. \tag{3.75}$$

*Then, the associated exponential attractor $n\mapsto\mathcal{M}_U(n)$ is also a heteroclinic orbit joining the "autonomous" exponential attractors $\mathcal{M}_{S_1}$ and $\mathcal{M}_{S_2}$ associated with the limit semigroups:*

$$\lim_{n\to-\infty}\operatorname{dist}^{symm}_{H_1}(\mathcal{M}_U(n),\mathcal{M}_{S_1}) = \lim_{n\to+\infty}\operatorname{dist}^{symm}_{H_1}(\mathcal{M}_U(n),\mathcal{M}_{S_2}) = 0.$$

Indeed, according to (3.57), we have

$$\operatorname{dist}^{symm}_{H_1}(\mathcal{M}_U(n),\mathcal{M}_{S_i}) \le C \sup_{l\in(-\infty,n)}\left\{e^{-\beta(n-l)}\|U(l)-S_i\|^{\kappa}_{\mathbb{S}}\right\}. \tag{3.76}$$

Passing to the limit $n\to\pm\infty$ and using (3.75) and the obvious fact that $\|U(n)-S_i\|_{\mathbb{S}}\le 2\|B\|_{H_1}$ is uniformly bounded, we deduce the assertion of the corollary.

REMARK 3.9. If the dynamical process $U$ is quasiperiodic or almost periodic with respect to time (in the sense of Bochner-Amerio), then it is not difficult to verify, using (3.57), that the associated exponential attractor, considered as a set-valued function $n\mapsto\mathcal{M}_U(n)$, will also be quasiperiodic or almost periodic with the same frequency basis (see [42] for analogous results for nonautonomous regular attractors).

THEOREM 3.7. *Let the spaces $H$ and $H_1$ and the set $B$ be the same as in Definition 3.3. We assume, in addition, that the set $B$ can be covered by a finite number of $\delta$-balls in the space $H_1$ with centers $V_0\subset B$. Then, for every dynamical process $U$ such that $U(n)\in\mathbb{S}_{\delta,\varepsilon,K}(B)$ for every $n\in\mathbb{Z}$, there exists an exponential attractor $n\mapsto\mathcal{M}_U(n)\subset B$, $n\in\mathbb{Z}$, which satisfies all the assertions of Theorem 3.6 (with constants $C_i$, $i=1,\dots,4$, $\alpha$, and $\kappa$ depending only on $H$, $H_1$, $B$, $\delta$, $\varepsilon$, and $K$).*

THEOREM 3.8. *Let the spaces $H$ and $H_1$ and the set $B$ be the same as in Definition 3.3. We assume, in addition, that the set $B$ can be covered by a finite number of $\delta$-balls in the space $H$ with centers $V_0\subset\mathcal{O}_\delta(B)$. Then, for every dynamical process $U$ such that $U(n)\in\mathbb{S}'_{\delta,\varepsilon,K}(B)$ for every $n\in\mathbb{Z}$, there exists an exponential attractor $n\mapsto\mathcal{M}_U(n)\subset\mathcal{O}_\delta(B)$, $n\in\mathbb{Z}$, which satisfies all the assertions of Theorem 3.6, in which the space $H_1$ is replaced by $H$.*

The proofs of these theorems are completely analogous to that of Theorem 3.6 (and to those of Theorems 3.4 and 3.5 in the autonomous case) and we leave them to the reader.

REMARK 3.10. The idea of replacing the initial neighborhood $\mathcal{O}_\delta(B)$ by the special "neighborhood" $\mathcal{O}'_\delta(B)$ (see (3.47)) described in Remark 3.4 obviously remains valid in the nonautonomous case as well.

## 3.4. Exponential attractors for a nonautonomous reaction-diffusion system

In this section of Chapter 3, we apply the results obtained above to the following reaction-diffusion problem in a bounded domain $\Omega \subset\subset \mathbb{R}^n$:

$$\partial_t u = a\Delta_x u - f(u) + g(t), \quad u\big|_{t=\tau} = u_\tau, \quad u\big|_{\partial\Omega} = 0. \tag{3.77}$$

Here, $u = (u_1, \dots, u_k)$ is an unknown vector-valued function, $a$ is a given constant diffusion matrix with positive symmetric part, $a + a^* > 0$, and $f \in C^2(\mathbb{R}^k, \mathbb{R}^k)$ is a given nonlinear interaction function which satisfies the following standard dissipativity and growth assumptions:

$$\begin{cases} (1) \ \ f(u) \cdot u \geq -C, \\ (2) \ \ f'(u) \geq -K, \\ (3) \ \ |f(u)| \leq C(1 + |u|^p), \quad p < p_{\max} := \frac{n+2}{n-2}, \end{cases} \tag{3.78}$$

where $u \cdot v$ denotes the standard inner product in $\mathbb{R}^k$ and $f'(u) \geq -K$ means that $f'(u)v \cdot v \geq -K|v|^2$ for all $u$, $v \in \mathbb{R}^k$. Finally, we assume that the external forces $g(t) := g(t, x)$ belong to the space $L^\infty(\mathbb{R}, L^2(\Omega))$ and satisfy

$$\|g\|_{L^\infty(\mathbb{R}, L^2(\Omega))} \leq M \tag{3.79}$$

for some given (possibly large) constant $M$.

It is well known (see, e.g., [19] or [20], [37]) that under the above assumptions, equation (3.77) possesses, for every $\tau \in \mathbb{R}$ and $u_\tau \in H_0^1(\Omega)$, a unique solution $u(t)$, $t \geq \tau$, which satisfies the following dissipative estimate:

$$\|u(t)\|_{H_0^1(\Omega)} \leq Q(\|u_\tau\|_{H_0^1(\Omega)})e^{-\alpha(t-\tau)} + C_K, \quad t \geq \tau, \tag{3.80}$$

where $\alpha$ is a positive constant, $Q$ is a monotonic function depending only on $a$ and $f$, and where the positive constant $C_K$ depends also on $M$ (but is independent of the concrete choice of $g$). Thus, equation (3.77) defines a dynamical process $\{U_g(t, \tau),\ \tau \in \mathbb{R},\ t \geq \tau\}$ in the phase space $H_0^1(\Omega)$ by

$$U_g(t, \tau)u_\tau := u(t), \quad \text{where } u(t) \text{ solves (3.77) with } u(\tau) = u_\tau. \tag{3.81}$$

Moreover, the following Lipschitz continuity and smoothing properties for the difference of two solutions $u_1(t)$ and $u_2(t)$ have been verified in [37]:

$$\begin{aligned} &\|u_1(\tau + t) - u_2(\tau + t)\|_{H_0^1(\Omega)} \\ &\qquad \leq Ce^{Kt}\|u_1(\tau) - u_2(\tau)\|_{H_0^1(\Omega)}, \quad t \in \mathbb{R}_+, \ \tau \in \mathbb{R}, \end{aligned} \tag{3.82}$$

and

$$\begin{aligned} &\|u_1(\tau + t) - u_2(\tau + t)\|_{H_0^1(\Omega)} \\ &\qquad \leq C(1 + t^{-1/2})e^{Kt}\|u_1(\tau) - u_2(\tau)\|_{L^2(\Omega)}, \quad t \in \mathbb{R}_+, \ \tau \in \mathbb{R}, \end{aligned} \tag{3.83}$$

where the constants $C$ and $K$ depend on $M$ and on the $H^1$-norms of $u_1(\tau)$ and $u_2(\tau)$ but are independent of $t$ and $\tau$. Furthermore, the following natural Lipschitz continuity holds for solutions of equation (3.77) with different external forces $g_1$ and $g_2$ satisfying (3.79) (see, e.g., [19] and [37]):

(3.84)
$$\|U_{g_1}(\tau+t,\tau)u_\tau - U_{g_2}(\tau+t,\tau)u_\tau\|^2_{H^1_0(\Omega)} \le Ce^{Kt}\int_\tau^{\tau+t}\|g_1(s)-g_2(s)\|^2_{L^2(\Omega)}\,ds,$$

where $C$ and $K$ depend on $M$ and the $H^1$-norm of $u_\tau$ but are independent of $t$ and $\tau$. We also recall that the trajectories of (3.77) possess a further smoothing property of the form

$$\|U_g(t+\tau,\tau)u_\tau\|_{H^{1+\delta}(\Omega)} \le Q_T(\|u_\tau\|_{H^1_0(\Omega)}),\quad t\ge T>0, \tag{3.85}$$

where $0<\delta<1$ and the monotonic function $Q_T$ depends on $M$, $\delta$, and $T$ but is independent of $t$ and $\tau$ (see [19] and [37] for details).

The main aim of this section is to show that the above estimates are sufficient to construct a robust family of nonautonomous exponential attractors for problem (3.77). To be more precise, the main result of this section is the following theorem.

THEOREM 3.9. *Let the nonlinear function $f$, the diffusion matrix $a$, and the external forces $g$ satisfy the above assumptions. Then, for every external force $g$ satisfying* (3.79)*, there exists an exponential attractor $t\mapsto \mathcal{M}_g(t)$ of the dynamical process* (3.81) *which satisfies the following properties:*

(1) *The sets $\mathcal{M}_g(t)$ are compact finite-dimensional subsets of $H^1_0(\Omega)$:*
$$\dim_F(\mathcal{M}_g(t),H^1_0(\Omega))\le C_1,\quad t\in\mathbb{R}, \tag{3.86}$$
*where the constant $C_1$ only depends on $M$ (and is independent of $t$ and $g$).*

(2) *These sets are semiinvariant with respect to $U_g(t,\tau)$ and translation-invariant with respect to time-shifts:*
$$\text{(i)}\ \ U_g(t,\tau)\mathcal{M}_g(\tau)\subset\mathcal{M}_g(t),\quad \text{(ii)}\ \ \mathcal{M}_g(t+s)=\mathcal{M}_{T_s g}(t), \tag{3.87}$$
*where $t,s,\tau\in\mathbb{R}$, $t\ge\tau$, and $\{T_h,\ h\in\mathbb{R}\}$ is the group of temporal shifts, $(T_h g)(t):=g(t+h)$.*

(3) *They satisfy a uniform exponential attraction property as follows: there exist a positive constant $\alpha$ and a monotonic function $Q$ (both depending only on $M$) such that, for every bounded subset $B$ of $H^1_0(\Omega)$, we have*
$$\operatorname{dist}_{H^1_0(\Omega)}(U_g(\tau+t,\tau)B,\mathcal{M}_g(\tau+t))\le Q(\|B\|_{H^1_0(\Omega)})e^{-\alpha t} \tag{3.88}$$
*for all $\tau\in\mathbb{R}$ and $t\ge 0$.*

(4) *The map $g\mapsto\mathcal{M}_g(t)$ is Hölder continuous in the following sense:*
$$\operatorname{dist}^{symm}_{H^1_0}(\mathcal{M}_{g_1}(t),\mathcal{M}_{g_2}(t))\le C_2\left(\int_{-\infty}^t e^{-\beta(t-s)}\|g_1(s)-g_2(s)\|^2_{L^2(\Omega)}\,ds\right)^{\kappa}, \tag{3.89}$$
*where the positive constants $C$, $\beta$, and $\kappa$ only depend on $M$ and are independent of $g_1$, $g_2$, and $t$. In particular, the function $t\mapsto\mathcal{M}_g(t)$ is uniformly Hölder continuous in the Hausdorff metric, i.e.,*
$$\operatorname{dist}^{symm}_{H^1_0(\Omega)}(\mathcal{M}_g(t+s),\mathcal{M}_g(t))\le C_3|s|^{\kappa_1}, \tag{3.90}$$
*where $C_3$ and $\kappa_1$ are also independent of $g$, $t$, and $s$.*

PROOF. We first construct a family of exponential attractors for the discrete dynamical processes associated with equation (3.77). To this end, we note that, according to estimate (3.80), the ball $\mathbb{B} = B_R := \{v \in H_0^1(\Omega),\ \|v\|_{H_0^1(\Omega)} \le R\}$, where $R$ is a sufficiently large number depending only on $M$ given in (3.79), is a uniform absorbing set for all the processes $U_g(t,\tau)$ generated by equation (3.77). Thus, it only remains to construct the required exponential attractors for initial data belonging to this ball. Moreover, it also follows from the above estimates that there exists $T = T(M)$ such that

$$U_g(\tau+T,\tau)\mathcal{O}_1(\mathbb{B}) \subset \mathbb{B}$$

for all $\tau \in \mathbb{R}$ and all $g$ satisfying (3.79). This embedding, together with estimate (3.83), proves that, for a sufficiently large $K = K(M)$, we have

$$U_g(\tau+T,\tau) \in \mathbb{S}_{1,K}(\mathbb{B}) \text{ with } H := L^2(\Omega) \text{ and } H_1 := H_0^1(\Omega) \tag{3.91}$$

for all $\tau$ and $g$. Thus, we can apply Theorem 3.6 to the family of discrete dynamical processes $U_g^\tau(m,l) := U_g(\tau+mT,\tau+lT)$, $m,l \in \mathbb{Z}$, $m \ge l$. According to this theorem, these processes possess exponential attractors $l \mapsto \mathcal{M}_g(l,\tau)$, $l \in \mathbb{Z}$, which satisfy the following properties:

(1) These sets are compact subsets of $H_0^1(\Omega)$ whose fractal dimension is uniformly bounded:

$$\mathbb{H}_\varepsilon(\mathcal{M}_g(l,\tau), H_0^1(\Omega)) \le C_1 \log_2 \frac{1}{\varepsilon} + C_2, \tag{3.92}$$

where the positive constants $C_1$ and $C_2$ only depend on $M$.

(2) They are semiinvariant with respect to the discrete processes:

$$U_g(\tau+mT,\tau+lT)\mathcal{M}_g(l,\tau) \subset \mathcal{M}_g(m,\tau).$$

(3) They enjoy the following uniform exponential attraction property:

$$\operatorname{dist}_{H_0^1(\Omega)}(U_g(\tau+mT,\tau+lT)\mathbb{B}, \mathcal{M}_g(m,\tau)) \le C_3 e^{-\alpha(m-l)}, \tag{3.93}$$

where the positive constants $C_3$ and $\alpha$ only depend on $M$ and are independent of $m,l \in \mathbb{Z}$, $m \ge l$, $\tau \in \mathbb{R}$, and $g$ satisfying (3.79).

For different $g_1$ and $g_2$ satisfying (3.79), we have

$$\begin{aligned}
&\operatorname{dist}^{symm}_{H_0^1(\Omega)}(\mathcal{M}_{g_1}(0,\tau), \mathcal{M}_{g_2}(0,\tau)) \\
&\quad \le C_4 \sup_{n\in\mathbb{N}} \{e^{-\beta n} \|U_{g_1}(\tau-(n-1)T,\tau-nT) - U_{g_2}(\tau-(n-1)T,\tau-nT)\|_{\mathbb{S}}^{\kappa}\} \\
&\quad \le C_5 \left( \sup_{n\in\mathbb{N}} e^{-2\beta n/\kappa} \int_{\tau-nT}^{\tau-(n-1)T} \|g_1(s)-g_2(s)\|^2_{L^2(\Omega)}\,ds \right)^{\kappa/2} \\
&\quad \le C_6 \left( \int_{-\infty}^{\tau} e^{-\beta'(\tau-s)} \|g_1(s)-g_2(s)\|^2_{L^2(\Omega)}\,ds \right)^{\kappa'},
\end{aligned} \tag{3.94}$$

where all the constants are positive and only depend on $M$ (here, we have used estimates (3.57) and (3.84)). Moreover, analogously to (3.74), we have the following translation invariance:

$$(1)\ \mathcal{M}_g(l,\tau) = \mathcal{M}_g(0,lT+\tau), \quad (2)\ \mathcal{M}_{T_s g}(l,\tau) = \mathcal{M}_g(l,\tau+s). \tag{3.95}$$

We now verify the Hölder continuity of the function $\tau \mapsto \mathcal{M}_g(0,\tau)$. To this end, we need the following lemma which gives the Hölder continuity of the processes $U_g(t,\tau)$ with respect to time.

LEMMA 3.5. *Let the assumptions of Lemma 3.4 be fulfilled for the equation (3.77). Then, for every $u_\tau \in H_0^1(\Omega)$,*

$$\|U_g(\tau+s+t,\tau)u_\tau - U_g(\tau+t,\tau)u_\tau\|_{L^2(\Omega)} \le C|s|^{1/2}, \tag{3.96}$$

*where the constant $C$ depends on $M$ and on the $H^1$-norm of $u_\tau$ but is independent of $t \ge 0$, $\tau \in \mathbb{R}$, and $s \ge 0$. Moreover, for every $T > 0$, we also have*

$$\begin{aligned} &\|U_g(t+\tau+s,\tau+s)u_\tau - U_g(t+\tau,\tau)u_\tau\|_{H_0^1(\Omega)} \\ &\qquad \le C_T e^{Kt}|s|^\gamma, \quad t \ge T, \quad 0 \le s \le T/2, \end{aligned} \tag{3.97}$$

*where $\gamma$ is a positive number and the positive constant $C_T$ depends on $T$ and on the $H^1$-norm of $u_\tau$ but is independent of $t$, $\tau$, and $s$.*

PROOF. It follows from estimate (3.80) and equation (3.77) that the function $u(t) := U_g(t,\tau)u_\tau$, $t \ge \tau$, satisfies

$$u \in L^\infty([\tau,+\infty), H_0^1(\Omega)), \quad \partial_t u \in L^\infty([\tau,+\infty), H^{-1}(\Omega)), \tag{3.98}$$

and is uniformly bounded in these spaces, which immediately implies the Hölder continuity (3.96). In order to verify (3.97), we note that, due to (3.96) and (3.83), for every $v \in H_0^1(\Omega)$, we have

$$\begin{aligned} &\|U_g(t+s+\tau,\tau+s)v - U_g(\tau+t,\tau)v\|_{L^2(\Omega)} \\ &\qquad \le \|U_g(t+s+\tau,t+\tau)(U_g(t+\tau,\tau+s)v) - U_g(t+\tau,\tau+s)v\|_{L^2(\Omega)} \\ &\qquad\quad + \|U_g(t+\tau,\tau+s)v - U_g(t+\tau,\tau+s)(U_g(\tau+s,\tau)v)\|_{L^2(\Omega)} \\ &\qquad \le C_T|s|^{1/2} + C_T e^{K(t-s)}\|v - U_g(\tau+s,\tau)v\|_{L^2(\Omega)} \le C_T' e^{Kt}|s|^{1/2}, \end{aligned} \tag{3.99}$$

where all the constants depend on $T$, $M$, and the $H^1$-norm of $v$ but are independent of $t$, $s$ and $\tau$. Now using the smoothing property (3.85), we verify that

$$U_g(t+s+\tau,\tau+s)v \quad \text{and} \quad U_g(t+\tau,\tau)v$$

are uniformly bounded in $H^{1+\delta}(\Omega)$, with positive $\delta$. The obvious interpolation inequality

$$\|\cdot\|_{H_0^1(\Omega)} \le C\|\cdot\|_{L^2(\Omega)}^{\delta/(1+\delta)}\|\cdot\|_{H^{1+\delta}(\Omega)}^{1/(1+\delta)}$$

now finishes the proofs of estimate (3.97) and Lemma 3.5. □

We are now ready to verify the Hölder continuity of the attractors $\mathcal{M}_g(0,\tau)$ with respect to $\tau$. Indeed, according to (3.57) and (3.97), we have

(3.100)

$$\begin{aligned}\operatorname{dist}^{symm}_{H^1_0(\Omega)}&(\mathcal{M}_g(0,\tau+s),\mathcal{M}_g(0,\tau))\\ &\le C_7\sup_{l\in\mathbb{N}} e^{-\beta l}\|U_g(\tau-(l-1)T,\tau-lT)-U_g(\tau+s-(l-1)T,\tau+s-lT)\|_{\mathbb{S}}\\ &\le C_8|s|^\gamma.\end{aligned}$$

We then define the required exponential attractors for continuous time by the following natural formula:

$$\mathcal{M}_g(\tau):=\bigcup_{s\in[0,T]}U_g(\tau,\tau-T-s)\mathcal{M}_g(0,\tau-T-s). \tag{3.101}$$

We claim that the exponential attractor $\tau\mapsto\mathcal{M}_g(\tau)$ thus defined satisfies all the assertions of Theorem 3.9. We first verify semiinvariance (3.87)(i). To this end, we first note that, due to the semiinvariance of $\mathcal{M}_g(0,\tau)$ with respect to the discrete processes $U_g(\tau+lT,\tau+mT)$ (i.e., $U_g(\tau+lT,\tau)\mathcal{M}_g(0,\tau)\subset\mathcal{M}_g(l,\tau)=\mathcal{M}_g(0,\tau+lT)$), it is sufficient to verify (3.87)(i) for $t-\tau:=\alpha\in[0,T]$ only. Then we have

$$\begin{aligned}U_g(t+\alpha,t)\mathcal{M}_g(t)&=\bigcup_{s\in[0,T]}U_g(t+\alpha,t-T-s)\mathcal{M}_g(0,t-T-s)\\ &=\bigcup_{s\in[\alpha,T]}U_g(t+\alpha,t-T-s)\mathcal{M}_g(0,t-T-s)\\ &\quad\cup\bigcup_{s\in[0,\alpha]}U_g(t+\alpha,t-T-s)\mathcal{M}_g(0,t-T-s)\\ &=\bigcup_{s'\in[0,T-\alpha]}U_g(t+\alpha,t+\alpha-T-s')\mathcal{M}_g(0,t+\alpha-T-s')\\ &\quad\cup\bigcup_{s'\in[T-\alpha,T]}U_g(t+\alpha,t+\alpha-T-s')\\ &\quad\circ U_g(t+\alpha-T-s',t+\alpha-2T-s')\mathcal{M}_g(0,t+\alpha-2T-s')\\ &\subset\bigcup_{s'\in[0,T-\alpha]}U_g(t+\alpha,t+\alpha-T-s')\mathcal{M}_g(0,t+\alpha-T-s')\\ &\quad\cup\bigcup_{s'\in[T-\alpha,T]}U_g(t+\alpha,t+\alpha-T-s')\mathcal{M}_g(0,t+\alpha-T-s')\\ &=\mathcal{M}_g(t+\alpha).\end{aligned}$$

Thus, semiinvariance is verified. The translation invariance with respect to time-shifts (3.87)(ii) is an immediate consequence of (3.95)(2) and the obvious translation identity $U_{T_sg}(t,\tau)=U_g(t+s,\tau+s)$. Estimate (3.88) follows in a standard way from (3.93) and (3.80) and from the fact that the ball $\mathbb{B}$ is chosen as a uniformly absorbing set for all the processes $U_g(t,\tau)$. The Hölder continuity (3.89) follows easily from its analogue (3.94) for discrete times, definition (3.101), and the Lipschitz continuity (3.83) and (3.84), and the Hölder continuity (3.90) follows immediately from its discrete analogue (3.100) and the Hölder continuity (3.97) (we have introduced the additional time shift $T$ in (3.101) in order to be able to apply (3.97)). We also note that the sets $\mathcal{M}_g(t)$ are closed in $H^1_0(\Omega)$, since the set-valued function $\tau\mapsto\mathcal{M}_g(0,\tau)$

is (Hölder) continuous and all the maps $U_g(t,t-T-s)$ are also continuous. Thus, it only remains to verify the finiteness of the fractal dimension of $\mathcal{M}_g(t)$. In order to prove this, we first note that, according to the Hölder continuities (3.100) and (3.97), we have

(3.102)
$$\operatorname{dist}^{symm}_{H^1_0(\Omega)}(U_g(t,t-T-s_1)\mathcal{M}_g(0,t-T-s_1),U_g(t,t-T-s_2)\mathcal{M}_g(0,t-T-s_2))\\ \le C|s_1-s_2|^{\kappa''}$$

for all $s_1,s_2\in[0,T]$ and $t\in\mathbb{R}$ and for some positive constants $C$ and $\kappa''$. Thus, for a given $\varepsilon>0$, the minimal number $N_\varepsilon(\mathcal{M}_g(t),H^1_0(\Omega))$ of $\varepsilon$-balls which are necessary to cover $\mathcal{M}_g(t)$ can be estimated as follows:

(3.103)
$$N_\varepsilon\left(\mathcal{M}_g(t),H^1_0(\Omega)\right)\\ \le \sum_{l=0}^{[(\frac{2C}{\varepsilon})^{1/\kappa''}]} N_{\varepsilon/2}\left(U_g(t,t-T-l(\varepsilon/2C)^{1/\kappa''})\mathcal{M}_g(0,t-T-l(\varepsilon/2C)^{1/\kappa''}),H^1_0(\Omega)\right).$$

Moreover, since the maps $U_g(t,t-T-s)$ are uniformly Lipschitz continuous, estimate (3.92) implies that

(3.104)
$$\mathbb{H}_\varepsilon\left(U_g(t,t-T-l(\varepsilon/2C)^{1/\kappa''})\mathcal{M}_g(0,t-T-l(\varepsilon/2C)^{1/\kappa''}),H^1_0(\Omega)\right)\\ \le \mathbb{H}_{\varepsilon/L}\left(\mathcal{M}_g(0,t-T-l(\varepsilon/2C)^{1/\kappa''}),H^1_0(\Omega)\right)\le C_1'/\kappa''\log_2\frac{1}{\varepsilon}+C_2'.$$

Combining (3.103) and (3.104), we finally obtain

$$\mathbb{H}_\varepsilon(\mathcal{M}_g(t),H^1_0(\Omega))\le\frac{C}{\kappa''}\log_2\frac{1}{\varepsilon}+C' \tag{3.105}$$

for some constants $C$ and $C'$ which are independent of $t$. Theorem 3.9 is proven. □

To conclude, we compare the nonautonomous exponential attractor $t\mapsto\mathcal{M}_g(t)$ obtained above with the so-called infinite-dimensional (uniform) exponential attractor constructed in [37]. To this end, we first briefly recall (for the convenience of the reader) the most important objects related to the uniform attractor approach for nonautonomous dynamical systems, using our reaction-diffusion system as a model example (see [19] for detailed expositions).

Let $g\in L^\infty(\mathbb{R},L^2(\Omega))$ be some external forces. The hull $\mathcal{H}(g)\subset L^\infty(\mathbb{R},L^2(\Omega))$ is defined as the following set:

$$\mathcal{H}(g):=\left[T_sg,\ s\in\mathbb{R}\right]_{L^\infty_{loc}(\mathbb{R},L^2(\Omega))}, \tag{3.106}$$

where the closure is taken in the *local* topology of $L^\infty(\mathbb{R},L^2(\Omega))$. Obviously, the group of temporal shifts $\{T_h,\ h\in\mathbb{R}\}$ acts on the hull of $g$,

$$T_h\mathcal{H}(g)=\mathcal{H}(g),\quad h\in\mathbb{R}, \tag{3.107}$$

which is usually endowed with the local topology $L^\infty_{loc}(\mathbb{R},L^2(\Omega))$. Using the standard skew-product technique (see [19]), we can embed, for every external forces $g$

satisfying (3.79), the associated dynamical process $U_g(t,\tau)$ into the autonomous dynamical system $\mathbb{S}_t$ acting on the extended phase space $\Phi := H_0^1(\Omega) \times \mathcal{H}(g)$ via

$$\mathbb{S}_t(u_0, h) := (U_h(t,0)u_0, T_t h), \quad u_0 \in H_0^1(\Omega), \quad h \in \mathcal{H}(g), \quad t \geq 0. \tag{3.108}$$

It is well known that (3.108) is indeed a semigroup. If this semigroup possesses the global attractor $\mathbb{A} = \mathbb{A}(g) \subset \Phi$, then its projection $\mathcal{A}^{un}(g) := \Pi_1 \mathbb{A}(g)$ onto the first component of the Cartesian product is called the uniform attractor associated with problem (3.77).

It is also known that the uniform attractor $\mathcal{A}^{un}(g)$ exists under the relatively weak assumption that the hull $\mathcal{H}(g)$ is compact in $L^\infty_{loc}(\mathbb{R}, L^2(\Omega))$, but unfortunately, for more or less general external forces $g$, its Hausdorff and fractal dimensions are infinite. Instead, the following estimate for its Kolmogorov $\varepsilon$-entropy holds; see [21] and [37].

PROPOSITION 3.1. *Let the above assumptions hold and let, in addition, the hull $\mathcal{H}(g)$ of the initial external forces be compact. Then, equation* (3.1) *possesses the uniform attractor $\mathcal{A}^{un}(g)$, and its Kolmogorov $\varepsilon$-entropy can be estimated in terms of the Kolmogorov $\varepsilon$-entropy of the hull $\mathcal{H}(g)$ as follows:*

$$\begin{aligned} \mathbb{H}_\varepsilon(\mathcal{A}^{un}(g), H_0^1(\Omega)) \leq & C_1 \log_2 \frac{1}{\varepsilon} \\ & + \mathbb{H}_{\varepsilon/L}\left(\mathcal{H}(g)\big|_{[0,T_0\log_2\frac{1}{\varepsilon}]\times\Omega}, L^\infty\left(\left[0, T_0 \log_2 \frac{1}{\varepsilon}\right], L^2(\Omega)\right)\right) \end{aligned} \tag{3.109}$$

*for some positive constants $C_1$, $L$, and $T_0$ depending on $a$ and $f$.*

Since an exponential attractor always contains the global attractor, a uniform exponential attractor $\mathcal{M}^{un}(g)$ should necessarily be infinite dimensional if the uniform attractor $\mathcal{A}^{un}(g)$ has infinite dimension. Thus, following [19], it is natural to use Kolmogorov $\varepsilon$-entropy in order to control the "size" of an exponential attractor. To be more precise, the following object has been introduced in [37].

DEFINITION 3.5. A set $\mathcal{M}^{un}(g)$ is a (uniform) exponential attractor of equation (3.77) if the following properties are satisfied:

(1) Entropy estimate: $\mathcal{M}^{un}(g)$ is a compact subset of the phase space $H_0^1(\Omega)$ which satisfies estimate (3.109) (possibly for larger constants $C_1$, $L$, and $T_0$).

(2) Semiinvariance: For every $u_0 \in \mathcal{M}^{un}(g)$, there exists $h \in \mathcal{H}(g)$ such that $U_h(t,0)u_0 \subset \mathcal{M}^{un}(g)$ for all $t \geq 0$.

(3) Uniform exponential attraction property: There exists a positive constant $\alpha$ and a monotonic function $Q$ such that, for every $h \in \mathcal{H}(g)$ and every bounded subset $B \subset H_0^1(\Omega)$, we have

$$\operatorname{dist}_{H_0^1(\Omega)}(U_h(t+\tau,\tau)B, \mathcal{M}^{un}(g)) \leq Q(\|B\|_{H_0^1(\Omega)})e^{-\alpha t} \tag{3.110}$$

for all $\tau \in \mathbb{R}$ and $t \geq 0$.

The existence of an (infinite-dimensional) uniform exponential attractor for problem (3.77) has been verified directly in [37]. We show (see Corollary 3.3 below) that a uniform exponential attractor $\mathcal{M}^{un}(g)$ can be easily constructed if the (nonautonomous) exponential attractor $t \mapsto \mathcal{M}_g(t)$ has already been constructed.

COROLLARY 3.3. *Let the assumptions of Theorem* 3.9 *hold and let, in addition, the hull $\mathcal{H}(g)$ of some external forces satisfying* (3.79) *be compact. Then, there* *exists a uniform exponential attractor $\mathcal{M}^{un}(g)$ for problem (3.77) which can be constructed as follows:*

$$\mathcal{M}^{un}(g) := \left[\bigcup_{t\in\mathbb{R}} \mathcal{M}_g(t)\right]_{H_0^1(\Omega)} = \bigcup_{h\in\mathcal{H}(g)} \mathcal{M}_h(0). \tag{3.111}$$

Indeed, the second equality in (3.111) is an immediate consequence of the Hölder continuity (3.99) and the definition of the hull $\mathcal{H}(g)$. The entropy estimate (3.109) for $\mathcal{M}^{un}(g)$ is also a standard consequence of this Hölder continuity and of the uniform entropy estimates (3.105) for the nonautonomous attractors. The semiinvariance of $\mathcal{M}^{un}(g)$ follows from equality (3.111) and from the semiinvariance of $\mathcal{M}_h(t)$, and the exponential attraction property (3.110) is an immediate consequence of the analogous property (3.88) for nonautonomous exponential attractors.

REMARK 3.11. We now consider exponential attractors for continuous time. It is not difficult to extend the results of Sections 3.1, 3.2, and 3.3 to continuous time. Indeed, let us consider dynamical processes $\{U(t,s),\ t,s\in\mathbb{R},\ t\geq s\}$ with continuous time satisfying

$$U(t,\tau)\circ U(\tau,s) = U(t,s), \quad U(t,t) = \mathrm{Id}, \tag{3.112}$$

for all $t,s,\tau\in\mathbb{R}$ such that $t\geq s\geq\tau$. The main idea of our extension is to consider, for every $t\in\mathbb{R}$, the discrete dynamical process $U_t^d(m,n) := U(t+m,t+n)$, $m,n\in\mathbb{Z}$, $m\geq n$, construct "discrete" exponential attractors $\mathcal{M}_U^d(t) := \mathcal{M}_{U_t^d}(0)$ by using the results of the previous section, and then extend these attractors to continuous time by the following natural formula:

$$\mathcal{M}_U(t) := \bigcup_{s\in[0,1]} U(t,t-s)\mathcal{M}_U^d(t-s). \tag{3.113}$$

The next theorem is the analogue of Theorem 3.6 for continuous time.

THEOREM 3.10. *Let the spaces $H$ and $H_1$ and the set $B$ be the same as in Definition 3.3 and let the dynamical process $U$ (with continuous time) satisfy the following assumptions:*

(1) *For every $t\in\mathbb{R}$, the map $U(t) := U(t+1,t)$ belongs to the class $\mathbb{S}_{\delta,K}(B)$.*
(2) *For every $t\in\mathbb{R}$ and $s\in[0,1]$, the map $U(t+s,t)$ is uniformly Lipschitz continuous on $B$, i.e.,*

$$\|U(t+s,t)h_1 - U(t+s,t)h_2\|_{H_1} \leq K_1\|h_1-h_2\|_{H_1} \tag{3.114}$$

*for all $h_1,h_2\in B$.*

(3) *For every $t\in\mathbb{R}$, the operators $U(t+s,t)$ are uniformly Hölder continuous with respect to $s$, i.e., for every $h\in B$ and every $s\in[0,1]$,*

$$\|U(t+s,t)h - h\|_{H_1} \leq Cs^{\beta} \tag{3.115}$$

*for some positive constants $C$ and $\beta\in(0,1]$ which are independent of $h$, $s$, and $t$.*

*Then, there exists an exponential attractor $t \mapsto \mathcal{M}_U(t)$, $t \in \mathbb{R}$, associated with the process $U$, which enjoys the following properties:*

(1) *Finite dimensionality: The estimate*

$$\dim_F(\mathcal{M}_U(t), H_1) \leq C_1 \tag{3.116}$$

*holds uniformly with respect to $t \in \mathbb{R}$.*

(2) *Semiinvariance: $U(t,s)\mathcal{M}_U(s) \subset \mathcal{M}_U(t)$, $t, s \in \mathbb{R}$, $t \geq s$.*

(3) *Uniform exponential attraction:*

$$\operatorname{dist}_{H_1}(U(t+s,s)B, \mathcal{M}_U(t+s)) \leq C_2 e^{-\gamma t}, \tag{3.117}$$

*where $C_2$ and $\gamma > 0$ are independent of $t \in \mathbb{R}$ and $s \geq 0$.*

(4) *m Hölder continuity with respect to perturbations: For all dynamical processes $U_1$ and $U_2$ satisfying assumptions (1)–(3) of Theorem 3.10, we have*

$$\begin{aligned} &\operatorname{dist}_{H_1}^{symm}(\mathcal{M}_{U_1}(t), \mathcal{M}_{U_2}(t)) \\ &\leq C_3 \sup_{s\in(-\infty,t-1)} \left\{ e^{-\alpha s} \|U_1(s) - U_2(s)\|_{\mathbb{S}}^{\kappa} \right\} \\ &\quad + C_3 \sup_{\tau\in[0,1]} \|U_1(t-\tau,t) - U_2(t-\tau,t)\|_{\mathbb{S}}, \end{aligned} \tag{3.118}$$

*where $U_i(t) := U_i(t+1,t)$ and the positive constants $C_i$, $i = 1,2,3$, $\gamma$, $\alpha$, and $\kappa$ are independent of the concrete choice of the process $U$ satisfying assumptions (1)–(3) of the theorem.*

## 3.5. Pull-back exponential attractor

In this section we continue to deal with a time-dependent attractor, more precisely pull-back attractor, which is also a useful object for studying long-time dynamics of solutions of nonlinear nonautonomous PDE and stochastic equations (see [60]). In contrast to previous studies our construction of the exponential pull-back attractor allow us to consider the time-dependent bounded absorbing set. Indeed, let us consider the nonautonomous problem

$$\begin{cases} \dfrac{dU}{dt} = F(t,U), & s < t < \infty, \\ U(s) = U_s, & -\infty < s < \infty, \end{cases} \tag{3.119}$$

in a Banach space $X$. Assuming that (3.119) is well-posed for $U_s \in X$, we have the family of solving operators

$$U(t,s)U_s : U_s \mapsto U(t), \quad -\infty < s \leq t < \infty.$$

The family of operators has the properties

$$U(s,s) = I, \quad -\infty < s < \infty, \tag{3.120}$$

$$U(t,r) \circ U(r,s) = U(t,s), \quad -\infty < s \leq r \leq t < \infty. \tag{3.121}$$

It is then said that $U(t,s)$ forms an evolution operator, or a process on the phase space $X$. We especially emphasize that the theory of attractors for nonautonomous systems is less understood than that for autonomous systems. We have essentially three approaches.

The first one, initiated by A. Haraux (see [54]) and further studied and developed by V. V. Chepyzhov and M. I. Vishik (see, e.g., [20], [19]) is based on the notion of a uniform attractor. The major drawback of this approach is that it leads, for general (translation-compact; see [20]) time dependences, to an artificial infinite dimensionality of the uniform attractor. This can already be seen for the following simple linear equation:

$$\frac{\partial u}{\partial t} - \Delta u = h(t), \quad u\big|_{\partial\Omega} = 0,$$

in a bounded smooth domain $\Omega$, whose dynamics is simple; namely, one has one exponentially attracting trajectory. However, the uniform attractor has infinite dimension and infinite topological entropy (see [19]). However, for periodic and quasiperiodic time dependences, one has in general finite-dimensional uniform attractors (i.e., if the same is true for the corresponding autonomous system, see [18], [46]). Furthermore, one can derive sharp upper and lower bounds on the dimension of the uniform attractor, so that this approach is quite relevant in that case. We can note that, as in the autonomous case, an exponential attractor in this setting always contains the uniform attractor and, again, one has, for general time dependences, an artificial infinite dimensionality.

The second approach was developed in Section 3.3 and dealt with so-called time-dependent forward exponential attractors.

The third approach is based on the notion of a pull-back attractor (see, e.g., [15], [60] and the references therein). In that case, one has a time-dependent attractor $\{\mathcal{A}(t)\}_{t\in\mathbb{R}}$, contrary to the uniform attractor which is time-independent.

DEFINITION 3.6. A family $\{\mathcal{A}(t)\}_{t\in\mathbb{R}}$ is a pull-back attractor for the evolution operator $U(t,s)$ on $X$ if:

(i) Each $\mathcal{A}(t)$ is a compact set of $X$.
(ii) It is invariant, i.e., $U(t,s)\mathcal{A}(s) = \mathcal{A}(t)$ for all $-\infty < s \le t < \infty$.
(iii) It satisfies the following pull-back attraction property:

$$\forall B \subset X \text{ bounded}, \lim_{s\to+\infty} \operatorname{dist}_X(U(t,t-s)B, \mathcal{A}(t)) = 0.$$

One can prove that, in general, $\mathcal{A}(t)$ has finite dimension for every $t \in R$. Furthermore, if the system is autonomous, then one recovers the global attractor. Now, the attraction property essentially means that, at time $t$, the attractor $\mathcal{A}(t)$ attracts the bounded sets of initial data coming from the past (i.e., from $-\infty$). However, in (iii), the rate of attraction is not uniform in $t$, so that the forward convergence is not true in general (see, nevertheless, [14], [16] for cases where the forward convergence can be proven).

We want to introduce a version of the exponential attractor for nonautonomous equations as a time-dependent set satisfying certain natural assumptions. Our definition is stated as follows.

DEFINITION 3.7. A family $\{\mathcal{M}(t)\}_{t\in\mathbb{R}}$ is an exponential attractor for the evolution operator $U(t,s)$ on $X$ if:

(i) Each $\mathcal{M}(t)$ is a compact set of $X$ and its fractal dimension is finite and uniformly bounded; i.e., $\sup_{t\in\mathbb{R}} \dim \mathcal{M}(t) < \infty$.
(ii) It is positively invariant; i.e., $U(t,s)\mathcal{M}(s) \subset \mathcal{M}(t)$ for all $-\infty < s \le t < \infty$.

(iii) There exist an exponent $\alpha > 0$ and two monotonic functions $Q$ and $\tau$ such that

$$\forall B \subset X \text{ bounded, } \operatorname{dist}_X(U(t,s)B, \mathcal{M}(t)) \leq Q(\|B\|_X)e^{-\alpha(t-s)},$$
$$s \in \mathbb{R},\ s + \tau(\|B\|_X) \leq t < \infty.$$

The first purpose is then to show the construction of exponential attractors for nonautonomous systems. To this end, we will assume the existence of a family of bounded sets $\mathcal{X}(t)$, $t \in \mathbb{R}$, which is positively invariant and absorbs all bounded sets, and we will generalize (3.6)–(3.7) into the form

$$\|U(\tau^* + s, s)U_0 - U(\tau^* + s, s)V_0\|_Z \leq c\|U_0 - V_0\|_X, \quad U_0,\ V_0 \in \mathcal{X}(s),$$
$$\text{for all } s \in \mathbb{R}, \tag{3.122}$$

where $\tau^* > 0$ is some fixed constant. (Actually, our assumption will be of the more general form; see (3.123) and (3.124).) This condition together with some minor ones in fact enables us to generalize the method of construction for autonomous systems (due to [35]) to nonautonomous. Our exponential attractor $\mathcal{M}(t)$ then depends on $t$ continuously if $t \neq n\tau^*$, $n \in \mathbb{Z}$, and is right continuous at $t = n\tau^*$, $n \in \mathbb{Z}$. Left discontinuity of $\mathcal{M}(t)$ at time $n\tau^*$ comes completely from a technical reason. We notice in applications that (3.122) is actually verified for any $\tau^*$ contained in some interval $(\tau_0, \tau_1)$, where $0 < \tau_0 < \tau_1$, which means that, even if $\mathcal{M}(t)$ is left discontinuous at $n\tau^*$, it is possible to choose another $\overline{\tau}^*$ in order to construct another exponential attractor $\overline{\mathcal{M}}(t)$ which is now continuous at $n\tau^*$.

The second purpose is to apply this construction to a nonautonomous chemotaxis system. For autonomous chemotaxis systems, we have already constructed exponential attractors in the papers [3], [83] (cf. also Chapter 12 of [95]). In [4] we estimated their fractal dimensions from below and showed that if the chemotaxis parameter becomes large, then the fractal dimensions also increase and finally tend to infinity. Meanwhile, in [45] we proved that the exponential attractor can depend continuously on the chemotaxis parameter. In this book, we will consider a time-dependent sensitivity function. Under reasonable assumptions on the function, our general result will be applied to constructing exponential attractors as before. Our result seems to be in good agreement with the former ones in the sense that the dimension of $\mathcal{M}(t)$ is uniformly bounded and is continuous with respect to the variable $t$.

**Construction of exponential attractors**

Let $X$ be a Banach space with norm $\|\cdot\|_X$. Let $\mathcal{K}$ be a subset of $X$ which is a metric space equipped with the distance $d(U,V) = \|U - V\|_X$. We consider a family of nonlinear operators $U(t,s)$ acting on $\mathcal{K}$ defined for

$$(t,s) \in \Delta = \{(t,s);\ -\infty < s \leq t < \infty\}.$$

We assume that $U(t,s)$ has the properties (3.120) and (3.121) on $\mathcal{K}$. A family of $U(t,s)$ having these properties is called an evolution operator, or a process on the space $\mathcal{K}$. We assume also that $U(t,s)$ is continuous in the sense that

$$\text{the mapping } \quad G\colon \Delta \times \mathcal{K} \to \mathcal{K},\ ((t,s), U_0) \mapsto U(t,s)U_0 \quad \text{is continuous.}$$

Such an evolution operator is said simply to be continuous on $\mathcal{K}$. When $U(t,s)$ is a continuous evolution operator on $\mathcal{K}$, the triplet $(U(t,s), \mathcal{K}, X)$ is called a nonautonomous dynamical system, and $\mathcal{K}$ and $X$ are called the phase space and the universal space, respectively. The trace of a function $U(\cdot, s)U_0$ for $t \in [s, \infty)$ in the space $\mathcal{K}$ is called a trajectory starting from $U_0 \in \mathcal{K}$ at initial time $s \in \mathbb{R}$.

We now restate the definition of exponential attractors. (Note that in Definition 3.7, $\mathcal{K}$ coincides with $X$.)

DEFINITION 3.8. A family $\{\mathcal{M}(t)\}_{t\in\mathbb{R}}$ of subsets of $\mathcal{K}$ is called an exponential attractor for $(U(t,s), \mathcal{K}, X)$ if:

(i) Each $\mathcal{M}(t)$ is a compact set of $X$ and its fractal dimension is finite and uniformly bounded; i.e., $\sup_{t\in\mathbb{R}} \dim \mathcal{M}(t) < \infty$.
(ii) It is positively invariant; i.e., $U(t,s)\mathcal{M}(s) \subset \mathcal{M}(t)$ for all $(t,s) \in \Delta$.
(iii) There exist an exponent $\alpha > 0$ and two monotonic functions $Q$ and $\tau$ such that

$$\forall B \subset \mathcal{K} \text{ bounded, } \operatorname{dist}_X(U(t,s)B, \mathcal{M}(t)) \le Q(\|B\|_X)e^{-\alpha(t-s)},$$
$$s \in \mathbb{R},\ s + \tau(\|B\|_X) \le t < \infty.$$

In order to construct exponential attractors, we have to assume the existence of a family $\{\mathcal{X}(t)\}_{t\in\mathbb{R}}$ of bounded closed subsets of $\mathcal{K}$ with the following properties:

(1) The diameter $\|\mathcal{X}(t)\|_X$ of $\mathcal{X}(t)$ is uniformly bounded; i.e., $\sup_{t\in\mathbb{R}} \|\mathcal{X}(t)\|_X = R < \infty$.
(2) It is positively invariant; i.e., $U(t,s)\mathcal{X}(s) \subset \mathcal{X}(t)$ for all $(t,s) \in \Delta$.
(3) It is absorbing in the sense that there is a monotonic function $\sigma$ such that

$$\forall B \subset \mathcal{K} \text{ bounded, } U(t,s)B \subset \mathcal{X}(t), \quad s \in \mathbb{R},\ s + \sigma(\|B\|_X) \le t < \infty.$$

(4) There is $\tau^* > 0$ such that, for every $s \in \mathbb{R}$, $U(\tau^* + s, s)$ is a compact perturbation of contraction on $\mathcal{X}(s)$ in the sense that

$$\begin{aligned} &\|U(\tau^*+s,s)U_0 - U(\tau^*+s,s)V_0\|_X \\ &\quad \le \delta\|U_0 - V_0\|_X + \|K(s)U_0 - K(s)V_0\|_X, \quad U_0, V_0 \in \mathcal{X}(s), \end{aligned} \tag{3.123}$$

where $\delta$ is a constant such that $0 \le \delta < \frac{1}{2}$ and where $K(s)$ is an operator from $\mathcal{X}(s)$ into another Banach space $Z$ which is embedded compactly in $X$ and satisfies a Lipschitz condition

$$\|K(s)U_0 - K(s)V_0\|_Z \le L_1\|U_0 - V_0\|_X, \quad U_0, V_0 \in \mathcal{X}(s), \tag{3.124}$$

with some constant $L_1 > 0$ independent of $s$.
(5) For any $s \in \mathbb{R}$ and any $\tau \in [0, \tau^*]$,

$$\|U(\tau+s,s)U_0 - U(\tau+s,s)V_0\|_X \le L_2\|U_0 - V_0\|_X, \quad U_0, V_0 \in \mathcal{X}(s), \tag{3.125}$$

with some constant $L_2 > 0$ independent of $s$ and $\tau$.

THEOREM 3.11. *Let $(U(t,s), \mathcal{K}, X)$ be a nonautonomous dynamical system in $X$. Assume that the conditions (1)–(5) are satisfied. Then, one can construct an exponential attractor $\{\mathcal{M}(t)\}_{t\in\mathbb{R}}$ for $(U(t,s), \mathcal{K}, X)$.*

PROOF. For $m, n \in \mathbb{Z}$ with $m \le n$, let $U^*(n,m) = U(n\tau^*, m\tau^*)$. Let us first consider a discrete nonautonomous dynamical system $(U^*(n,m), \mathcal{K}, X)$. In the first three steps of the proof, we will construct an exponential attractor for the discrete system $(U^*(n,m), \mathcal{K}, X)$.

For $n \in \mathbb{Z}$, let $\mathcal{X}^*(n) = \mathcal{X}(n\tau^*)$. A family $\mathcal{M}^*(n)$, $n \in \mathbb{Z}$, of compact sets of $X$ is called an exponential attractor for $(U^*(n,m), \mathcal{K}, X)$ if $\mathcal{M}(n)$ satisfies:

(1) The fractal dimension of $\mathcal{M}^*(n)$ is uniformly bounded for $n$; i.e.,

$$\sup_{n\in\mathbb{Z}} \dim \mathcal{M}^*(n) \leq c_1^*.$$

(2) $U^*(n,m)\mathcal{M}^*(m) \subset \mathcal{M}^*(n)$ for all $-\infty < m \leq n < \infty$.

(3) For some $0 < a < 1$, it holds true that

$$\mathrm{dist}_X(U^*(n,m)\mathcal{X}^*(m), \mathcal{M}^*(n)) \leq Ra^{n-m}, \quad -\infty < m \leq n < \infty.$$

The last step will be devoted to handling the continuous case.

*Step 1.* Let $\theta$ be any number such that $0 < \theta < \frac{1-2\delta}{2L_1}$ and let $a_\theta = 2(\delta + \theta L_1)$. Clearly, $0 < a_\theta < 1$. The purpose of this step is to construct, for any $-\infty < m \leq n < \infty$, a covering of $U^*(n,m)\mathcal{X}^*(m)$ by $N_\theta^{n-m}$-closed balls of $X$ with uniform radius $Ra_\theta^{n-m}$ centered at points in $U^*(n,m)\mathcal{X}^*(m)$, where $N_\theta$ is a minimal number of closed balls of $X$ with radius $\theta$ which cover the closed unit ball $\overline{B}^Z(0;1)$ of $Z$ centered at 0. That is, for $-\infty < m \leq n < \infty$,

$$U^*(n,m)\mathcal{X}^*(m) \subset \bigcup_{i=1}^{N_\theta^{n-m}} \overline{B}(W_{n,m,i}; Ra^{n-m}), \quad \text{where } a = a_\theta, \tag{3.126}$$

with $W_{n,m,i} \in U^*(n,m)\mathcal{X}^*(m)$, $1 \leq i \leq N_\theta^{n-m}$.

Let us construct the covering (3.126) by induction on $n$ ($m$ being fixed). If $n = m$, then we can take $W_{m,m,1} \in \mathcal{X}^*(m)$ arbitrarily. (Remember condition (1) for $\mathcal{X}(t)$.) Assume that we have the covering (3.126) for $n \geq m$. Then,

$$\begin{aligned} U^*(n+1,m)\mathcal{X}^*(m) &= U^*(n+1,n)U^*(n,m)\mathcal{X}^*(m) \\ &\subset \bigcup_{i=1}^{N_\theta^{n-1-m}} U^*(n+1,n)\Big(\overline{B}(W_{n,m,i};\, Ra^{n-m}) \cap U^*(n,m)\mathcal{X}^*(m)\Big). \end{aligned}$$

So, it suffices to cover each set

$$U^*(n+1,n)\big(\overline{B}(W_{n,m,i}; Ra^{n-m}) \cap U^*(n,m)\mathcal{X}^*(m)\big)$$

by $N_\theta$-closed balls with the radius $Ra^{n+1-m}$ centered at $U^*(n+1,m)\mathcal{X}^*(m)$. Using (3.124) with $s = n\tau^*$, we see that

$$K(n\tau^*)\Big(\overline{B}(W_{n,m,i};\, Ra^{n-m}) \cap U^*(n,m)\mathcal{X}^*(m)\Big) \subset \overline{B}^Z(K(n\tau^*)W_{n,m,i};\, L_1 Ra^{n-m}).$$

Then, by the compactness of closed bounded balls of $Z$ in $X$, the last ball can be covered by $N_\theta$-closed balls of $X$ in such a way that

$$\overline{B}^Z(K(n\tau^*)W_{n,m,i};\, L_1Ra^{n-m}) \subset \bigcup_{j=1}^{N_\theta} \overline{B}(\widetilde{V}_{n,m,i,j};\, \theta L_1 Ra^{n-m})$$

with centers $\widetilde{V}_{n,m,i,j} \in X$, $1 \leq j \leq N_\theta$, and radius $\theta L_1 Ra^{n-m}$. Therefore,

$$K(n\tau^*)\Big(\overline{B}(W_{n,m,i};\, Ra^{n-m}) \cap U^*(n,m)\mathcal{X}^*(m)\Big) \subset \bigcup_{j=1}^{N_\theta} \overline{B}(\widetilde{V}_{n,m,i,j};\, \theta L_1 Ra^{n-m}). \tag{3.127}$$

We are here allowed to assume that

$$K(n\tau^*)\Big(\overline{B}(W_{n,m,i};\ Ra^{n-m}) \cap U^*(n,m)\mathcal{X}^*(m)\Big) \cap \overline{B}(\widetilde{V}_{n,m,i,j};\ \theta L_1 Ra^{n-m}) \neq \emptyset$$

for every $j$, since, if not for some $j$, we can exclude these balls from the covering. So, we can choose for each $j$ a point $V_{n,m,i,j}$ such that

$$\begin{aligned} V_{n,m,i,j} &\in \overline{B}(W_{n,m,i};\ Ra^{n-m}) \cap U^*(n,m)\mathcal{X}^*(m), \\ K(n\tau^*)V_{n,m,i,j} &\in \overline{B}(\widetilde{V}_{n,m,i,j};\ \theta L_1 Ra^{n-m}). \end{aligned} \tag{3.128}$$

Therefore, from (3.127) we deduce that

$$\begin{aligned} &K(n\tau^*)\Big(\overline{B}(W_{n,m,i};\ Ra^{n-m}) \cap U^*(n,m)\mathcal{X}^*(m)\Big) \\ &\subset \bigcup_{j=1}^{N_\theta} \overline{B}(K(n\tau^*)V_{n,m,i,j};\ 2\theta L_1 Ra^{n-m}). \end{aligned}$$

Now let $U \in \overline{B}(W_{n,m,i};\ Ra^{n-m}) \cap U^*(n,m)\mathcal{X}^*(m)$. Then, there is some $j$ such that $K(n\tau^*)U \in \overline{B}(K(n\tau^*)V_{n,m,i,j}; 2\theta L_1 Ra^{n-m})$. As a consequence, it follows from (3.123) that

$$\begin{aligned} \|U^*(n+1,n)U - U^*(n+1,n)V_{n,m,i,j}\|_X &\le \delta\|U - V_{n,m,i,j}\|_X \\ &\quad + \|K(n\tau^*)U - K(n\tau^*)V_{n,m,i,j}\|_X \\ &\le \delta\|U - V_{n,m,i,j}\|_X + 2\theta L_1 Ra^{n-m}. \end{aligned}$$

In addition, by (3.128),

$$\|U - V_{n,m,i,j}\|_X \le \|U - W_{n,m,i}\|_X + \|W_{n,m,i,j} - V_{n,m,i,j}\|_X \le 2Ra^{n-m},$$

so that $\|U^*(n+1,n)U - U^*(n+1,n)V_{n,m,i,j}\|_X \le 2(\delta + \theta L_1)Ra^{n-m} = Ra^{n+1-m}$. Hence, it holds that

$$\begin{aligned} &U^*(n+1,n)\Big(\overline{B}(W_{n,m,i};\ Ra^{n-m}) \cap U^*(n,m)\mathcal{X}^*(m)\Big) \\ &\subset \bigcup_{j=1}^{N_\theta} \overline{B}(U^*(n+1,n)V_{n,m,i,j};\ Ra^{n+1-m}). \end{aligned} \tag{3.129}$$

We observe from (3.128) that $U^*(n+1,n)V_{n,m,i,j} \in U^*(n+1,m)\mathcal{X}^*(m)$.

A covering of the form (3.129) can, of course, be constructed for all other balls. Therefore, the desired covering (3.126) for $n+1$ is obtained by locating central points as

$$\begin{aligned} \{W_{n+1,m,i};\ 1 \le i \le N_\theta^{n+1-m}\} &= \{U^*(n+1,n)V_{n,m,i,j};\ 1 \le i \le N_\theta^{n-m},\ 1 \le j \le N_\theta\} \\ &\subset U^*(n+1,m)\mathcal{X}^*(m). \end{aligned}$$

*Step 2.* For $-\infty < m \le n < \infty$, we let

$$E_m(n) = \{U^*(n,m+k)W_{m+k,m,i_k};\ 0 \le k \le n-m,\ 1 \le i_k \le N_\theta^k\}.$$

It is clear by definition that

$$E_m(n) \subset U^*(n,m)\mathcal{X}^*(m) \subset \mathcal{X}^*(n).$$

In addition, for $n \le p < \infty$,

$$U^*(p,n)E_m(n) \subset E_m(p). \tag{3.130}$$

We then set, for each $-\infty < n < \infty$,

$$\mathcal{M}^*(n) = \overline{\bigcup_{m=-\infty}^{n} E_m(n)}. \tag{3.131}$$

This family $\mathcal{M}^*(n)$, $-\infty < n < \infty$, will indeed give an exponential attractor for $(U^*(n,m), \mathcal{K}, X)$.

Let us estimate in this step the fractal dimension of $\mathcal{M}^*(n)$. Let $n$ be fixed and let $0 < \varepsilon < 1$ be any number. Let $m_\varepsilon$ $(\le n)$ be the largest integer such that $Ra^{n-m} \le \varepsilon$, i.e., $m_\varepsilon \le \frac{\log(R^{-1}a^{-n}\varepsilon)}{-\log a}$. For all $m$ such that $-\infty < m \le m_\varepsilon$, we have

$$E_m(n) \subset U^*(n,m)\mathcal{X}^*(m) \subset U^*(n,m_\varepsilon)U^*(m_\varepsilon,m)\mathcal{X}^*(m) \subset U^*(n,m_\varepsilon)\mathcal{X}^*(m_\varepsilon).$$

Therefore, by (3.126), we deduce that the set $\overline{\bigcup_{m=-\infty}^{m_\varepsilon} E_m(n)}$ is covered by $N_\theta^{n-m_\varepsilon}$-closed balls with radius $\varepsilon$. Meanwhile, for each $m_\varepsilon < m \le n$, $E_m(n)$ is a finite set. Hence, $\mathcal{M}^*(n)$ is a precompact set of $X$ and actually is a compact set of $X$. Denote by $N(\varepsilon)$ the minimal number of balls with radius $\varepsilon$ which can cover $\mathcal{M}^*(n)$. Then,

$$\begin{aligned} N(\varepsilon) &\le N_\theta^{n-m_\varepsilon} + \sum_{m=m_\varepsilon+1}^{n} \#E_m(n) \\ &= N_\theta^{n-m_\varepsilon} + \sum_{m=m_\varepsilon+1}^{n} \sum_{k=0}^{n-m} N_\theta^k \le (n-m_\varepsilon)N_\theta^{n-m_\varepsilon}. \end{aligned}$$

Since $\varepsilon < Ra^{n-m_\varepsilon-1}$, it follows that

$$\frac{\log N(\varepsilon)}{-\log\varepsilon} \le \frac{(n-m_\varepsilon)\log N_\theta + \log(n-m_\varepsilon)}{-(n-m_\varepsilon-1)\log a - \log R}.$$

Letting $\varepsilon \to 0$, we conclude that $\dim \mathcal{M}^*(n) \le \frac{\log N_\theta}{-\log a}$.

*Step 3.* It is seen by (3.130) that

$$\begin{aligned} U^*(p,n)\mathcal{M}^*(n) &= U^*(p,n)\overline{\bigcup_{m=-\infty}^{n} E_m(n)} \\ &\subset \overline{U^*(p,n)\bigcup_{m=-\infty}^{n} E_m(n)} \\ &\subset \overline{\bigcup_{m=-\infty}^{n} E_m(p)} \subset \mathcal{M}^*(p). \end{aligned} \tag{3.132}$$

Meanwhile, it is seen by (3.126) that $\mathrm{dist}_X(U^*(n,m)\mathcal{X}^*(m), \mathcal{M}^*(n)) \le Ra^{n-m}$ since $W_{n,m,i} \in \mathcal{M}^*(n)$ for $1 \le i \le N_\theta^{n-m}$.

We have thus verified that $\mathcal{M}^*(n)$ is an exponential attractor for $(U^*(n,m), \mathcal{K}, X)$.

*Step 4.* Let us now consider the continuous dynamical system $(U(t,s), \mathcal{K}, X)$. For $-\infty < t < \infty$, let $n$ be the integer such that $n\tau^* \le t < (n+1)\tau^*$. We then set

$$\mathcal{M}(t) = U(t, n\tau^*)\mathcal{M}^*(n), \quad n\tau^* \le t < (n+1)\tau^*.$$

Since $U(t, n\tau^*)$ is a continuous mapping from $\mathcal{K}$ into $X$, the image $\mathcal{M}(t)$ of a compact set $\mathcal{M}^*(n)$ by $U(t, n\tau^*)$ is also a compact set of $X$. Similarly, since $U(t, n\tau^*)$

is Lipschitz continuous due to (3.125), the fractal dimension of $\mathcal{M}(t)$ is finite and does not exceed dim $\mathcal{M}^*(n)$; namely, dim $\mathcal{M}(t) \leq \frac{\log N_\theta}{-\log a}$ for any $t$.

For $-\infty < s < t < \infty$, let $m\tau^* \leq s < (m+1)\tau^*$ and $n\tau^* \leq t < (n+1)\tau^*$ with integers $m \leq n$. Then, by (3.132),

$$\begin{aligned} U(t,s)\mathcal{M}(s) &= U(t,n\tau^*)U(n\tau^*,s)U(s,m\tau^*)\mathcal{M}^*(m) \\ &= U(t,n\tau^*)U^*(n,m)\mathcal{M}^*(m) \\ &\subset U(t,n\tau^*)\mathcal{M}^*(n) \\ &= \mathcal{M}(t). \end{aligned}$$

Let $U_s \in \mathcal{X}(s)$. We write

$$\begin{aligned} &d(U(t,s)U_s, \mathcal{M}(t)) \\ &\quad = d(U(t,n\tau^*)U(n\tau^*,(m+1)\tau^*)U((m+1)\tau^*,s)U_s, U(t,n\tau^*)\mathcal{M}^*(n)). \end{aligned}$$

Noting that $U_{m+1} = U((m+1)\tau^*,s)U_s \in \mathcal{X}^*(m+1)$, we obtain that

$$d(U(t,s)U_s,\mathcal{M}(t)) \leq L_2 d(U^*(n,m+1)U_{m+1},\mathcal{M}^*(n)) \leq L_2 R a^{n-m-1}.$$

Hence it holds true that

$$\mathrm{dist}_X(U(t,s)\mathcal{X}(s),\mathcal{M}(t)) \leq L_2 R a^{-2} e^{-\alpha(t-s)}, \quad -\infty < s \leq t < \infty,$$

with $\alpha = \frac{-\log a}{\tau^*}$.

We have thus verified that the family of sets $\mathcal{M}(t)$, $-\infty < t < \infty$, enjoys the desired properties. □

**Continuous dependence of $\mathcal{M}(t)$ in $t$**

We are concerned with the continuity of $\mathcal{M}(t)$ with respect to the variable $t$. We make the following assumptions. For each fixed $-\infty < t < \infty$,

$$\lim_{t' \searrow t} \sup_{U_t \in \mathcal{X}(t)} \|[U(t',t) - 1]U_t\|_X = 0. \tag{3.133}$$

For each fixed $-\infty < t < \infty$,

$$\lim_{t' \nearrow t} \sup_{U_{t'} \in \mathcal{X}(t')} \|[U(t,t') - 1]U_{t'}\|_X = 0. \tag{3.134}$$

THEOREM 3.12. *Let $\{\mathcal{M}(t)\}_{t\in\mathbb{R}}$ be the exponential attractor constructed in Theorem 3.11. Let $U(t,s)$ satisfy (3.133) and (3.134), too. Then, $\mathcal{M}(t)$ is right continuous at any $t \in \mathbb{R}$; i.e., $\lim_{t'\searrow t} d(\mathcal{M}(t'),\mathcal{M}(t)) = 0$. If $t \neq n\tau^*$ for any $n \in \mathbb{Z}$, then $\mathcal{M}(t)$ is left continuous, too, i.e., $\lim_{t'\nearrow t} d(\mathcal{M}(t'),\mathcal{M}(t)) = 0$. If $t = n\tau^*$ with some $n \in \mathbb{Z}$, then $\mathcal{M}(t)$ is at least left outer continuous, i.e., $\lim_{t'\nearrow t} \mathrm{dist}_X(\mathcal{M}(t'),\mathcal{M}(t)) = 0$.*

PROOF. Let $n\tau^* \leq t < t' < (n+1)\tau^*$. Then, $\mathcal{M}(t') = U(t',t)\mathcal{M}(t)$. For any $U_{t'} \in \mathcal{M}(t')$, there is a point $U_t \in \mathcal{M}(t)$ such that $U_{t'} = U(t',t)U_t$. Therefore,

$$d(U_{t'},\mathcal{M}(t)) \leq d(U(t',t)U_t, U_t) \leq \sup_{U_t\in\mathcal{M}(t)} \|[U(t',t)-1]U_t\|_X.$$

Consequently,

$$d(\mathcal{M}(t'),\mathcal{M}(t)) \leq \sup_{U_t\in\mathcal{X}(t)} \|[U(t',t)-1]U_t\|_X.$$

In the meantime, let $U_t \in \mathcal{M}(t)$. Then,

$$d(U_t, \mathcal{M}(t')) \le d(U_t, U(t',t)U_t) \le \sup_{U_t \in \mathcal{M}(t)} \|[U(t',t) - 1]U_t\|_X.$$

Consequently,

$$d(\mathcal{M}(t), \mathcal{M}(t')) \le \sup_{U_t \in \mathcal{X}(t)} \|[U(t',t) - 1]U_t\|_X.$$

Therefore, (3.133) implies $\lim_{t' \to t} d(\mathcal{M}(t'), \mathcal{M}(t)) = 0$.

Let $t \neq n\tau^*$ for any $n \in \mathbb{Z}$. Indeed, let $n\tau^* < t' < t < (n+1)\tau^*$. Then, we have $\mathcal{M}(t) = U(t,t')\mathcal{M}(t')$. By the same arguments as above, we can conclude from (3.134) that $\lim_{t' \nearrow t} d(\mathcal{M}(t'), \mathcal{M}(t)) = 0$.

Let $t = n\tau^*$ with $n \in \mathbb{Z}$. Let $(n-1)\tau^* < t' < t = n\tau^*$. Since $U(t,t')\mathcal{M}(t') \subset \mathcal{M}(t)$, we deduce from (3.134) that $\lim_{t' \nearrow t} \mathrm{dist}_X(\mathcal{M}(t'), \mathcal{M}(t)) = 0$. $\square$

## 3.6. Nonautonomous chemotaxis system

In this section we apply the result of Section 3.5 to a class of nonautonomous chemotaxis growth systems.

Indeed, consider the initial boundary value problem for nonautonomous chemotaxis growth equations

$$\begin{cases} \dfrac{\partial u}{\partial t} = a\Delta u - \nabla \cdot [u \nabla \chi(t,\rho)] + f(t,u) & \text{in} \quad \Omega \times (s,\infty), \\ \dfrac{\partial \rho}{\partial t} = b\Delta\rho - c\rho + \nu u & \text{in} \quad \Omega \times (s,\infty), \\ \dfrac{\partial u}{\partial n} = \dfrac{\partial \rho}{\partial n} = 0 & \text{on} \quad \partial\Omega \times (s,\infty), \\ u(x,s) = u_s(x), \;\; \rho(x,s) = \rho_s(x) & \text{in} \quad \Omega, \end{cases} \tag{3.135}$$

in a bounded domain $\Omega \subset \mathbb{R}^2$ with initial time $s \in \mathbb{R}$.

We assume that $\Omega$ is a two-dimensional bounded domain with sufficiently smooth boundary $\partial\Omega$, say of $\mathcal{C}^4$ class. For each $t$, the sensitivity function $\chi(t,\rho)$ is a $\mathcal{C}^3$ function for $0 \le \rho < \infty$ satisfying

$$\left| \frac{\partial^i \chi}{\partial \rho^i}(t,\rho) \right| \le C_1, \quad -\infty < t < \infty, \; 0 \le \rho < \infty, \; i = 1,2,3, \tag{3.136}$$

with some constant $C_1 > 0$. The partial derivatives also satisfy uniform Lipschitz conditions

$$\left| \frac{\partial^i \chi}{\partial \rho^i}(s,\rho) - \frac{\partial^i \chi}{\partial \rho^i}(t,\rho) \right| \le C_2|t-s|, \quad -\infty < s,\, t < \infty, \, 0 \le \rho < \infty, \, i = 1,2,3, \tag{3.137}$$

with some constant $C_2 > 0$. The growth function $f(t,u)$ is a continuous function for $(t,u) \in \mathbb{R} \times \mathbb{R}_+$ satisfying

$$c_1 u - c_2 u^2 \le f(t,u) \le c_3 u - c_4 u^2 \tag{3.138}$$

with some positive constants $c_i > 0\,(i = 1, 2, 3, 4)$. We also assume a Lipschitz condition of the form

$$|f(s,u) - f(t,v)| \le C_3(u+v+1) \times [(u+v+1)|t-s| + |u-v|],$$
$$-\infty < s,\, t < \infty,\, 0 \le u,\, v < \infty, \tag{3.139}$$

with some constant $C_3 > 0$.

We will treat this problem in the product space

$$X = \left\{ U = \begin{pmatrix} u \\ \rho \end{pmatrix};\ u \in L^2(\Omega) \quad \text{and} \quad \rho \in H^2_N(\Omega) \right\}. \tag{3.140}$$

As for the space of initial functions, we set

$$\mathcal{K} = \left\{ U_s = \begin{pmatrix} u_s \\ \rho_s \end{pmatrix};\ 0 \le u_s \in L^2(\Omega) \quad \text{and} \quad 0 \le \rho_s \in H^2_N(\Omega) \right\}. \tag{3.141}$$

**Local solutions**

We want to appeal to the theory of nonlinear abstract parabolic evolution equations (see [95]). Problem (3.135) is formulated as the Cauchy problem for a nonautonomous semilinear evolution equation

$$\begin{cases} \dfrac{dU}{dt} + AU = F(t,U), \quad s < t < \infty, \\ U(s) = U_s, \end{cases} \tag{3.142}$$

in the product space $X$ given by (3.140). Here, $A$ is a matrix linear operator of $X$ given by

$$A = \begin{pmatrix} A_1 & 0 \\ -\nu & A_2 \end{pmatrix},$$

where $A_1$ (resp. $A_2$) is a realization of the elliptic operator $-a\Delta + 1$ (resp. $-b\Delta + c$) in $L^2(\Omega)$ under the Neumann boundary conditions on $\partial\Omega$ and is a positive definite selfadjoint operator of $L^2(\Omega)$ with domain $\mathcal{D}(A_1) = \mathcal{D}(A_2) = H^2_N(\Omega)$. But, since the underlying space for the equation of $\rho$ is the space $H^2_N(\Omega)$ (see (3.140)), $A_2$ is actually an operator from $\mathcal{D}(A_2^2)$ into $\mathcal{D}(A_2)$. The nonlinear operator $F(t,U)$ is given by

$$F(t,U) = \begin{pmatrix} -\nabla \cdot [u\nabla\chi(t, \operatorname{Re}\rho)] + f(t, \operatorname{Re} u) + u \\ 0 \end{pmatrix}, \quad U = \begin{pmatrix} u \\ \rho \end{pmatrix}.$$

The initial value is given by $U_s = {}^t(u_s, \rho_s) \in \mathcal{K}$.

We shall use the standard techniques of reducing the nonautonomous problems to autonomous ones by introducing a new unknown function $\tau = \tau(t)$. Namely, we rewrite (3.142) in the form

$$\begin{cases} \dfrac{d}{dt}\begin{pmatrix} \tau \\ U \end{pmatrix} + \begin{pmatrix} 1 & 0 \\ 0 & A \end{pmatrix}\begin{pmatrix} \tau \\ U \end{pmatrix} = \begin{pmatrix} \tau + 1 \\ F(\operatorname{Re}\tau, U) \end{pmatrix}, \quad s < t < \infty, \\ \begin{pmatrix} \tau \\ U \end{pmatrix}(s) = \begin{pmatrix} s \\ U_s \end{pmatrix}, \end{cases}$$

in the product space $\mathbb{X}$ of $\mathbb{C}$ and $X$. Then, we have the Cauchy problem of the form

$$\begin{cases} \dfrac{d\widetilde{U}}{dt} + \widetilde{A}\widetilde{U} = \widetilde{F}(\widetilde{U}), \quad s < t < \infty, \\ \widetilde{U}(s) = \widetilde{U}_s. \end{cases} \tag{3.143}$$

Here, $\widetilde{U} = {}^t(\tau, U) \in \mathbb{X}$ and $\widetilde{A} = \operatorname{diag}\{1, A\}$ is a matrix operator of $\mathbb{X}$. The nonlinear operator $\widetilde{F}$ is defined by

$$\widetilde{F}(\widetilde{U}) = \begin{pmatrix} \tau + 1 \\ F(\operatorname{Re}\tau, U) \end{pmatrix}, \quad \widetilde{U} = \begin{pmatrix} \tau \\ U \end{pmatrix}. \tag{3.144}$$

Finally, the initial value is given by $\widetilde{U}_s = {}^t(s, U_s) \in \mathbb{R} \times \mathcal{K}$.

Let us now apply the general results for semilinear abstract parabolic evolution equations, say Theorem 4.4 of [95], in order to construct local solutions to (3.143). In fact, we can verify that $\widetilde{F}(U)$ defined by (3.144) fulfills condition (4.21) of [95] with $\frac{1}{2} < \eta < 1$ by the analogous arguments in Section 6 of [3] or Section 12.1.2 of [95]. As a result, we conclude that, for any initial value $\widetilde{U}_s \in \mathbb{R} \times \mathcal{K}$, (3.143) has a unique local solution in the function space:

$$\widetilde{U} \in \mathcal{C}((s, s + T_{U_s}]; \mathcal{D}(\widetilde{A})) \cap \mathcal{C}([s, s + T_{U_s}]; \mathbb{X}) \cap \mathcal{C}^1((s, s + T_{U_s}]; \mathbb{X});$$

here $T_{U_s} > 0$ is determined by the norm $\|U_s\|_X$ alone. In addition,

$$(t - s)\|\widetilde{A}\widetilde{U}(t)\|_{\mathbb{X}} + \|\widetilde{U}(t)\|_{\mathbb{X}} \le C_{U_s}, \quad s < t \le s + T_{U_s}. \tag{3.145}$$

By definition, the local solution to (3.143) and hence to the original problem (3.142) (equally, (3.135)) is given by

$$U(t) = \operatorname{pr}_2 \widetilde{U}(t), \quad s < t \le s + T_{U_s}, \tag{3.146}$$

where $\operatorname{pr}_2 : \mathbb{X} \to X$ is the projection from $\mathbb{X}$ onto $X$. Inequality (3.145) then yields that

$$(t - s)\|AU(t)\|_X + \|U(t)\|_X \le C_{U_s}, \quad s < t \le s + T_{U_s}. \tag{3.147}$$

It is easy to verify that $u_s \ge 0$ and $\rho_s \ge 0$ imply that the local solution to (3.142) also satisfies $u(t) \ge 0$ and $\rho(t) \ge 0$ for every $s < t \le s + T_{U_s}$; see Theorem 3.5 of [83] or Section 12.1.3 of [95].

Let $0 < R < \infty$. Let $\mathcal{K}_R = \mathcal{K} \cap \overline{B}^X(0; R)$, where $\overline{B}^X(0; R)$ denotes the closed ball of $X$ centered at 0 with radius $R$. For each $U_s \in \mathcal{K}_R$, (3.143) has a unique local solution on an interval $[s, s + T_R]$, where $T_R > 0$ is determined by $R$ alone. We can then verify the Lipschitz continuity of the local solutions with respect to the initial data. Thanks to Theorem 4.5 of [95], we have

$$\begin{aligned} &(t - s)^\eta \|A^\eta[U_1(t) - U_2(t)]\|_X + \|U_1(t) - U_2(t)\|_X \\ &\qquad \le C_R \|U_s^1 - U_s^2\|_X, \quad s < t \le s + T_R, \end{aligned} \tag{3.148}$$

where $U_1(t)$ (resp. $U_2(t)$) is a local solution to (3.142) for initial function $U_s^1 \in \mathcal{K}_R$ (resp. $U_s^2 \in \mathcal{K}_R$).

**Global solutions**

We consider problem (3.142). For any $U_s \in \mathcal{K}$, we have already constructed a local solution on an interval $[s, T_{U_s}]$. Let $U = {}^t(u, \rho)$ be any extension of this local solution in the function space:

$$0 \leq u \in \mathcal{C}((s, s+T_U]; H^2_N(\Omega)) \cap \mathcal{C}([s, s+T_U]; L^2(\Omega)) \cap \mathcal{C}^1((s, s+T_U]; L^2(\Omega)),$$
$$0 \leq \rho \in \mathcal{C}((s, s+T_U]; H^4_{N^2}(\Omega)) \cap \mathcal{C}([s, s+T_U]; H^2_N(\Omega)) \cap \mathcal{C}^1((s, s+T_U]; H^2_N(\Omega)),$$

$U$ being defined on $[s, s+T_U]$. Then, repeating the similar arguments in Section 4 of [83] or Section 12.3.2 of [95], we can establish a priori estimates

$$\|U(t)\|_X \leq p(\|U_s\|_X); \quad s \leq t \leq s+T_U; \tag{3.149}$$

here $p(\cdot)$ denotes some specific continuous increasing function which is independent of $U(\cdot)$.

This a priori estimate shows that the local solution on $[s, s+T_{U_s}]$ mentioned above can be extended on an interval $[s, s+T_{U_s}+\tau]$, $\tau > 0$ being dependent only on $p(\|U_s\|_X)$ and independent of $s+T_{U_s}$. We will repeat such a procedure. In each step the time $\tau > 0$ is determined by $p(\|U_s\|_X)$ alone. Hence, we can construct a unique global solution of (3.142) in the function space:

$$0 \leq u \in \mathcal{C}((s, \infty); H^2_N(\Omega)) \cap \mathcal{C}([s, \infty); L^2(\Omega)) \cap \mathcal{C}^1((s, \infty); L^2(\Omega)),$$
$$0 \leq \rho \in \mathcal{C}((s, \infty); H^4_{N^2}(\Omega)) \cap \mathcal{C}([s, \infty); H^2_N(\Omega)) \cap \mathcal{C}^1((s, \infty); H^2_N(\Omega)).$$

Moreover, as shown by Proposition 5.1 of [83] or (12.38) of [95], the global solution satisfies a dissipative estimate

$$\|U(t)\|_X \leq p\big((t-s+1)^{-1}\|U_0\|_X + 1\big), \quad s < t < \infty.$$

This joined with the local estimate (3.147) provides a stronger dissipative estimate of the form

$$\|AU(t)\|_X \leq p\big((t-s)^{-1}\|U_0\|_X + 1\big), \quad s < t < \infty. \tag{3.150}$$

**Nonautonomous dynamical system**

Let $s \in \mathbb{R}$. For $U_s \in \mathcal{K}$, let $U(\cdot, s; U_s)$ be the global solution of (3.142). We then set

$$U(t,s)U_s = U(t, s; U_s) \quad \text{for } (t,s) \in \Delta.$$

This $U(t,s)$ defines an evolution operator acting on $\mathcal{K}$. It is indeed clear that $U(s,s) = I$ for $s \in \mathbb{R}$ and $U(t,s) = U(t,r) \circ U(r,s)$ for $(t,r)$, $(r,s) \in \Delta$.

Let us prove that $U(t,s)$ is a continuous evolution operator on $\mathcal{K}$.

Proposition 3.2. *Let $0 < R < \infty$ and $0 < T < \infty$ be arbitrarily fixed. For any $(t,s) \in \Delta$ such that $0 \leq t-s \leq T$, $U(t,s)$ satisfies*

$$\|U(t,s)U_0 - U(t,s)V_0\|_X \leq L_{R,T}\|U_0 - V_0\|_X, \quad U_0, V_0 \in \mathcal{K}_R, \tag{3.151}$$

*$L_{R,T} > 0$ being determined by $R$ and $T$ alone.*

Proof. We notice from (3.149) that $\|U(t,s)U_0\|_X \leq p(R)$ for any $0 \leq t-s < \infty$ provided $U_0 \in \mathcal{K}_R$.

In the meantime, by applying (3.148) with radius $p(R)$, we see that

$$\|U(t,s)U_1 - U(t,s)V_1\|_X \leq C_{p(R)}\|U_1 - V_1\|_X, \quad U_1, V_1 \in \mathcal{K}_{p(R)},$$

provided that $0 \le t-s \le T_{p(R)}$. Since $R \le p(R)$, i.e., $\mathcal{K}_R \subset \mathcal{K}_{p(R)}$, this means that the desired estimate (3.151) holds for $0 \le t-s \le T_{p(R)}$.

Next let $T_{p(R)} \le t-s \le 2T_{p(R)}$. Then,

$$\begin{aligned}
&\|U(t,s)U_0 - U(t,s)V_0\|_X \\
&\quad = \|U(t,t-T_{p(R)})U(t-T_{p(R)},s)U_0 - U(t,t-T_{p(R)})U(t-T_{p(R)},s)V_0\|_X \\
&\quad = \|U(t,t-T_{p(R)})U_1 - U(t,t-T_{p(R)})V_1\|_X \\
&\quad \le C_{p(R)}\|U_1 - V_1\|_X \\
&\quad \le C_{p(R)}^2\|U_0 - V_0\|_X.
\end{aligned}$$

That is, the desired estimate holds for $T_{p(R)} \le t-s \le 2T_{p(R)}$. Repeating this arguments, we see that

$$\|U(t,s)U_0 - U(t,s)V_0\|_X \le C_{p(R)}^n\|U_0 - V_0\|_X$$

for $(n-1)T_{p(R)} \le t-s \le nT_{p(R)}$, where $n = 1, 2, 3, \ldots$.

Hence, the proposition is proved. □

PROPOSITION 3.3. *Let $U_0 \in \mathcal{K}$ be arbitrarily fixed. Then, $U(t,s)U_0$ is a continuous function for $(t,s) \in \Delta$ with values in $X$.*

PROOF. Let $\widetilde{U}_s = {}^t(s, U_0)$. Let $\widetilde{U}(\cdot)$ be the global solution of (3.143) with the initial value $\widetilde{U}_s$. Then, $\widetilde{U}(t)$ is given by

$$\widetilde{U}(t) = e^{-(t-s)\widetilde{A}}\widetilde{U}_s + \int_s^t e^{-(t-\tau)\widetilde{A}}\widetilde{F}(\widetilde{U}(\tau))d\tau, \quad s < t < \infty.$$

In view of (3.146), we observe that $U(t,s)U_0$ satisfies the integral equation

$$U(t,s)U_0 = e^{-(t-s)A}U_0 + \int_s^t e^{-(t-\tau)A}F(\tau, U(\tau,s)U_0)d\tau, \quad s < t < \infty. \tag{3.152}$$

We can then verify without difficulty that $U(t,s)U_0$ is continuous for $(t,s)$ with values in $X$. □

These two propositions yield that the mapping $G : \Delta \times \mathcal{K} \to X$, where $G(t,s;U_0) = U(t,s)U_0$, is continuous. Hence, $(U(t,s), \mathcal{K}, X)$ generates a nonautonomous dynamical system determined from (3.142).

**Exponential attractors**

We now proceed with constructing an exponential attractor. It indeed suffices to show that there exists a family of closed bounded subsets $\mathcal{X}(t)$ of $X$ having the properties (1)~(5) of Definition 3.8.

In view of the dissipative estimate (3.150), we consider a subset

$$\mathcal{B} = \mathcal{K} \cap \overline{B}^{\mathcal{D}(A)}(0; p(2)),$$

where $p(\cdot)$ is the same continuous increasing function as in (3.150). This $\mathcal{B}$ is a compact set of $X$ and is a bounded subset of $\mathcal{D}(A)$. From (3.150) we observe that, for any bounded set $B$ of $\mathcal{K}$, there exists a time $t_B > 0$ such that $U(t,s)B \subset \mathcal{B}$ for every $t \ge t_B + s$; here $t_B$ is independent of $s$.

Here, for each $t \in \mathbb{R}$, we set

$$\mathcal{X}(t) = \bigcup_{-\infty < s \le t} U(t,s)\mathcal{B}. \tag{3.153}$$

Since $\mathcal{B}$ is a bounded subset of $\mathcal{K}$, $\mathcal{B}$ itself is absorbed by $\mathcal{B}$, i.e., $U(t,s)\mathcal{B} \subset \mathcal{B}$ for any $(t,s) \in \Delta$ such that $t \geq t_{\mathcal{B}} + s$. This means that $\mathcal{X}(t)$ is written by

$$\mathcal{X}(t) = \bigcup_{t-t_{\mathcal{B}} \leq s \leq t} U(t,s)\mathcal{B}, \tag{3.154}$$

too.

Let us see that $\mathcal{X}(t)$, $t \in \mathbb{R}$, fulfills all the desired conditions. It is clear that $\mathcal{B} \subset \mathcal{X}(t) \subset \mathcal{K}$. In addition, $\mathcal{X}(t)$ is considered as the image of a mapping $g\colon [t - t_{\mathcal{B}}, t] \times \mathcal{B} \to \mathcal{K}$ such that $g(s, U_0) = U(t,s)U_0$. Since $[t - t_{\mathcal{B}}, t] \times \mathcal{B}$ is compact and $g$ is continuous, its image $g([t - t_{\mathcal{B}}] \times \mathcal{B}) = \mathcal{X}(t)$ is also compact. Hence, condition (1) is fulfilled. Moreover, we have the following result.

PROPOSITION 3.4. *The union $\bigcup_{t\in\mathbb{R}} \mathcal{X}(t)$ is a bounded subset of $\mathcal{D}(A)$. Consequently, the union is a relatively compact set of $X$.*

PROOF. To prove this, we have to go back to the abstract problem (3.143). Let the initial data $U_s$ satisfy $U_s \in \mathcal{D}(A)$ such that $\|AU_s\|_X \leq p(2)$ and consequently $\widetilde{U}_s \in \mathcal{D}(\widetilde{A})$ with $\|\widetilde{A}\widetilde{U}_s\|_{\mathbb{X}} \leq p(2)$. Let $\widetilde{U}(t)$ be the global solution of (3.143). We want to use the estimates obtained in Theorem 4.2 of [95] with $\gamma = \eta$ to conclude that

$$\begin{cases} \|\widetilde{A}^\eta \widetilde{U}(t)\|_X \leq C, & s \leq t \leq s+T, \\ \|\widetilde{A}^\eta[\widetilde{U}(t) - \widetilde{U}(\tau)]\|_X \leq C(t-\tau)^{1-\eta}(\tau - s)^{-\eta}, & s \leq \tau < t \leq s+T, \end{cases}$$

for the solution with the initial data $U_s$ with some $T > 0$ and $C > 0$ depending only on $p(2)$. Therefore, by (3.146),

$$\begin{cases} \|A^\eta U(t,s)U_s\|_X \leq C, & s \leq t \leq s+T, \\ \|A^\eta[U(t,s)U_s - U(\tau,s)U_s]\|_X \leq C(t-\tau)^{1-\eta}(\tau - s)^{-\eta}, & s \leq \tau < t \leq s+T. \end{cases} \tag{3.155}$$

Let $-\infty < t < \infty$ and $t - t_{\mathcal{B}} \leq s \leq t$ and let $U_s \in \mathcal{B}$. By definition, $\|AU_s\|_X \leq p(2)$. As seen in (3.152), $U(t,s)U_s$ satisfies the integral equation

$$U(t,s)U_s = e^{-(t-s)A}U_s + \int_s^t e^{-(t-\tau)A} F(\tau, U(\tau,s)U_s)d\tau.$$

Therefore,

$$AU(t,s)U_s = e^{-(t-s)A}AU_s + \int_s^t Ae^{-(t-\tau)A}[F(\tau, U(\tau,s)U_s) - F(t, U(t,s)U_s)]d\tau$$
$$+ \int_s^t Ae^{-(t-\tau)A}F(t, U(t,s)U_s)d\tau.$$

Also,

$$\int_s^t Ae^{-(t-\tau)A}F(t, U(t,s)U_s)d\tau = (1 - e^{-tA})F(t, U(t,s)U_s).$$

Using (3.155), we easily obtain that

$$\|AU(t,s)U_s\|_X \leq C, \quad s \leq t \leq s+T,$$

the constant $C$ being determined by $p(2)$.

We have thus verified that the union $\bigcup_{t-T\leq s \leq t} AU(t,s)\mathcal{B}$ is uniformly bounded in $X$ with respect to $t$. Hence, the proof is complete if $T \geq t_{\mathcal{B}}$.

Let $T < t_{\mathcal{B}}$. For $(t, s)$ such that $T \le t - s \le t_{\mathcal{B}}$, we utilize the global estimate (3.147) to conclude that

$$\|AU(t,s)U_s\|_X \le p(T^{-1}\|U_s\|_X + 1), \quad T + s \le t \le t_{\mathcal{B}} + s.$$

This means that the union $\bigcup_{t-t_{\mathcal{B}} \le s \le t-T} U(t,s)\mathcal{B}$ is also uniformly bounded in $X$ with respect to $t$. Hence, the proof is complete even in this case. □

Let us verify condition (2). By (3.153),

$$\mathcal{X}(s) = \bigcup_{-\infty < r \le s} U(s,r)\mathcal{B}.$$

For each $-\infty < r \le s$, it follows that $U(t,s) \circ U(s,r)\mathcal{B} = U(t,r)\mathcal{B} \subset \mathcal{X}(t)$. Hence, $U(t,s)\mathcal{X}(s) \subset \mathcal{X}(t)$.

Consider any bounded subset $B$ of $\mathcal{K}$. Then, there exists a time $t_B$ such that $U(t,s)B \subset \mathcal{B}$ for every $t \ge t_B + s$. Since $\mathcal{B} \subset \mathcal{X}(t)$, this means that condition (3) is valid.

We set $Z = \mathcal{D}(A^\eta)$, where $\eta > 0$ is the exponent appearing in (3.148). By Proposition 3.4, there exists $R > 0$ such that $\bigcup_{t\in\mathbb{R}} \mathcal{X}(t) \subset \mathcal{K}_R$. Then, (3.148) shows that the Lipschitz condition of (4) is valid provided $\tau^* = T_R$. The estimate also provides the Lipschitz condition of (5).

We have thus verified that all the conditions (1)~(5) are fulfilled. Hence, Theorem 3.11 yields the existence of an exponential attractor $\mathcal{M}(t)$, $-\infty < t < \infty$, for $(U(t,s), \mathcal{K}, X)$.

Let us finally verify that $U(t,s)$ satisfies (3.133) and (3.134). For $(t,s) \in \Delta$, we see from (3.142) that

$$U(t,s)U_s - U_s = \int_s^t [-AU(\tau,s)U_s + F(\tau, U(\tau,s)U_s)]d\tau, \quad U_s \in \mathcal{X}(s).$$

Therefore,

$$\|[U(t,s) - 1]U_s\|_X \le C(t-s) \sup_{s\le\tau\le t} \|AU(\tau,s)U_s\|_X, \quad U_s \in \mathcal{X}(s).$$

Then, Proposition 3.4 gives that

$$\sup_{U_s\in\mathcal{X}(s)} \|[U(t,s) - 1]U_s\|_X \le C(t-s), \quad -\infty < s \le t < \infty.$$

This means that both (3.133) and (3.134) are fulfilled.

CHAPTER 4

# Porous medium equation in homogeneous media: Long-time dynamics

As we already mentioned in Chapter 2, the long-time behavior of many dissipative systems generated by evolution PDEs of mathematical physics can be described in terms of the so-called attractors.

In particular, in the case of dissipative PDEs in bounded domains, this attractor usually has finite Hausdorff and fractal dimension; see [9], [93], [32] and some model examples of mathematical physics studied in Chapter 3. Consequently, in spite of the infinite dimensionality of the initial phase space, the reduced dynamics on the attractor occurred (in a sense) finite dimensionally and can be studied by the methods of the classical (finite-dimensional) theory of dynamical systems.

In contrast to that, infinite-dimensional global/uniform attractors are typical for dissipative PDEs in *unbounded* domains or/and for the *nonautonomous* equations. In order to study such attractors, one usually uses the concept of Kolmogorov $\varepsilon$-entropy; see the recently published book by M. A. Efendiev [32] for a systematic study of this issue and the details.

We, however, note that the above results have been obtained mainly for evolution PDEs with more or less regular structure (e.g., uniformly parabolic or uniformly hyperbolic). In contrast to this, very little is known about the equations with degeneration which also play a significant role in the life sciences; see [31] and the references therein. Indeed, although the attractors for such equations have been considered in a number of papers (see [31] and the references therein), the questions related to the finite or infinite dimensionality of these attractors have not been studied yet.

The main aim of this chapter is to give a detailed study of the fractal dimension and Kolmogorov entropy of attractors of the following degenerate porous media equation:

$$\partial_t u = \Delta_x(f(u)) - g(u) + h \tag{4.1}$$

in a bounded domain $\Omega \subset \mathbb{R}^n$ (equipped by Dirichlet boundary conditions). Here $f(u) \sim u|u|^{p-1}$ has a degeneration at $u = 0$ ($p > 1$), the function $g$ satisfies the standard dissipativity assumptions, and $h = h(x)$ is a given external force (see Section 4.1 for the rigorous conditions).

This chapter is organized as follows. In Section 4.1, we briefly recall some basic results on the existence, uniqueness, and regularity of solutions of equation (4.1).

A natural class of equations of the form (4.1) whose global attractors are finite dimensional is considered in Section 4.2. The finite dimensionality is proven under the additional assumption

$$g'(0) > 0 \tag{4.2}$$

and is strongly based on the global Hölder continuity of solutions of equation (4.1); see Theorem 4.2 of Section 4.1.

An exponential attractor (in the sense of Chapter 3) for problem (4.1) under assumption (4.2) is constructed in Section 4.3.

Finally, in Section 4.4, we show that the global attractor is usually infinite dimensional if condition (4.2) is violated and, thus, the sign of $g'(0)$ occurs crucial for the finite or infinite dimensionality of the global attractor. Namely, we consider here the particular case of (4.1) of the following form:

$$\partial_t u = \Delta_x(u|u|^{p-1}) + u - g(u) \tag{4.3}$$

with $p > 1$ and $g(u)$ vanishing near $u = 0$. Under these assumptions, we prove (see Theorem 4.5) that the associated global attractor is infinite dimensional.

Moreover, we also study the Kolmogorov $\varepsilon$-entropy $\mathbb{H}_\varepsilon(\mathcal{A})$ of this attractor (which, by definition, is a logarithm from the minimal number of $\varepsilon$-balls which cover the compact set $\mathcal{A}$). To be more precise, we establish the following lower bounds for that quantity:

$$\mathbb{H}_\varepsilon(\mathcal{A}) \geq C\left(\frac{1}{\varepsilon}\right)^{n(p-1)/2}$$

where $C$ is some positive number independent of $\varepsilon > 0$.

Thus, porous media equations of the form (4.1) give natural examples of dissipative equations of mathematical physics in *bounded* domains with infinite-dimensional attractors. It is also worth noting that, although the infinite-dimensional global attractors are typical for the regular (nondegenerate) equations in unbounded domains, even in that case the asymptotics of their Kolmogorov $\varepsilon$-entropy were always logarithmical (like $(\log_2 \frac{1}{\varepsilon})^{n+1}$; see [32] for a systematic study of this issues). To the best of our knowledge, it is the first example of a global attractor whose $\varepsilon$-entropy has a polynomial (with respect to $\varepsilon^{-1}$) asymptotics.

As we will see in the next chapters, the method of study of the dimension of global attractors of degenerate parabolic equations developed in Chapter 4 seems to be of a general nature and can be applied to other classes of degenerate equations.

## 4.1. A priori estimates and regularity of solutions

In this section, we briefly recall the known results on the regularity of solutions of porous media equations which will be systematically used in the next sections; see, e.g., [13], [27], [47], [58], [59] for more details.

In a bounded domain $\Omega \subset\subset \mathbb{R}^n$ with a sufficiently smooth boundary, we consider the porous media equation in the following form:

$$\begin{cases} \partial_t u = \Delta_x f(u) - g(u) + h, \\ u\big|_{\partial\Omega} = 0, \quad u\big|_{t=0} = u_0, \end{cases} \tag{4.4}$$

where $u = u(t, x)$ is an unknown function, $\Delta_x$ is a Laplacian with respect to the variable $x = (x_1, \dots, x_n)$, $f$ and $g$ are given functions, and $h = h(x)$ is a given external force.

We assume that the function $f \in C^2(\mathbb{R})$ has a polynomial degeneration at $u = 0$ and is nondegenerate for $u \neq 0$. To be more precise, we assume that

$$C_1|u|^{p-1} \leq f'(u) \leq C_2|u|^{p-1}, \quad f(0) = 0, \tag{4.5}$$

for some positive constants $C_i$ and $p > 1$. It is also assumed that the function $g$ satisfies the dissipativity condition

$$g'(u) \geq -C + \kappa|u|^{q-1}, \tag{4.6}$$

for some $q > 1$, $\kappa > 0$, and that the external force $h$ belongs to $L^\infty(\Omega)$.

As usual, in order to prove the existence of a solution of problem (4.4), one considers the nondegenerate analogue of (4.1)

$$\begin{cases} \partial_t u = \Delta_x f(u) + \varepsilon \Delta_x u - g(u) + h, \\ u\big|_{\partial\Omega} = 0, \quad u\big|_{t=0} = u_0, \end{cases} \tag{4.7}$$

which obviously has a unique solution for every $\varepsilon > 0$ and sufficiently smooth $u_0$ (see, e.g., [68]) and then passes to the limit $\varepsilon \to 0$. Following this scheme, we first derive estimates that are uniform with respect to $\varepsilon$ for equation (4.7). We start from the standard $L^1 - L^\infty$-estimates.

LEMMA 4.1. *Let the above assumptions hold and let $u$ be a solution of equation (4.1). Then the following estimates hold:*

$$\begin{cases} (1)\ \|u(t)\|_{L^1(\Omega)} \leq \|u(0)\|_{L^1(\Omega)} e^{-\alpha t} + C(1 + \|h\|_{L^1(\Omega)}), \\ (2)\ \|u(t)\|_{L^\infty(\Omega)} \leq \|u(0)\|_{L^\infty(\Omega)} e^{-\alpha t} + C(1 + \|h\|_{L^\infty(\Omega)}) \end{cases} \tag{4.8}$$

*where the positive constants $\alpha$ and $C$ depend only on the function $g$ and are independent of $\varepsilon$, $t$, $u_0$, and $u$. Moreover, the following $L^1 - L^\infty$-smoothing property holds:*

$$\|u(t)\|_{L^\infty(\Omega)} \leq Q(t^{-1} + \|h\|_{L^\infty(\Omega)}) \quad \text{for all } t > 0,\ u_0 \in L^1(\Omega), \tag{4.9}$$

*where the monotonic function $Q$ is independent of $\varepsilon$, $t$, and $u$.*

PROOF. Indeed, taking the $L^2(\Omega)$ inner product of equation (4.1) with the function $\operatorname{sgn} u = \operatorname{sgn}(f(u) + \varepsilon u)$ and using the Kato inequality $(\Delta_x v, \operatorname{sgn} v) \leq 0$ and the dissipativity assumption (4.6), we deduce that

$$\partial_t \|u(t)\|_{L^1(\Omega)} + \kappa \|u(t)\|^q_{L^q(\Omega)} \leq C + \|h\|_{L^1(\Omega)}. \tag{4.10}$$

Since $\kappa \|u\|^q_{L^q(\Omega)} \geq \|u(t)\|_{L^1(\Omega)} - C$, the the Gronwall inequality applied to (4.10) implies the first estimate of (4.8). In order to deduce the second estimate of (4.8), we use the comparison principle for second-order parabolic equations and deduce that

$$y_-(t) \leq u(t,x) \leq y_+(t) \tag{4.11}$$

where $y_\pm(t)$ solve the ODEs

$$y'_\pm(t) + g(y_\pm(t)) = \pm\|h\|_{L^\infty(\Omega)}, \quad y_\pm(0) = \pm\|u_0\|_{L^\infty(\Omega)}. \tag{4.12}$$

It remains to note that, due to the dissipativity assumption (4.6), the solutions $y_\pm(t)$ satisfy the analogue of estimate (4.8)(2) which together with (4.11) finishes the proof of estimate (4.8)(2). Finally, in order to verify (4.9), it remains to recall that, due to our assumptions, $g$ has a superlinear growth rate as $u \to \infty$. Consequently, the solutions $y_\pm(t)$ satisfy estimate (4.9) which together with estimate (4.11) imply estimate (4.9) for the solution $u$ and this finishes the proof of the lemma. □

The next lemma gives some kind of energy estimates for equation (4.7).

LEMMA 4.2. *Let the above assumptions hold and let $u$ be a solution of (4.7). Then, for every $\delta > 0$, the following estimate holds:*

$$\|u\|_{W^{1/p-\delta,2p}([t,t+1]\times\Omega)} + \|\partial_t f(u)\|_{L^2([t,t+1]\times\Omega)} + \|\nabla_x f(u)\|_{L^\infty([t,t+1],L^2(\Omega))} \le Q(t^{-1} + \|h\|_{L^\infty(\Omega)}) \tag{4.13}$$

*where the monotonic function $Q$ is independent of $\varepsilon$, $t > 0$, and $u$ (here and below, $W^{s,p}$ denotes the Sobolev space of distributions whose derivatives up to order $s$ belong to $L^p$; see, e.g., [68]).*

PROOF. Without loss of generality, we can assume that $t \le 1$. Then, taking the $L^2(\Omega)$ inner product of (4.7) with $f_\varepsilon(u) := f(u) + \varepsilon u$ and integrating over $[\delta, 2]$, $\delta > 0$, we get

$$(F_\varepsilon(u(2)), 1) - (F_\varepsilon(u(\delta)), 1) + \int_\delta^2 \|\nabla_x f_\varepsilon(u(t))\|^2_{L^2(\Omega)}\, dt + \int_\delta^2 (f_\varepsilon(u(t), g(u(t))) - (h, f_\varepsilon(u(t))))\, dt = 0 \tag{4.14}$$

(here and below we denote by $(\cdot,\cdot)$ the standard inner product in $L^2(\Omega)$ and $F_\varepsilon(u) = \int_0^v f_\varepsilon(v)\, dv$). Together with the $L^\infty$-estimate (4.9) this estimate gives

$$\int_\delta^2 \|\nabla_x f_\varepsilon(u)\|^2_{L^2(\Omega)}\, dt \le Q(\delta^{-1} + \|h\|_{L^\infty(\Omega)}). \tag{4.15}$$

Let us now multiply equation (4.7) by $(t-\delta)\partial_t f_\varepsilon(u)$ and integrate over $[\delta, T] \times \Omega$, $\delta \le T \le 2$. Then, we have

$$\begin{aligned}\int_\delta^2 (t-\varepsilon) f'_\varepsilon(u(t))|\partial_t u(t)|^2\, dt \\ + (T-\delta)(\frac{1}{2}\|\nabla_x f_\varepsilon(u(T))\|^2_{L^2(\Omega)} + (F_{\varepsilon,g}(u(T)), 1) - (f_\varepsilon(u(T)), h)) \\ = \int_\delta^T \frac{1}{2}\|\nabla_x f_\varepsilon(u(t))\|^2_{L^2(\Omega)} + (F_{\varepsilon,g}(u(t)), 1) - (f_\varepsilon(u(t)), h)\, dt\end{aligned} \tag{4.16}$$

where $F_{\varepsilon,g}(u) := \int_0^u f'_\varepsilon(v) g(v)\, dv$. This estimate, together with the $L^\infty$-estimate (4.9) and estimate (4.16), implies that

$$\int_{2\delta}^2 f'(u(t))|\partial_t u(t)|^2\, dt + \|\nabla_x f(u)\|^2_{L^\infty([2\delta,2],L^2(\Omega))} \le Q(\delta^{-1} + \|h\|_{L^\infty(\Omega)}) \tag{4.17}$$

for the appropriate monotonic function $Q$. This estimate, together with the $L^\infty$-estimate, implies, in turn, that

$$\|\partial_t f(u)\|_{L^2([t,t+1]\times\Omega)} + \|\nabla_x f(u)\|_{L^\infty([t,t+1],L^2(\Omega))} \le Q(t^{-1} + \|h\|_{L^\infty(\Omega)}). \tag{4.18}$$

Thus, it only remains to estimate the first term on the left-hand side of (4.13). To this end, we note that, according to (4.18), we have $v = f(u) \in W^{1,2}([t, t+1]\times\Omega) \subset W^{1-\delta,2}([t, t+1] \times \Omega)$. Then, due to Lemma 1.2, we have

$$\|u\|_{W^{1/p-\delta,2p}([t,t+1]\times\Omega)} \le C_{f,\delta}\|v\|^{1/p}_{W^{1,2}([t,t+1]\times\Omega)}$$

where the constant $C_{f,\delta}$ depends only on $f$ and $\delta > 0$. Lemma 4.2 is proven. □

The next lemma gives the uniform Lipschitz continuity of solutions in $L^1(\Omega)$.

LEMMA 4.3. *Let the above assumptions hold and let $u_1(t)$ and $u_2(t)$ be two solutions of equation* (4.7). *Then, the following estimate holds:*

$$\|u_1(t) - u_2(t)\|_{L^1(\Omega)} \le e^{Kt}\|u_1(0) - u_2(0)\|_{L^1(\Omega)} \tag{4.19}$$

*where* $K := \max_{v\in\mathbb{R}}\{-g'(v)\}$.

PROOF. Indeed, let $v(t) := u_1(t) - u_2(t)$. Then, this function satisfies the following linear equation:

$$\partial_t v = \Delta_x(l_1(t)v) + \varepsilon\Delta_x v - l_2(t)v, \quad v\big|_{\partial\Omega} = 0, \quad v\big|_{t=0} = u_1(0) - u_2(0) \tag{4.20}$$

where

$$l_1(t) := \int_0^1 f'(su_1(t) + (1-s)u_2(t))\,ds \ge 0$$

and

$$l_2(t) := \int_0^1 g'(su_1(t) + (1-s)u_2(t))\,ds \ge -K.$$

Now multiplying equation (4.20) by $\operatorname{sgn} v = \operatorname{sgn}((l_1(t)+\varepsilon)v)$ and again using the Kato inequality, we arrive at

$$\partial_t\|v(t)\|_{L^1(\Omega)} - K\|v(t)\|_{L^1(\Omega)} \le 0. \tag{4.21}$$

Applying the Gronwall inequality to this relation, we finish the proof of Lemma 4.3. □

We are now ready to verify the existence and uniqueness of a solution for the initial degenerate problem (4.1). To this end, we first formulate the definition of a weak solution of that problem.

DEFINITION 4.1. We say that a function $u$ is a weak solution of (4.1) if $u \in C([0,T], L^1(\Omega))$, $u \in L^\infty([t,T]\times\Omega)$, and $f(u) \in L^2([t,T], W_0^{1,2}(\Omega))$, for every $t > 0$ and it satisfies (4.1) in the sense of distributions.

The following theorem can be considered as the main result of the section.

THEOREM 4.1. *Let the above assumptions hold. Then, for every $u_0 \in L^1(\Omega)$, there exists a unique weak solution of problem* (4.1) *and this solution satisfies all of the estimates formulated in Lemmas* 4.1–4.3.

PROOF. We first establish the existence of a solution and assume additionally that $u_0$ is smooth enough. Let us consider a sequence $u_{\varepsilon_n}(t)$ of solutions of the auxiliary problem (4.7) with $\varepsilon_n \to 0$. Then, this sequence satisfies estimates (4.8) and (4.13) uniformly with respect to $n$. Moreover, since $u_0$ is smooth, estimate (4.13) holds for $t = 0$ as well. In particular,

$$\|u_{\varepsilon_n}\|_{W^{1/p-\delta,2p}([t,t+1]\times\Omega)} \le C \tag{4.22}$$

uniformly with respect to $t$ and $n$. Thus, without loss of generality, we can assume that $u_{\varepsilon_n} \to u$ strongly in $C([0,T], L^1(\Omega))$ (due to the compactness of the embedding $W^{1/p-\delta,2p}([0,T]\times\Omega) \subset C([0,T], L^1(\Omega))$ if $\delta$ is small enough). Now passing in a standard way (see, e.g., [9]) to the limit $n \to \infty$ in equations (4.7), we verify that $u_0$ satisfies the initial equation (4.4) (in the sense of distributions), and passing to the limit $n \to \infty$ in the uniform estimates of Lemmas 4.1–4.3, we verify that the solution thus constructed satisfies estimates (4.8), (4.9), (4.13), and (4.19). In particular, these estimates show that $u$ is a weak solution in the sense of Definition 4.1. Thus, for smooth initial data $u_0$ the existence of a solution is verified. In order to relax the smoothness assumption, it remains to recall that the solutions constructed satisfy (4.19) with the constants which are independent of the initial data. Consequently, approximating in $L^1(\Omega)$ the nonsmooth initial data $u_0 \in L^1(\Omega)$ by a sequence of the smooth one $u_0^n$, constructing the associated solutions $u^n(t)$, and passing to the limit $n \to \infty$, we obtain a weak solution $u(t)$ for every $u_0 \in L^1(\Omega)$. Obviously, this solution will also satisfy all of the estimates of Lemmas 4.1–4.3. Thus, existence is verified.

Let us now prove uniqueness. Indeed, let $u_1(t)$ and $u_2(t)$ be two weak solutions of equation (4.4) and let $v(t) := u_1(t) - u_2(t)$. Then, this function satisfies the equation

$$\partial_t v = \Delta_x(l_1(t)v) - l_2(t)v \tag{4.23}$$

where $l_i(t)$ are the same as in (4.20). It would be natural (analogously to the proof of Lemma 4.3) to multiply equation (4.23) by $\operatorname{sgn}(v)$ and to use the Kato inequality, which would immediately give estimate (4.19) and finish the proof of uniqueness, but, unfortunately, in contrast to the situation in Lemma 4.3, we do not now have enough regularity for the expression $(\Delta_x(l(t)v), \operatorname{sgn}(v))$ to make sense. Thus, we need to proceed in a little more accurate way. To this end, we assume, in addition, that $u_i \in L^\infty([0,T] \times \Omega)$ and we introduce the following "regularized" version of the conjugate equation for (4.23):

$$-\partial_t w = l_1(t)\Delta_x w + \varepsilon \Delta_x w, \quad w\big|_{t=T} = w_T, \quad w\big|_{\partial\Omega} = 0, \tag{4.24}$$

which we will consider in the space $W^{(1,2),2}([0,T]\times\Omega)$ (here and below, we denote by $W^{(1,2),q}$ the anisotropic Sobolev space consisting of distributions whose $t$-derivatives up to order one and $x$-derivatives up to order two belong to $L^q$; see [68]).

The next lemma gives the solvability result for that equation.

LEMMA 4.4. *Let the above assumptions hold. Then, for every $w_T \in W_0^{1,2}(\Omega)$ and every $\varepsilon > 0$, equation (4.24) possesses a unique solution $w \in W^{(1,2),2}([0,T]\times\Omega)$ and the following estimate holds:*

$$\|\nabla_x w(t)\|^2_{L^2(\Omega)} + 2\varepsilon \int_0^T \|\Delta_x w(t)\|^2_{L^2(\Omega)} \le \|\nabla_x w(T)\|^2_{L^2(\Omega)}, \quad t \in [0,T]. \tag{4.25}$$

*Moreover, if in addition, $C_1 \le w_T(x) \le C_2$, then*

$$C_1 \le w(t,x) \le C_2, \quad t \in [0,T]. \tag{4.26}$$

PROOF. Indeed, according to our assumption, $l_i \in L^\infty([0,T] \times \Omega)$ and, moreover, due to (4.5), $l_1(t) \ge 0$. Therefore, equation (4.24) is nondegenerate. The a priori estimate (4.25) can be obtained by multiplying (4.24) by $\Delta_x w$ and integrating over $[0,T] \times \Omega$, and the $L^2$-estimate for the derivative then follows from (4.25) and equation (4.24). Thus, the a priori estimate in $W^{(1,2),2}([0,T] \times \Omega)$ is obtained. The existence of a solution can be easily verified by, e.g., the Galerkin method; see [9]. Finally, estimate (4.26) is just a maximum principle for the linear second-order parabolic equation (4.24). (As mathematicians, we cannot apply the classical maximum principle directly to equation (4.24) since the function $l_1(t,x)$ is only from $L^\infty$ (and not smooth), but approximating it by the smooth ones, say, in $L^2([0,T] \times \Omega)$, we may apply the maximum principle for the solutions of the associated smooth equations and then pass to the limit in a standard way.) Lemma 4.4 is proven. □

We are now ready to finish the proof of the uniqueness for weak solutions of (4.4). To this end, we multiply equation (4.23) by the solution $w(t)$ of the "conjugate" equation (4.24) (with some $w_T$) and integrate over $[\delta, T]\times\Omega$. Then, after integration by parts, we have

(4.27)
$$(v(T), w(T)) - (v(0), w(0)) + \varepsilon \int_0^T (\Delta_x w(t), v(t))\, dt + \int_0^T (l_2(t)v(t), w(t))\, dt = 0.$$

We now approximate the function $w_T^0 := \operatorname{sgn}(v(T))$ in the $L^2(\Omega)$ metric by $w_T^n \in W_0^{1,2}(\Omega)$ in such a way that $-1 \le w_T^n \le 1$ and we construct the appropriate solutions $w^n(t)$ of equation (4.24). Then, due to (4.26), $-1 \le w^n(t,x) \le 1$ and, consequently, (4.27) reads as

$$(v(T), w_T^n) + \varepsilon \int_0^T (\Delta_x w^n(t), v(t))\, dt \le \|v(0)\|_{L^1(\Omega)} + L_2 \int_0^T \|v(t)\|_{L^1(\Omega)}\, dt \tag{4.28}$$

where $L_2 = \|l_2(t,x)\|_{L^\infty([0,T]\times\Omega)}$. We now pass to the limit $\varepsilon \to 0$ (with a fixed $n$) in the inequality (4.28) using (4.25) and

$$\varepsilon \int_0^T (\Delta_x w^n(t), v(t))\, dt \le \varepsilon^{1/4}(\varepsilon^{1/2}\|\Delta_x w^n\|^2_{L^2([0,T]\times\Omega)} + \|v\|^2_{L^2([0,T]\times\Omega)}).$$

Then, we have

$$(v(T), w_T^n) \le \|v(0)\|_{L^1(\Omega)} + L_2 \int_0^T \|v(t)\|_{L^1(\Omega)}\, dt. \tag{4.29}$$

Finally, passing to the limit $n \to \infty$ in (4.29), we get

$$\|v(T)\|_{L^1(\Omega)} \le \|v(0)\|_{L^1(\Omega)} + L_2 \int_0^T \|v(t)\|_{L^1(\Omega)}\, dt. \tag{4.30}$$

Since $T > 0$ is arbitrary, the Gronwall inequality applied to (4.30) implies that

$$\|v(t)\|_{L^1(\Omega)} \le e^{L_2 t}\|v(0)\|_{L^1(\Omega)}.$$

Thus, we have proven that every weak solution $u(t)$ of (4.1) is unique under the additional assumption $u \in L^\infty([0,T]\times\Omega)$. Therefore, every such solution coincides with the solution obtained by passing to the limit $\varepsilon \to 0$ in the nondegenerate equations (4.7). This implies, in turn, that all such solutions should satisfy estimate (4.19).

Let us now consider the general case of two weak solutions $u_1$ and $u_2$ which do not belong to $L^\infty([0,T]\times\Omega)$. Then, due to the definition of a weak solution, $u_i \in L^\infty([\delta,T]\times\Omega)$ for every $\delta > 0$ and, consequently, due to (4.19), we have

$$\|u_1(t) - u_2(t)\|_{L^1(\Omega)} \le e^{K(t-\delta)}\|u_1(\delta) - u_2(\delta)\|_{L^1(\Omega)}. \tag{4.31}$$

Now passing to the limit $\delta \to 0$ in (4.31) and taking into account that $u_i \in C([0,T], L^1(\Omega))$, we obtain estimate (4.19) for any two weak solutions of (4.4). Theorem 4.1 is proven. □

REMARK 4.1. In Theorem 4.1, we have proven, in particular, that every weak solution of (4.4) can be approximated by smooth solutions of the nondegenerate problem (4.7). This allows us in the sequel to use the Kato inequality for deriving more delicate estimates without concerning ourselves with regularity. Indeed, all these estimates can be easily justified by this approximating procedure.

We also note that the rather strong dissipativity condition (which guarantees, in particular, the superlinear growth rate of the nonlinearity $g$) has been posed only in order to avoid the technicalities in proving the $L^1 - L^\infty$-smoothing property for the solutions of (4.4) and can be relaxed to the standard dissipativity condition:

$$\limsup_{|u|\to\infty} \frac{g(u)}{u} > 0.$$

We conclude this section by formulating the result on the Hölder continuity of solutions of degenerate parabolic equations, which is crucial for our study of the dimension of the attractor.

THEOREM 4.2. *Let the above assumptions hold and let $u$ be a weak solution of* (4.4). *Then, there exists a positive constant $\alpha$ such that*

$$\|u\|_{C^\alpha([t,t+1]\times\Omega)} \le Q(t^{-1} + \|h\|_{L^\infty(\Omega)}) \tag{4.32}$$

*where $t > 0$ and for some monotonic function $Q$.*

In the multidimensional case $n \ge 2$, the Hölder continuity of (4.32) is a rather delicate fact and its proof is based on the proper modification of the De Giorgi technique; see [30], [27], [58]. In contrast to that, in the one-dimensional case, it can be easily derived from the standard energy estimates. For the convenience of the reader, we give below the proof for the one-dimensional case.

PROOF (THE ONE-DIMENSIONAL CASE). Indeed, according to Lemma 4.2 and Theorem 4.1, any weak solution $u$ satisfies

$$\|f(u)\|_{L^\infty([t,t+1],W^{1,2}(\Omega))\cap W^{1,2}([t,t+1],L^2(\Omega))} \le Q(t^{-1} + \|h\|_{L^\infty(\Omega)}). \tag{4.33}$$

Moreover, due to interpolation (see [68]), we have

$$\|v\|_{C^\alpha([t,t+1],W^{1-2\alpha,2}(\Omega))} \le C_\alpha \|v\|_{L^\infty([t,t+1],W^{1,2}(\Omega))\cap W^{1,2}([t,t+1],L^2(\Omega))} \tag{4.34}$$

for $0 \le \alpha < 1/2$. In the one-dimensional case, we have the embedding $W^{1-2\alpha,2}(\Omega) \subset C^{1/2-2\alpha}(\Omega)$. Taking $\alpha = 1/6$, we finally derive

$$\|f(u)\|_{C^{1/6}([t,t+1]\times\Omega)} \le Q(t^{-1} + \|h\|_{L^\infty(\Omega)}) \tag{4.35}$$

for some monotonic function $Q$. Lemma 1.2 together with (4.35) implies (4.32) with $\alpha = 1/(6p)$. Theorem 4.2 for the one-dimensional case is proven. □

## 4.2. Finite-dimensional global attractor

In the previous section proved that equation (4.1) generates a uniformly Lipschitz continuous semigroup $S(t)$ on the phase space $\Phi = L^1(\Omega)$ via

$$S(t)u_0 = u(t), \quad u_0 \in L^1(\Omega), \quad t > 0, \tag{4.36}$$

where $u(t)$ is a unique weak solution of (4.1) (see Theorem 4.1). The present section is devoted to the study of the long-time behavior of the trajectories of that semigroup in terms of finite-dimensional global attractors (see Definition 2.1). The case where the limit dynamics is infinite dimensional will be considered in Section 4.4.

The next lemma states the existence of such an attractor.

LEMMA 4.5. *Let the assumptions of Section* 4.1 *hold. Then, semigroup* $S(t)$ *associated with equation* (4.1) *possesses a global attractor* $\mathcal{A}$ *in the phase space* $L^1(\Omega)$ *which is globally bounded in* $C^\alpha(\Omega)$ *(for some sufficiently small* $\alpha$*) and has the following structure:*

$$\mathcal{A} = \mathcal{K}\big|_{t=0} \tag{4.37}$$

*where* $\mathcal{K}$ *is a set of all bounded solutions of* (4.1) *defined for all* $t$. *Moreover, this set satisfies*

$$\|\mathcal{K}\|_{C^\alpha(\mathbb{R}\times\Omega)} \le Q(\|h\|_{L^\infty(\Omega)}) \tag{4.38}$$

*for some monotone function* $Q$.

PROOF. As usual, in order to verify the existence of a global attractor, one needs to verify two properties:

(1) The maps $S(t) : \Phi \to \Phi$ are continuous for every fixed $t$.

(2) The semigroup $S(t)$ possesses a (pre)compact absorbing set in $\Phi$; see [9], [93].

In our case, the first property is obvious, since, due to Lemma 4.3, the semigroup $S(t)$ is even globally Lipschitz continuous in $\Phi$. Moreover, the existence of an absorbing set, bounded in $C^\alpha(\Omega)$, is an immediate corollary of Theorem 4.2. Thus, due to the abstract theorem on the existence of the attractor, this semigroup possesses a global attractor $\mathcal{A}$, bounded in $C^\alpha(\Omega)$. Formula (4.37) is also a corollary of that theorem and (4.38) follows from Theorem 4.2. Lemma 4.5 is proven. □

The next theorem, which establishes the finite dimensionality of the global attractor under the additional assumption that equation (4.4) is asymptotically stable near $u = 0$, can be considered as the main result of the section.

THEOREM 4.3. *Let the assumptions of Section* 4.1 *hold and let, in addition,*

$$g'(0) > 0. \tag{4.39}$$

*Then the fractal dimension of* $\mathcal{A}$ *in* $C(\Omega)$ *is finite:*

$$\dim_f(\mathcal{A}, C(\Omega)) < \infty. \tag{4.40}$$

PROOF. As usual, see [19], [35], [36], [93]. In order to prove the finite dimensionality of the attractor, we need to consider an arbitrary finite $\varepsilon$-net $V_\varepsilon$ in $\mathcal{A}$ in the metric of $L^1(\Omega)$ (with a sufficiently small positive $\varepsilon$) and to construct, using this net, a $\kappa\varepsilon$-net $V_{\kappa\varepsilon}$ (with $\kappa < 1$) in $\mathcal{A}$ satisfying

$$\#V_{\kappa\varepsilon} \le L\#V_\varepsilon \tag{4.41}$$

where the constants $\kappa$ and $L$ are independent of $\varepsilon$ and of the initial covering $V_\varepsilon$. Then, iterating this procedure, we can prove the finite dimensionality of the attractor.

Let $V_\varepsilon = \{u_0^i\}_{i=1}^{N_\varepsilon}$, $V_\varepsilon \subset \mathcal{A}$ an arbitrary $\varepsilon$-net in $\mathcal{A}$ (with $N_\varepsilon = \#V_\varepsilon$). Then, in order to construct the required $\kappa\varepsilon$-net, it is sufficient to construct, for every $u_0 \in \mathcal{A}$, the $\kappa\varepsilon$-net $V_{\kappa\varepsilon}(u_0)$ in the image $S(T)(B(\varepsilon, u_0, L^1) \cap \mathcal{A})$ (for some positive $T$) of the $\varepsilon$-ball centered at $u_0$ intersected with the attractor (here and below we denote by $B(R, x, X)$ an $R$-ball in the space $X$ centered at $x \in X$) satisfying

$$\#V_\varepsilon(u_0) \le L. \tag{4.42}$$

Then, obviously, the set $V_{\kappa\varepsilon} := \bigcup_{u_0 \in V_\varepsilon} V_{\kappa\varepsilon}(u_0)$ gives a $\kappa\varepsilon$-net in $S(T)\mathcal{A}$ satisfying (4.41). Finally, since $S(T)\mathcal{A} = \mathcal{A}$, the required $\kappa\varepsilon$-net in $\mathcal{A}$ would be constructed.

Thus, we only need to construct the $\kappa\varepsilon$-net in the set $S(T)(B(\varepsilon, u_0, L^1(\Omega)) \cap \mathcal{A})$ for all sufficiently small $\varepsilon$, $u_0 \in \mathcal{A}$, and some $T > 0$ satisfying (4.42) with the constant $L$ independent of $\varepsilon$ and $u_0$. So, let $u_0 \in \mathcal{A}$ and let $\varepsilon \ll 1$ be fixed.

Let us introduce, for every $\theta > 0$, the following sets:

$$\begin{aligned} L(\theta) = L(\theta, u_0) &:= \{x \in \Omega, \ \ |u_0(x)| > \theta\}, \\ S(\theta) = S(\theta, u_0) &:= \{x \in \Omega, \ \ |u_0(x)| < \theta\}. \end{aligned} \tag{4.43}$$

Then, obviously, $S(\theta_1) \subset S(\theta_2)$ and $L(\theta_2) \subset L(\theta_1)$ if $\theta_1 \le \theta_2$. Moreover, since $u_0 \in C^\alpha(\Omega)$ and $\|u_0\|_{C^\alpha} \le M$, these sets are open,

$$\partial S(\theta) = \partial L(\theta) = \{x \in \Omega, \ u_0(x) = \theta\}, \quad \Omega = S(\theta) \cup L(\theta) \cup \partial L(\theta),$$

and, for every $\delta > 0$,

$$d[\partial S(\theta + \delta), \partial S(\theta)] \ge C_\delta \tag{4.44}$$

where the constant $C_\delta$ depends only on $\delta$, $\alpha$, and $M$ and is independent of $\theta$ and of the concrete choice of $u_0 \in \mathcal{A}$. Here and below we denote by $d(X, Y)$ the standard metric distance between sets in $\mathbb{R}^n$:

$$d[X, Y] := \inf_{x \in X} \inf_{y \in Y} \|x - y\|.$$

Let us now fix $\theta > 0$ and $\beta > 0$ in such a way that

$$g'(u) > 3\beta > 0, \quad \forall |u| < 5\theta \tag{4.45}$$

(it is possible to do this due to assumption (4.39)) and the cut-off function $\phi \in C^\infty(\mathbb{R}^n)$, $\phi \ge 0$, such that

$$\phi(x) = \begin{cases} 1, & x \in S(4\theta), \\ 0, & x \in L(5\theta). \end{cases} \tag{4.46}$$

Due to (4.44) and Proposition 1.1, this cut-off function $\phi$ can be chosen in such a way that

$$\|\phi\|_{C^k(\Omega)} \le C_k \tag{4.47}$$

where the constants $C_k$ depend only on $M$, $\alpha$, and $k$ and are independent of $u_0$.

We also recall that the trajectory $u(t) := S_t u_0$ belongs to $C^\alpha$ with respect to $t$ and $x$. Consequently, there exists time $T > 0$ (also depending only on $M$, $\beta$, and $\theta$) such that

$$\begin{aligned} g'(u(t,x)) > 2\beta, \quad x \in S(5\theta), \ t \in [0, T], \\ |u(t,x)| > \theta/2, \quad x \in L(\theta), \ t \in [0, T]. \end{aligned} \tag{4.48}$$

On the other hand, due to the interpolation inequality

$$\|w\|_{C(\Omega)} \le C \|w\|^\gamma_{L^1(\Omega)} \|w\|^{1-\gamma}_{C^\alpha(\Omega)} \tag{4.49}$$

(for some $0 < \gamma < 1$) and Hölder continuity, we obtain that

$$|v(t,x) - u(t,x)| \le C_1 \varepsilon^\gamma, \quad x \in \Omega, \ t \in [0, T], \tag{4.50}$$

for every solution $v(t)$ such that $v(0) \in \mathcal{A} \cap B(\varepsilon, u_0, L^1)$. Thus, assuming that $\varepsilon$ is small enough ($\varepsilon \le \varepsilon_0 \ll 1$ where $\varepsilon_0 > 0$ is independent of $u_0 \in \mathcal{A}$), we may improve (4.48) in the following way:

$$
\begin{aligned}
&g'(v(t,x)) > \beta, \quad x \in S(5\theta), \ t \in [0,T],\\
&|v(t,x)| > \theta/4, \quad x \in L(\theta), \ t \in [0,T],
\end{aligned}
\tag{4.51}
$$

uniformly with respect to $v_0 \in \mathcal{A} \cap B(\varepsilon, u_0, L^1)$.

In order to construct the required $\kappa\varepsilon$-net in $S(T)(\mathcal{A} \cap B(\varepsilon, u_0, L^1))$, we need to derive some smoothing property for differences of solutions. To this end, we consider the difference $w(t) := u_1(t) - u_2(t)$ of two solutions satisfying $u_i(0) \in B(\varepsilon, u_0, L^1)$. Then the function $w(t)$ solves the following equation:

$$
\partial_t w(t) = \Delta_x(l_1(t)w) - l_2(t)w, \quad w\big|_{t=0} = u_1(0) - u_2(0), \quad t \in [0,T],
\tag{4.52}
$$

where the $l_i(t)$ are the same as in (4.20).

Let us first consider the case of domains $L(\theta)$ where the equation (4.4) is, in a sense, nondegenerate. To this end, we need the following lemma, which is similar to the classical interior regularity estimates for the linear parabolic equation (4.52).

LEMMA 4.6. *Let $u_0 \in \mathcal{A}$ be arbitrary, let the sets $L(\theta, u_0)$ be defined via (4.43). Assume also that $u_1(t)$ and $u_2(t)$ are two solutions of (4.4) such that $u_i(0) \in \mathcal{A} \cap B(\varepsilon, u_0, L^1(\Omega))$. Then, the following estimate holds for every $t_0 \in (0,T)$:*

$$
\|u_1 - u_2\|_{C^\alpha([t_0,T]\times L(3\theta))} \le C_{t_0}\|u_1(0) - u_2(0)\|_{L^1(\Omega)}
\tag{4.53}
$$

*where the constant $C_{t_0}$ depends on $t_0$ and is independent of $\varepsilon$, $u_0$, $u_1$, and $u_2$.*

PROOF. We first prove that, for every $r > 2$, the functions $u_1$ and $u_2$ satisfy the following estimate:

$$
\|u_i\|_{W^{(1,2),r}([t_0/2,T]\times L(2\theta))} \le C_r, \quad i = 1,2,
\tag{4.54}
$$

where the constant $C_r$ depend on $r$ but is independent of the concrete choice of $u_0$, $\varepsilon$ and of the trajectories $u_1$ and $u_2$ (starting from $\mathcal{A} \cap B(\varepsilon, u_0, L^1)$). Indeed, let us verify it for $u = u_1$ (for $u = u_2$ it can be verified analogously). To this end, we introduce a new dependent variable $v(t,x) := f(u(t,x))$. Then, since $f(u)$ is nondegenerate if $|u| > \theta > 0$, one can easily verify that the function $v$ solves the following equation:

$$
\partial_t v = a\Delta_x v + h_u, \quad (t,x) \in [0,T] \times L(\theta),
\tag{4.55}
$$

where $a(t,x) := f'(u(t,x))$ and $h_u(t,x) := f'(u(t,x))[h(x) - g(u(t,x))]$. Moreover, due to (4.38), the coefficient $a$ is uniformly (with respect to $u \in \mathcal{K}$) Hölder continuous and the function $h_u$ is uniformly bounded in $L^\infty$. Furthermore, due to the second inequality of (4.51) and assumption (4.5), we have

$$
a(t,x) \ge C_1, \quad (t,x) \in [0,T] \times L(\theta),
\tag{4.56}
$$

where the constant $C_1$ is also independent of the choice of $u_0$ and $u$. Thus, we can apply the standard $L^r$-interior regularity estimate for the solution of the linear nondegenerate equation (4.55); see Proposition 1.3 and Corollary 1.1. Due to (4.44) with $\delta = \theta$, this estimate implies

$$
\|v\|_{W^{(1,2),r}([t_0/2,T]\times L(2\theta))} \le C_r(\|h\|_{L^r([0,T]\times L(\theta))} + \|v\|_{L^1([0,T]\times L(\theta))}) \le C_r'.
$$

Returning back to the variable $u = f^{-1}(v)$ and using the fact that $f \in C^2$ (and the fact that it is nondegenerate outside of zero), we indeed deduce estimate (4.54).

We now return to equation (4.52), which will now be considered in the domain $[t_0/2, T]\times L(2\theta)$. To this end, we first need to study the regularity of the coefficient $l_1(t)$. Indeed, since $f \in C^2$ and estimate (4.38) holds,

$$\|l_1\|_{C^\alpha([0,T]\times\Omega)} \le C \tag{4.57}$$

where the constant $C$ is independent of $u_1$ and $u_2$. Moreover, due to (4.54), we have

$$\|\partial_t l_1\|_{L^r([t_0/2,T]\times L(2\theta))} \le C\sum_{i=1}^{2}\|\partial_t u_i\|_{L^r([t_0/2,T]\times L(2\theta))} \le C_r'' \tag{4.58}$$

and, finally, due to the second inequality of (4.51), we also have

$$l_1(t,x) \ge \kappa > 0, \quad (t,x)\in [t_0/2, T]\times L(2\theta), \tag{4.59}$$

where the constants $C$, $C_r''$, and $\kappa$ are independent of the concrete choice of $u_0$, $u_1$, and $u_2$.

Let us introduce a new dependent variable $Z(t) := l_1(t)w(t)$. Then, this function solves

$$\partial_t Z = a(t,x)\Delta_x Z + l(t,x)Z, \quad (t,x)\in [t_0/2,T]\times L(2\theta), \tag{4.60}$$

where $a(t,x) := l_1(t,x)$ and $l(t,x) := l_2(t,x) - \frac{\partial_t l_1(t,x)}{l_1(t,x)}$. Furthermore, estimates (4.57)–(4.59) (together with the obvious fact that $l_2$ is uniformly bounded in the $L^\infty$-norm) allow us to apply the $L^q$-interior regularity estimate for equation (4.60), which gives (see Proposition 1.3 and Corollary 1.1)

$$\|Z\|_{W^{(1,2),q}([t_0,T]\times L(3\theta))} \le C_q\|Z\|_{L^1([t_0/2,T]\times L(2\theta))} \le C_q'\|w\|_{L^1([0,T]\times\Omega)}. \tag{4.61}$$

Now fixing $q$ large enough to have the embedding $W^{(1,2),q}\subset C^\alpha$, returning to the initial variable $w$, and using (4.57), we have

$$\|w\|_{C^\alpha([t_0,T]\times L(3\theta))} \le C\|w\|_{L^1([0,T]\times\Omega)}. \tag{4.62}$$

Estimating the right-hand side of (4.61) by (4.19), we deduce (4.53) and finish the proof of Lemma 4.6. □

Let us consider now equation (4.52) on the set $S(4\theta)$ where, due to the first condition of (4.51), we have, in a sense, the contraction property for the differences of solutions. Indeed, let us multiply equation (4.52) by

$$\phi(x)\operatorname{sgn}(w(t,x)) = \phi(x)\operatorname{sgn}(\phi(x)l_1(t,x)w(t,x)) \tag{4.63}$$

(where $\phi$ is defined by (4.46)) and use that

$$\phi\Delta_x[l_1(t)w] = \Delta_x(\phi(x)l_1(t)w) - 2\nabla_x\phi.\nabla_x(l_1(t)w) - \Delta_x\phi l_1(t)w. \tag{4.64}$$

Then integrating over $x\in\Omega$ and using the Kato inequality, we derive that

$$\partial_t(\phi,|w|) \le (\Delta_x\phi, l_1(t)|w|) - (g(u_1)-g(u_2), \phi\operatorname{sgn}(u_1-u_2)). \tag{4.65}$$

Taking now in account the first inequality of (4.51) and the fact that $\Delta_x\phi(x) = 0$ for $x\in S(4\theta)$, we deduce from (4.65) that

$$\partial_t(\phi,|w(t)|) + \beta(\phi,|w(t)|) \le C\|w(t)\|_{L^1(L(4\theta))} \tag{4.66}$$

and consequently, due to the Gronwall inequality and estimate (4.19), we infer

$$\begin{aligned}\|u_1(T)-u_2(T)\|_{L^1(S(4\theta))} &\le e^{Kt_0-\beta(T-t_0)}\|u_1(0)-u_2(0)\|_{L^1(\Omega)}\\ &+C_{t_0}\|u_1-u_2\|_{L^1([t_0,T]\times L(4\theta))}\end{aligned}\tag{4.67}$$

where $t_0$ is an arbitrary time in the interval $(0,T)$.

Let us now fix $t_0$ in such a way that

$$e^{Kt_0-\beta(T-t_0)}<1-\delta<1.$$

In this case (4.60) really gives a contraction in $S(4\theta)$. Moreover, using that

$$\|w\|_{L^1(\Omega)}\le\|w\|_{L^1(S(4\theta))}+\|w\|_{L^1(L(7\theta/2))}$$

and that $\|w(T)\|_{L^1(L(7\theta/2))}\le C\|w\|_{C([t_0,T]\times L(7\theta/2))}$, we derive from (4.53) and (4.67) the following basic inequalities:

$$\begin{cases}\|u_1-u_2\|_{C^\alpha([t_0,T]\times L(3\theta))}\le & P\|u_1(0)-u_2(0)\|_{L^1(\Omega)},\\ \|u_1(T)-u_2(T)\|_{L^1(\Omega)}\le & (1-\delta)\|u_1(0)-u_2(0)\|_{L^1(\Omega)}\\ & +P\|u_1-u_2\|_{C([t_0,T]\times L(7\theta/2))},\end{cases}\tag{4.68}$$

which is valid for all solutions $u_i$ such that $u_i(0)\in B(\varepsilon,u_0,L^1)\cap\mathcal{A}$ where the constants $T>0$, $\delta>0$, and $P$ are independent of the concrete choice of $\varepsilon\le\varepsilon_0$ and $u_0\in\mathcal{A}$.

Our next observation is the fact that the embedding $C^\alpha([t_0,T]\times L(3\theta,u_0))\subset C([t_0,T]\times L(7\theta/2,u_0))$ is compact. Moreover, since $L(7\theta/2,u_0)\subset L(3\theta,u_0)$ and $d[\partial L(3\theta,u_0),L(7\theta/2,u_0)]\ge C_\theta$ with the constant $C_\theta$ independent of $u_0\in\mathcal{A}$, this embedding is uniformly (with respect to $u_0\in\mathcal{A}$) compact. This means that there exists a monotone decreasing function $\mathbb{M}(\delta)$ such that

$$\mathbb{H}_\delta(B(1,0,C^\alpha([t_0,T]\times L(3\theta,u_0))),C([t_0,T]\times L(7\theta/2,u_0)))\le\mathbb{M}(\delta)\tag{4.69}$$

uniformly with respect to $u_0\in\mathcal{A}$ and $\delta>0$; see Proposition 1.4.

We are now ready to construct the required $\kappa\varepsilon$-net in the set $S(T)(B(\varepsilon,u_0,L^1)\cap\mathcal{A})$. To this end, we fix a minimal $\delta\varepsilon/(4P)$-net $V$ in the ball $B(P,u,C^\alpha([t_0,T]\times L(3\theta)))$, where $u(t):=S(t)u_0$, endowed by the metric of $C([t_0,T]\times L(7\theta/2))$. Then, due to (4.69), the number of points in that net can be estimated via

$$\begin{aligned}\#V&=N_{\varepsilon\delta/(4P)}(B(P\varepsilon,u,C^\alpha([t_0,T]\times L(3\theta))),C([t_0,T]\times L(7\theta/2)))\\ &=N_{\delta/(4P^2)}(B(1,0,C^\alpha([t_0,T]\times L(3\theta)),C([t_0,T]\times L(7\theta/2))))\\ &\le e^{\mathbb{M}(\delta/(4P^2))}\\ &:=L\end{aligned}\tag{4.70}$$

where $L$ is independent of $u_0$. Moreover, since we only need to control the trajectories $v(t)$ starting from $\mathcal{A}\cap B(\varepsilon,u_0,L^1)$ (all these trajectories are contained in the ball $B(P\varepsilon,u,C^\alpha([t_0,T]\times L(3\theta))$ due to the first estimate of (4.68)), then increasing the radii of the balls by a factor of two, we may construct the $\delta\varepsilon/(2P)$-net $\bar V=\{u^1,\dots,u^N\}$ in the set of these trajectories (in the metric of $C([t_0,T]\times L(7\theta/2))$) such that the functions $\{u^1,\dots,u^L\}$ are also the trajectories of (4.4) started from $\mathcal{A}\cap B(\varepsilon,u_0,L^1)$ and $\#\bar V\le L$. We claim that the set

$$V_{\kappa\varepsilon}(u_0):=\bar V\big|_{t=T}\tag{4.71}$$

is the required $\kappa\varepsilon$-net in $S(T)(B(\varepsilon, u_0, L^1) \cap \mathcal{A})$ with $\kappa = 1 - \delta/2 < 1$. Indeed, let $v(t)$ be an arbitrary trajectory starting from the $B(\varepsilon, u_0, L^1) \cap \mathcal{A}$. Then, due to our construction of the net $\bar{V}$, there exists a solution $u^i \in \bar{V}$ satisfying

$$\|u^i - v\|_{C([t_0,T]\times L(7\theta/2))} \le \delta\varepsilon/(2P). \tag{4.72}$$

Inserting this estimate into the second estimate of (4.68) and using that $\|u^i(0) - v(0)\|_{L^1(\Omega)} \le \varepsilon$, we infer

$$\|u^i(T) - v(T)\|_{L^1(\Omega)} \le (1-\delta)\varepsilon + \delta\varepsilon/2 = (1-\delta/2)\varepsilon.$$

Thus, (4.71) is indeed the required $\kappa\varepsilon$-net in $S(T)(B(\varepsilon, u_0, L^1) \cap \mathcal{A})$. Since an $\varepsilon$-ball of the attractor has been chosen arbitrarily, the recurrent formula (4.41) is verified for $\varepsilon \le \varepsilon_0$.

We are now ready to finish the proof of Theorem 4.3. Indeed, since the attractor $\mathcal{A}$ is compact in $L^1(\Omega)$,

$$\mathbb{H}_{\varepsilon_0}(\mathcal{A}, L^1(\Omega)) \le C_{\varepsilon_0} < \infty. \tag{4.73}$$

Moreover, starting from that $\varepsilon_0$-net and using the recurrent procedure described above, we prove that

$$\mathbb{H}_{\kappa^m\varepsilon_0}(\mathcal{A}, L^1(\Omega)) \le C_{\varepsilon_0} + m\log_2 L \tag{4.74}$$

for all $m \in \mathbb{N}$. It remains to recall (see Definition 1.1) that the fractal dimension of $\mathcal{A}$ can be expressed in terms of Kolmogorov $\varepsilon$-entropy, so that we obtain the estimate

$$\dim_f(\mathcal{A}, L^1(\Omega)) \le \frac{\log_2 L}{\log_2 1/\kappa} < \infty. \tag{4.75}$$

The finite dimensionality in $C(\Omega)$ is now an immediate corollary of Hölder continuity (4.38) and the interpolation inequality (4.49). Theorem 4.3 is proven. □

## 4.3. Exponential attractor

In the previous section, we proved the existence of a finite-dimensional global attractor $\mathcal{A}$ for problem (4.4). However, according to the definition of $\mathcal{A}$, we know only that $\operatorname{dist}_{L^1(\Omega)}(S(T)B, \mathcal{A})$ tends to zero as $t \to \infty$ (for every bounded subset $B$) and we do not have any information on the *rate* of convergence to the global attractor. Moreover, this rate of convergence can be arbitrarily slow and, to the best of our knowledge, there is not any way to control this rate of convergence in a more or less general situation (e.g., to express it in terms of physical parameters of the system considered). This leads to essential difficulties (as we already indicated in Chapter 3) in numerical simulations of global attractors and even makes them, in a sense, unobservable.

As we have seen in Chapter 3, in order to overcome this difficulty, the concept of the so-called *exponential* attractor has been suggested. In the present section, we construct the exponential attractor for the porous media equation (4.1). The main result of the section is formulated in the following theorem.

THEOREM 4.4. *Let the assumptions of Theorem 4.2 hold. Then, semigroup* (4.36) *generated in* $\Phi = L^1(\Omega)$ *by equation* (4.1) *possesses an exponential attractor* $\mathcal{M}$ *in the sense of the above definition. Moreover, this attractor is bounded in* $C^\alpha(\Omega)$, *for some* $\alpha > 0$.

PROOF. Let us introduce the set

$$\mathcal{C} := \big[\bigcup_{t\ge 1} S(t)\Phi\big]_{C^\alpha(\Omega)}$$

where $[\cdot]_V$ is a closure in the space $V$. Then, due to Theorem 4.2, we have

$$\|\mathcal{C}\|_{C^\alpha(\Omega)} \le M$$

and, due to the construction of $\mathcal{C}$, we also have

$$S(t)\mathcal{C} \subset \mathcal{C}, \quad t \ge 0.$$

Thus, instead of constructing an exponential attractor for $S(t)$ on the whole phase space $\Phi$, it is sufficient to construct it only for the restriction of that semigroup on a compact invariant subset $\mathcal{C}$. To this end, we will use the algorithm of constructing the $\varepsilon$-nets, developed in the proof of Theorem 4.2. To be more precise, let $V_{\varepsilon_0} \subset \mathcal{C}$ be the $\varepsilon_0$-net in the set $\mathcal{C}$ with sufficiently small $\varepsilon_0$. Then, arguing exactly as in the proof of Theorem 4.3, we can find positive numbers $T$, $L$, and $\kappa < 1$ such that, for every $u_0 \in \mathcal{C}$, the set $S(T)(B(\varepsilon_0, u_0, L^1) \cap \mathcal{C})$ possesses a $\kappa\varepsilon_0$-net with $L$-points. Thus, starting from the $\varepsilon_0$-net $V_{\varepsilon_0}$ of $\mathcal{C}$, we construct the $\kappa\varepsilon_0$-net $V_{\kappa\varepsilon_0} \subset S(T)\mathcal{C}$ of the set $S(T)\mathcal{C}$ such that

$$\#V_{\kappa\varepsilon_0} \le L\#V_{\varepsilon_0}.$$

Iterating this procedure, we then construct, for every $n \in \mathbb{N}$, $\kappa^n\varepsilon_0$-nets $V_{\kappa^n\varepsilon_0} \subset S(nT)\mathcal{C}$ in the set $S(nT)\mathcal{C}$ which satisfy

$$\#V_{\kappa^n\varepsilon_0} \le L^n\#V_{\varepsilon_0}. \tag{4.76}$$

These $\kappa^n\varepsilon_0$-nets in $S(nT)\mathcal{C}$ allow us to construct in a standard way the exponential attractor $\mathcal{M}_d$ for the *discrete* dynamical system, generated by the map $S = S(T) : \mathcal{C} \to \mathcal{C}$. This exponential attractor can be defined via the following expression:

$$\mathcal{M}_d := \big[\bigcup_{n=0}^{\infty}\bigcup_{m=0}^{\infty} S(m)V_{\kappa^n\varepsilon_0}\big]_{L^1(\Omega)}.$$

Indeed, semiinvariance and the exponential attraction property are obvious since

$$\operatorname{dist}(S(nT)\mathcal{C}, V_{\kappa^n\varepsilon_0}) \le \kappa^n\varepsilon_0, \quad n \in \mathbb{N}, \quad \kappa < 1. \tag{4.77}$$

The finiteness of a fractal dimension of $\mathcal{M}_d$ can be easily verified using (4.76), (4.77), and the fact that $V_{\kappa^n\varepsilon_0} \subset S(nT)\mathcal{C}$; see [36] for the details. Thus, since $\mathcal{M}_d$ is closed, it is indeed an exponential attractor for the map $S = S(T) : \mathcal{C} \to \mathcal{C}$. As usual, the required exponential attractor $\mathcal{M}$ for the semigroup $S(t)$ with continuous time can be defined via

$$\mathcal{M} = \bigcup_{t\in[T,2T]} S(t)\mathcal{M}_d.$$

Indeed, semiinvariance and the exponential attraction property follow immediately from the analogous properties of the discrete attractor $\mathcal{M}_d$ and the finiteness of a fractal dimension in $L^1(\Omega)$ can be easily verified using the facts that the dimension of $\mathcal{M}_d$ is finite and that the map $S(t)$ is uniformly Hölder continuous on $\mathcal{M}_d$; see [29], [36] for details. Thus, $\mathcal{M}$ is indeed the required exponential attractor and Theorem 4.4 is proven. □

REMARK 4.2. There exists a rather important exceptional class of dynamical systems whose global attractors are simultaneously the exponential ones. These are the so-called *regular* attractors (for details see Chapter 2 of this book and the references therein) which appear in *smooth* dynamical systems with the global Lyapunov function under the additional assumption that all of the equilibria are hyperbolic. In our case of the porous media equation (4.1), we obviously have the global Lyapunov function. Indeed, arguing as in Lemma 4.2, we can easily verify that the functional

$$\mathcal{G}(u) := \int_{x\in\Omega} \frac{1}{2}|\nabla_x f(u(x))|^2 + F_{0,g}(u(x)) - h(x)F_0(u(x))\,dx,$$

where $F_{0,g}$ and $F_0$ are the same as in Lemma 4.2, satisfies

$$\mathcal{G}(u(t)) - \mathcal{G}(u(0)) = -\int_0^T \int_{x\in\Omega} f'(u(t,x))|\partial_t u(t,x)|^2\,dx\,dt$$

and, consequently, gives a global Lyapunov function for (4.1).

Nevertheless, the regular attractor theory does not seem to be applicable here, since equation (4.1) is *degenerate* and we can obtain neither the differentiability of semigroup $S(t)$ with respect to the initial data nor the hyperbolicity of the equilibria.

REMARK 4.3. As we have already mentioned, the appropriate smoothing properties for differences of solutions play a crucial role in the modern theory of exponential attractors; see [29], [36]. The most simple abstract version (which gives an existence of an exponential attractor for the map $S$) of such a smoothing property is

$$\|Su_1 - Su_2\|_{H_1} \le K\|u_1 - u_2\|_H \tag{4.78}$$

where the constant $K$ is independent of $u_i$ belonging to a bounded invariant subset and $H_1$ and $H$ are two *Banach* spaces such that $H_1$ is compactly embedded in $H$; see Theorem 3.2.

Our proof (Theorem 4.4) of the existence of an exponential attractor can also be embedded in an abstract scheme, but, in contrast to (4.78), in our situation, the spaces $H_1$ and $H$ should *depend* on $u_1$ and $u_2$.

To be more precise, let $S$ be an abstract map acting on some Banach space $X$ and let $\mathcal{C}$ be a compact subset of $X$ such that

$$S\mathcal{C} \subset \mathcal{C}.$$

Let us also assume that, for every $u_0 \in \mathcal{C}$ and for every $\varepsilon \le \varepsilon_0$, there exists a pair of Banach spaces $H_1(u_0,\varepsilon)$ and $H(u_0,\varepsilon)$ such that $H_1$ is compactly embedded in $H$ and this embedding is *uniformly* (with respect to $\varepsilon$ and $u_0$) compact in the sense of Kolmogorov $\varepsilon$-entropy (compare with (4.69)) and a map $\mathcal{T}_{u_0,\varepsilon} : B(\varepsilon,u_0,X)\cap\mathcal{C} \to H_1(u_0,\varepsilon)$ such that, for every $u_1,u_2 \in B(\varepsilon,u_0,X)\cap\mathcal{C}$

$$\begin{cases} \|\mathcal{T}_{u_0,\varepsilon}u_1 - \mathcal{T}_{u_0,\varepsilon}u_2\|_{H_1} \le P\|u_1-u_2\|_X, \\ \|Su_1 - Su_2\|_X \le (1-\delta)\|u_1-u_2\|_X + P\|\mathcal{T}_{u_0,\varepsilon}u_1 - \mathcal{T}_{u_0,\varepsilon}u_2\|_H \end{cases}$$

(compare with (4.68)). Then, arguing exactly as in the proof of Theorems 4.3 and 4.4, we can verify the existence of an exponential attractor for the abstract map $S$.

## 4.4. Infinite-dimensional global attractor

We now show that the attractor $\mathcal{A}$ can be infinite dimensional if condition (4.39) is violated. To be more precise, we consider the following equation of the form (4.1):

$$\partial_t u = \Delta_x(u|u|^{p-1}) + u - g(u), \quad u\big|_{\partial\Omega} = 0, \tag{4.79}$$

where $p > 1$ and the function $g$ vanishes near zero and satisfies assumption (4.6) at infinity. As we will show, the associated attractor has infinite dimension. That is why we will study below its Kolmogorov $\varepsilon$-entropy. The following theorem which gives a natural lower bounds for the entropy of the attractor can be considered as the main result of this section.

THEOREM 4.5. *Let the above assumptions hold. Then the global attractor $\mathcal{A}$ associated with equation* (4.79) *is infinite dimensional and its $\varepsilon$-entropy possesses the following estimate:*

$$\mathbb{H}_\varepsilon(\mathcal{A}, L^\infty) \geq C\left(\frac{1}{\varepsilon}\right)^{n(p-1)/2}, \tag{4.80}$$

*for some positive constant $C$ independent of $\varepsilon$.*

PROOF. In order to prove the theorem, we will study as usual the so-called unstable set $M_+(0)$ of the equilibrium $u \equiv 0$ of equation (4.79). By definition,

$$M_+(0) = \{u_0 \in L^\infty(\Omega), \ \exists u \in \mathcal{K}, \ \lim_{t\to-\infty} \|u(t)\|_{L^\infty} = 0, \ u(0) = u_0\}. \tag{4.81}$$

Obviously, $M_+(0) \subset \mathcal{A}$. On the other hand, since the nonlinearity $g$ vanishes at the origin, it is sufficient to consider only the backward solutions of the following "linearized" problem:

$$\partial_t u = \Delta_x(u|u|^{p-1}) + u, \quad u(0) = u_0, \quad t \leq 0, \tag{4.82}$$

tending to zero as $t \to -\infty$ (all such solutions belonging to the sufficiently small ball in $L^\infty$ will also satisfy equation (4.79)). In order to solve equation (4.82), we change the unknown $v(t) := e^{-t}u(t)$. Then we arrive at

$$\partial_t v = e^{(p-1)t}\Delta_x(v|v|^{p-1}), \quad v(0) = u_0, \quad t \in (-\infty, 0). \tag{4.83}$$

Finally, making one more variable change $s := e^{(p-1)t}$, we obtain

$$\partial_s \tilde{v} = (p-1)\Delta_x(\tilde{v}|\tilde{v}|^{p-1}), \quad \tilde{v}(1) = u_0, \quad s \in (0, 1]. \tag{4.84}$$

Let $\mathcal{S}_t : L^\infty(\Omega) \to L^\infty(\Omega)$ be the solving operator of the following problem:

$$\partial_t w = (p-1)\Delta_x(w|w|^{p-1}), \quad w\big|_{t=0} = w_0, \quad t \geq 0. \tag{4.85}$$

Then, we have proven that the unstable set $M_+(0)$ contains the image of a sufficiently small ball $B(r_0) := B(r_0, 0, L^\infty)$:

$$\mathcal{S}_1 B(r_0) \subset M_+(0) \subset \mathcal{A}. \tag{4.86}$$

Thus, it is sufficient to estimate the $\varepsilon$-entropy of the set $\mathcal{S}_1 B(r_0)$. To this end, we recall that in contrast to the nondegenerate case, solutions of equation (4.85) possess the spatially localized solutions, i.e., there exists a nonzero solution $W(t,x) \geq 0$ of equation (4.85) such that for $W(0) \in B(r_0)$ we have

$$\operatorname{supp} W(s, \cdot) \subset K \subset\subset \Omega, \tag{4.87}$$

for all $s \in [0,1]$. For simplicity, we assume that $\|W(1)\|_{L^\infty} = 1$. On the other hand, if $W(s,x)$ solves (4.85), then the scaled function

$$W_\varepsilon(s,x) := \varepsilon W(s, \varepsilon^{(1-p)/2}x) \tag{4.88}$$

also solves (4.85) for every $\varepsilon \neq 0$ and

$$\operatorname{supp} W_\varepsilon(s,x) \subset K_\varepsilon := \varepsilon^{(p-1)/2}K. \tag{4.89}$$

Therefore, for every sufficiently small $\varepsilon$ it follows from (4.89) that there exists a finite set $R_\varepsilon := \{x_i\} \subset \Omega$ such that

$$\begin{aligned} &(1)\ \ (x+K_\varepsilon)\cap(y+K_\varepsilon)=\emptyset, \quad \forall x,y\in R_\varepsilon, \quad x\neq y,\\ &(2)\ \ \#R_\varepsilon \geq C\left(\frac{1}{\varepsilon}\right)^{n(p-1)/2},\\ &(3)\ \ x+K_\varepsilon \subset\subset \Omega, \quad \forall x\in R_\varepsilon. \end{aligned} \tag{4.90}$$

Consequently, for every $m \in \{0,1\}^{R_\varepsilon}$ the function

$$W_{m,\varepsilon}(s,x) := \sum_{i=1}^{\#R_\varepsilon} m_i W_\varepsilon(s, x - x_i) \tag{4.91}$$

solves (4.85) in $\Omega$. On the other hand, obviously,

$$\|W_{m^1,\varepsilon}(1,\cdot) - W_{m^2,\varepsilon}(1,\cdot)\|_{L^\infty} \geq \varepsilon \tag{4.92}$$

for $m^1 \neq m^2$. Since we have $2^{\#R_\varepsilon}$ different functions of that form,

$$\mathbb{H}_\varepsilon(\mathcal{A}, L^\infty) \geq \mathbb{H}_\varepsilon(\mathcal{S}_1 B(r_0), L^\infty) \geq \#R_\varepsilon \geq C\left(\frac{1}{\varepsilon}\right)^{n(p-1)/2}. \tag{4.93}$$

Theorem 4.5 is proven. □

REMARK 4.4. It is worth recalling the usual method of obtaining the lower bounds for attractor dimension based on the unstable manifolds theory. Namely, if we are able to find a (hyperbolic) equilibrium with large/infinite instability index, then due to this theory, the attractor contains a manifold of large/infinite dimension (which is equal to the instability index; see [9]). But this method is not applicable for the degenerate equations since the associated semigroups are usually *not differentiable.* Indeed, under the assumptions of Theorem 4.5 the formal linearization near zero equilibrium reads

$$\partial_t w = w$$

which, of course, has infinite instability index. But, in contrast to the nondegenerate case the backward solutions of that equation are not associated with the backward solutions of the whole nonlinear equation (due to the lack of regularity) and, consequently, do not give the infinite dimensionality of the associated unstable set. That is the reason we needed to develop above the alternative method based on the existence of a localized solution and scaling technique which is closely related to the degenerate nature of the problem considered.

The next corollary gives the lower bounds for the $\varepsilon$-entropy in the initial phase space $L^1(\Omega)$.

COROLLARY 4.1. *Let the assumptions of Theorem 4.5 hold. Then, the Kolmogorov $\varepsilon$-entropy of the attractor $\mathcal{A}$ in $L^1(\Omega)$ possesses the following estimate:*

$$\mathbb{H}_\varepsilon(\mathcal{A}, L^1(\Omega)) \geq C\left(\frac{1}{\varepsilon}\right)^{\frac{n(p-1)}{2+n(p-1)}} \tag{4.94}$$

*where the constant $C$ is independent of $\varepsilon$.*

PROOF. Indeed, according to (4.88),

$$\|W_\varepsilon(1,x)\|_{L^1(\Omega)} = C\varepsilon^{1+n(p-1)/2} \tag{4.95}$$

and, consequently, instead of (4.92), we now have

$$\|W_{m^1,\varepsilon}(1,\cdot) - W_{m^2,\varepsilon}(1,\cdot)\|_{L^1} \geq C\varepsilon^{1+n(p-1)/2}. \tag{4.96}$$

Therefore, the distance between any two functions of the form (4.91) is not less than $C\varepsilon^{1+n(p-1)/2}$. Since we have $2^{\#R_\varepsilon}$ such functions, estimate (4.94) is verified and Corollary 4.1 is proven. □

We note that, in contrast to the lower bounds for the entropy in the $L^\infty(\Omega)$-metric given in Theorem 4.5, estimate (4.94) seems to be very rough (in particular, the exponent on the right-hand side of it remains bounded as $p \to \infty$ or $n \to \infty$). Nevertheless, it allows us to establish the infinite dimensionality of global attractors for the essentially more general class of porous media equations. We illustrate this in the following example of the degenerate Chafee-Infante equation:

$$\partial_t u = \Delta_x(u^3) + u - u^3, \quad u\big|_{\partial\Omega} = 0. \tag{4.97}$$

COROLLARY 4.2. *Let $\mathcal{A}$ be the attractor of equation* (4.97). *Then, its Kolmogorov $\varepsilon$-entropy satisfies*

$$\mathbb{H}_\varepsilon(\mathcal{A}, L^1(\Omega)) \geq C\begin{cases}\varepsilon^{-1/2}, & n = 1,\\ \varepsilon^{-2/(n+1)}, & n \geq 2,\end{cases} \tag{4.98}$$

*for some $C > 0$ independent of $\varepsilon$.*

PROOF. Indeed, analogously to the proof of Theorem 4.5, replacing the dependent variable $u(t) = e^t v(t)$ in equation (4.97) and scaling time $s = e^{2t}$, we arrive at

$$\partial_s v = 2\Delta_x(v^3) - v^3, \quad s \in [0,1]. \tag{4.99}$$

Now let $W_\varepsilon(s,x) := \varepsilon W(s, \varepsilon^{-1}x)$, $\varepsilon \ll 1$, be the solutions of equation

$$\partial_s w = 2\Delta_x(w^3) \tag{4.100}$$

constructed in the proof of Theorem 4.5 and let us define, for every $m \in \{0,1\}^{R_\varepsilon}$, the functions $W_{m,\varepsilon}(s,x)$ via (4.91). We also recall that the $L^1$-norm of every solution of (4.100) with compact support preserves, consequently,

$$\|W_{m,\varepsilon}(s,\cdot)\|_{L^1(\Omega)} = C\varepsilon^{1+n}|m| \tag{4.101}$$

where $|m| = \sum m_i$. Let us now define the associated solutions $\bar{W}_{m,\varepsilon}(s,x)$ of (4.99) with $\bar{W}_{m,\varepsilon}(0,x) = W_{m,\varepsilon}(0,x)$. Then, the difference $Z(s) = \bar{W}_{m,\varepsilon}(s) - W_{m,\varepsilon}(s)$ satisfies

$$\partial_s Z = 2\Delta_x(\bar{W}^3_{m,\varepsilon} - W^3_{m,\varepsilon}) - (\bar{W}^3_{m,\varepsilon} - W^3_{m,\varepsilon}) - W^3_{m,\varepsilon}. \tag{4.102}$$

Multiplying (4.102) by $\operatorname{sgn} Z$, integrating, and using the Kato inequality together with (4.101), we obtain

$$\|\bar{W}_{m,\varepsilon}(1,\cdot) - W_{m,\varepsilon}(1,\cdot)\|_{L^1(\Omega)} \le \int_0^1 (W_{m,\varepsilon}(s))^3\,ds \le C\varepsilon^{3+n}|m|. \tag{4.103}$$

Thus, due to (4.96) with $p = 3$,

$$\begin{aligned} \|\bar{W}_{m^1,\varepsilon}(1) - \bar{W}_{m^2,\varepsilon}(1)\|_{L^1(\Omega)} &\ge \|W_{m^1,\varepsilon}(1) - W_{m^2,\varepsilon}(1)\|_{L^1(\Omega)} \\ &\quad - C\varepsilon^{3+n}(|m^1| + |m^2|) \\ &\ge C\varepsilon^{1+n}(1 - \varepsilon^2(|m^1| + |m^2|)) \end{aligned} \tag{4.104}$$

and, consequently, the functions $\bar{W}_{m^i,\varepsilon}$ are $\frac{1}{2}C\varepsilon^{1+n}$-separated if

$$|m^i| \le \frac{1}{4}\varepsilon^{-2}. \tag{4.105}$$

We recall that $\#R_\varepsilon \sim \varepsilon^{-n}$. Then, for $n = 1$, (4.105) is automatically satisfied for small $\varepsilon$ and so the number $N$ of $\frac{1}{2}C\varepsilon^2$-separated functions equals $2^{\#R_\varepsilon} \sim 2^{C\varepsilon^{-1}}$. In the case of $n \ge 2$, this number $N$, obviously, satisfies $N \ge 2^{\frac{1}{4}\varepsilon^{-2}}$. These estimates immediately imply (4.98) and this finishes the proof of Corollary 4.2. □

To conclude, we discuss also the upper bounds for the Kolmogorov $\varepsilon$-entropy of the attractors of porous media equations of the form of (4.79). To this end, we recall that the polynomial asymptotics of the Kolmogorov entropy (like $\varepsilon^{-k}$) are typical for the embeddings of Sobolev spaces, and, consequently, the upper bounds of the entropy in the same form can be obtained by studying the maximal smoothness of the attractor. In particular, Theorem 4.2 together with the standard asymptotics for the Kolmogorov entropy of the embedding $C^\alpha \subset C$ (see [61]) gives

$$\mathbb{H}_\varepsilon(\mathcal{A}, L^\infty) \le C\left(\frac{1}{\varepsilon}\right)^{n/\alpha}. \tag{4.106}$$

In particular, for $n = 1$ under the assumptions of Theorem 4.5, we have

$$C^{-1}\left(\frac{1}{\varepsilon}\right)^{(p-1)/2} \le \mathbb{H}_\varepsilon(\mathcal{A}, L^\infty) \le C\left(\frac{1}{\varepsilon}\right)^{6p}. \tag{4.107}$$

In turn, estimate (4.80) (and the scaling method, introduced in Theorem 4.5) gives the natural upper bounds for the smoothness of the attractor.

COROLLARY 4.3. *Let $W_\varepsilon(s,x)$ be as defined in (4.88). Then $W_\varepsilon(1,x) \notin C^{(\alpha)}(\Omega)$ for all $\alpha > 2/(p-1)$. Moreover, if the inequality*

$$\|u_0\|_{W^{1,1}(\Omega)} \le C \tag{4.108}$$

*holds uniformly with respect to $u_0$ belonging to the attractor $\mathcal{A}$ of (4.79), then, necessarily, $p \le 3$.*

PROOF. Indeed, analogously to the proof of Theorem 4.5, all functions $W_\varepsilon(1,x) = \varepsilon W(1, \varepsilon^{(1-p)/2}x)$ belong to the attractor. On the other hand,

$$\|W_\varepsilon(1)\|_{C^\alpha(\Omega)} = \varepsilon^{1-\alpha(p-1)/2}\|W(1)\|_{C^\alpha(\Omega)}. \tag{4.109}$$

Since the left-hand side of (4.109) should be bounded as $\varepsilon \to 0$, then, necessarily, $\alpha \le 2/(p-1)$.

Analogously,

$$\|W_\varepsilon(1)\|_{W^{1,1}(\Omega)} = \varepsilon^{(3-p)/2}\varepsilon^{n(p-1)/2}\|W(1)\|_{W^{1,1}(\Omega)}. \tag{4.110}$$

Let us now consider the function $W_{\vec{1},\varepsilon}(1,x)$ associated with (4.91) with all $m_i = 1$. Then, since $\#R_\varepsilon \sim \varepsilon^{n(1-p)/2}$, (4.110) implies that

$$\|W_{\vec{1},\varepsilon}(1)\|_{W^{1,1}(\Omega)} = C\varepsilon^{(3-p)/2}\|W(1)\|_{W^{1,1}(\Omega)}.$$

Thus, (4.108) indeed implies that $p \le 3$ and Corollary 4.3 is proven. ☐

CHAPTER 5

# Porous medium equation in heterogeneous media: Long-time dynamics

In Chapter 4 we studied the long-time behavior of solutions of a porous medium in a so-called homogeneous medium in terms of a global attractor.

In Chapter 4 we proved the following facts:

(a) the infinite dimensionality of the attractor,
(b) the polynomial asymptotics of its $\varepsilon$-Kolmogorov entropy,
(c) the difference of the asymptotics of the $\varepsilon$-Kolmogorov entropy depending on the choice of the underlying phase spaces,

which one cannot observe in the case of nondegenerate equations.

The main purpose of Chapter 5 is to extend the results of Chapter 4, that is, the study of the global attractors for a much wider class of porous medium equations in a heterogeneous medium. More precisely, we consider

$$(\mathrm{E})\begin{cases} \dfrac{\partial u}{\partial t} = \Delta\left(|u|^{p-1}u\right) - g(x,u) + h(x) & \text{in } \Omega \times [0,\infty), \qquad (\mathrm{E1}) \\ u = 0 & \text{in } \partial\Omega \times [0,\infty), \qquad (\mathrm{E2}) \\ u(x,0) = u_0(x) & x \in \Omega, \qquad (\mathrm{E3}) \end{cases}$$

where $\Omega$ is a bounded domain in $\mathbb{R}^N$ with smooth boundary $\partial\Omega$.

We especially emphasize that one cannot apply the direct approach developed in Chapter 4 for (E), so achieving our goal requires new ideas. To this end, we use some new comparison results and special scale transformations which will play a very important role in our arguments.

Chapter 5 is organized as follows. In Section 5.1, we briefly recall the results on the existence, uniqueness, and regularity of solutions of (E1)–(E3). The existence of the global attractor for the semigroup generated by (E1)–(E3), its infinite dimensionality, and the asymptotics of the Kolmogorov entropy are given in Section 5.2. In particular, as in Section 4.4, we show

(a) the infinite dimensionality of the attractor,
(b) the polynomial asymptotics of its $\varepsilon$-Kolmogorov entropy,
(c) the difference of the asymptotics of the $\varepsilon$-Kolmogorov entropy depending on the choice of the underlying phase spaces.

## 5.1. Existence of global solutions and a priori estimates

In this section we briefly investigate the solvability of the problem (E1)–(E3). For this purpose, we assume that $g(x,\theta)$ satisfies the following conditions:

(a) $g \in C(\overline{\Omega} \times \mathbb{R}^1)$ and $\frac{\partial g}{\partial \theta} \in C(\overline{\Omega} \times \mathbb{R}^1)$,
(b) $\frac{\partial g}{\partial \theta}(x, \theta) \geq -C + \kappa|\theta|^{q-1}$ for some $q > 1$, $\kappa > 0$ for all $(x, \theta) \in \overline{\Omega} \times \mathbb{R}^1$.

REMARK 5.1. We can relax the strong dissipativity assumption (b) assuming

$$\limsup_{|\theta| \to \infty} \frac{g(x, \theta)}{\theta} > 0$$

for all $(x, \theta) \in \overline{\Omega} \times \mathbb{R}^1$.

Then (E1)–(E3) admits a unique global weak solution in the following sense:

THEOREM 5.1. *Assume that conditions* (a) *and* (b) *are satisfied and* $h \in L^\infty(\Omega)$. *Then, for any* $u_0 \in L^1(\Omega)$, *there exists a unique weak solution* $u$ *of the problem* (E1)–(E3) *satisfying for every* $t > 0$

$$u \in C([0, T], L^1(\Omega)) \cap C([t, T], L^\infty(\Omega)),\ |u|^{p-1}u \in L^2([t, T], W_0^{1,2}(\Omega)). \tag{5.1}$$

*Furthermore,* $S(t) : u_0 \longmapsto u(t)$ *is continuous in the strong topology of* $L^1(\Omega)$ *for each fixed* $t > 0$.

PROOF. For the convenience of the reader we will briefly recall the basic steps of the proof, which can be done in the same manner as in Chapter 4 (see also [40]).

Step 1. We consider the nondegenerate approximation of (E1)–(E3), that is,

$$\begin{cases} \frac{\partial u}{\partial t} = \Delta_x \left(|u|^{p-1}u\right) + \varepsilon \Delta_x u - g(x, u) + h, \\ u|_{t=0} = u_0,\ u|_{\partial\Omega} = 0, \end{cases} \tag{5.2}$$

which, obviously, due to the assumptions on $g$ and $h$ has a unique solution for every $\varepsilon > 0$ and sufficiently smooth $u_0$.

Step 2. In this step, using the strong dissipativity assumption (b) on $g$, we obtain the following estimates:

$$||u(t)||_{L^1(\Omega)} \leq e^{-\alpha t}||u_0||_{L^1(\Omega)} + C\left(1 + ||h||_{L^1(\Omega)}\right), \tag{5.3}$$

$$||u(t)||_{L^\infty(\Omega)} \leq e^{-\alpha t}||u_0||_{L^\infty(\Omega)} + C\left(1 + ||h||_{L^\infty(\Omega)}\right), \tag{5.4}$$

where the positive constants $\alpha$ and $C$ depend only on the function $g$ and are independent of $\varepsilon, t, u_0$, and $h$. Moreover, the following $L^1 - L^\infty$-smoothing property holds:

$$||u(t)||_{L^\infty(\Omega)} \leq Q\left(\frac{1}{t} + ||h||_{L^\infty(\Omega)}\right) \text{ for all } t > 0,\ u_0 \in L^1(\Omega), \tag{5.5}$$

where the monotonic function $Q$ is independent of $\varepsilon, t$, and $u$.

Step 3. Let $u_{\varepsilon_n}(t, x)$ be a sequence of solutions of the (nondegenerate) approximated equation (5.2). Then the sequence $u_{\varepsilon_n}(t, x)$ satisfies estimates (5.3) and (5.4) of Step 2 uniformly in $n$. Moreover, since $u_0$ is smooth, it is easy to see that

$$||u_{\varepsilon_n}||_{W^{1/p-\delta,2p}([t,t+1]\times\Omega)} \leq C$$

holds uniformly in $t$, $t \geq 0$, and $n$. Due to the compactness (for $\delta > 0$ sufficiently small) of the embeddings

$$W^{1/p-\delta,2p}([t, t+1] \times \Omega) \subset\subset C([0, T], L^1(\Omega)),$$

we can assume there exists a subsequence $u_{\varepsilon_n} \to u$ strongly in $C([0,T], L^1(\Omega))$. Hence $u$ is a solution of (E1)–(E3). Thus for smooth initial data $u_0$ the existence of solution (5.2) is verified. For nonsmooth initial data $u_0 \in L^1(\Omega)$ a proof of the existence of a weak solution is given by a standard procedure; namely, we approximate in $L^1(\Omega)$ the nonsmooth initial data $u_0$ by a sequence of smooth ones $u_{0,n}$ and take into account that the solutions of (E1)–(E3) associated to $u_{0,n}$ (which we denote by $u_n(t)$) satisfy

$$\left\| u_n^{(1)}(t,x) - u_n^{(2)}(t,x) \right\|_{L^1(\Omega)} \le e^{Kt} \left\| u_{0,n}^{(1)}(x) - u_{0,n}^{(2)}(x) \right\|_{L^1(\Omega)} \tag{5.6}$$

where $K := \max_{\theta \in \mathbb{R}} \left\{ -\frac{\partial g}{\partial \theta}(x,\theta) \right\}$ and $u_n^{(1)}(t,x)$ and $u_n^{(2)}(t,x)$ are the solutions of (5.2) corresponding to the initial data $u_{0,n}^{(1)}(x)$ and $u_{0,n}^{(2)}(x)$, respectively. Consequently we can pass to the limit as $n \to \infty$, obtaining a weak solution $u(t,x)$ for every $u_0 \in L^1(\Omega)$. □

REMARK 5.2. Note that (5.6), that is,

$$\|u_1(t,x) - u_2(t,x)\|_{L^1(\Omega)} \le e^{Kt} \|u_1(0,x) - u_2(0,x)\|_{L^1(\Omega)},$$

where $K := \max_{\theta \in \mathbb{R}} \left\{ -\frac{\partial g}{\partial \theta}(x,\theta) \right\}$, is obtained by multiplying

(5.7)
$$\begin{cases} \frac{\partial v}{\partial t} = \Delta_x \left( |u_1|^{p-1} u_1 \right) - \Delta_x \left( |u_2|^{p-1} u_2 \right) - \left[ g(x, u_1(t,x)) - g(x, u_2(t,x)) \right], \\ v|_{t=0} = u_1(0,x) - u_2(0,x), \ v|_{\partial\Omega} = 0, \end{cases}$$

by sgn $v$ and using the "Kato inequality" (for a rigorous analysis see the proof of Theorem 4.1) as well as the strong dissipativity assumption (b) on $g(x,\theta)$; $v(t,x) := u_1(t,x) - u_2(t,x)$.

## 5.2. Infinite-dimensional global attractor

Our existence result for global attractors of (E1)–(E3) in the phase space $\Phi = L^1(\Omega)$ can be stated as follows.

THEOREM 5.2. *Let all the assumptions in Theorem* 5.1 *be satisfied. Then the semigroup* $S(t)$ *associated with* (E1)–(E3) *possesses a global attractor* $\mathcal{A}$ *in the phase space* $L^1(\Omega)$ *which is globally bounded in* $C^\alpha(\overline{\Omega})$ *with* $\alpha \in (0,1]$ *and has the following structure*: $\mathcal{A} := \mathcal{K}_0 := \{ u(0) \, ; \, \{u(t)\}_{t\in\mathbb{R}^1} \in \mathcal{K} \}$, *where* $\mathcal{K}$ *is the set of all bounded solutions of* (E1)–(E2) *defined on* $\mathbb{R}^1$, *i.e.,*

$$\mathcal{K} = \{\{u(t)\}_{t\in\mathbb{R}^1} \, ; \, u(t) \text{ is a solution of (E1)–(E2) on } \mathbb{R}^1, \ \sup_{t\in\mathbb{R}^1} \|u(t)\|_{L^1} < +\infty\}.$$

PROOF. In order to prove the existence of the global attractor $\mathcal{A}$ for (E1)–(E3), it suffices to show that the semigroup $S(t)$ associated with (E1)–(E3) is continuous in the topology of $L^1(\Omega)$ for each $t > 0$ and that there exists a pre-compact absorbing set $\mathcal{B}$ in $L^1(\Omega)$ such that for every $\xi \in L^1(\Omega)$, there exists $T = T(\xi) > 0$ such that $S(t)\,\xi \in \mathcal{B} \ \forall\, t \in [T, +\infty)$. (See Theorem 1.1 of [93].) For our case, the first property is ensured by Theorem 5.1 and the second by the inequality (4.32). The characterization of $\mathcal{A}$ in terms of $\mathcal{K}$ is derived by standard arguments. □

Next we present lower bounds for Kolmogorov $\varepsilon$-entropy of the attractor $\mathcal{A}$ in $\Phi = L^p(\Omega)$ $(1 \le p \le \infty)$. In compliance with Definition 1.1 we denote by $\mathcal{N}_\varepsilon(\mathcal{A}, \Phi)$ the minimal number of $\varepsilon$-balls in $\Phi$ that cover $\mathcal{A}$ (recall that $\mathcal{A}$ is a compact set in $\Phi$). From now on we assume that $p > 1$, $h \equiv 0$, and $g(\cdot, u)$ satisfies the following assumption:

$(\mathrm{I})_g$ There exists an open bounded subset $\omega$ of $\Omega$ and $\alpha > 0$ such that $g^\alpha(x,u) = g(x,u) + \alpha\, u$ satisfies the following:

$(\mathrm{I})_1$ There exist $a(s)$, $\rho \ge 0$, and $h(x,s,v)$ satisfying

$$s^{-\frac{p}{p-1}}\, g^\alpha(x, s^{\frac{1}{p-1}} v) = a(s)\,|v|^\rho\, h(x,s,v), \quad (x,s,v) \in \omega \times [0,1] \times \mathbb{R}^1,$$
$$h(x,s,v),\ h'_v(x,s,v) \in C(\overline{\omega} \times [0,1] \times \mathbb{R}^1), \quad h(x,s,0) = 0,$$
$$a \ge 0,\ a \in L^1(0,1),\ s\,a^2 \in L^1(0,1).$$

$(\mathrm{I})_2$ There exist $C > 0$ and $\delta > 0$ such that

$$|\,v\,|^\rho\,|\,h(x,s,v)\,| \le C\,|\,v\,|^{1+\delta}, \quad (x,s,v) \in \omega \times [0,1] \times [0,1].$$

Then our result on the infinite dimensionality of global attractors for (E1)–(E3) reads as follows.

THEOREM 5.3. *Let* $(\mathrm{I})_g$ *be satisfied and assume that* (E1)–(E3) *possess a global attractor* $\mathcal{A}$ *in the topology of* $L^1(\Omega)$. *Then the fractal dimension of* $\mathcal{A}$ *is infinite.*

The presentation of condition $(\mathrm{I})_g$ might seem to be somewhat obscure. In order to clarify the meaning of this condition, we show below that it covers a very large class of nonlinearity.

EXAMPLE 5.1. Let $g(x,u) = -\alpha\, u + b_1(x)\,|u|^{q_1-2}u - b_2(x)\,|u|^{q_2-2}u$, $\alpha > 0$, $2 < q_2 < q_1$, $b_1, b_2 \in C(\overline{\omega})$. Then $g(\cdot,u)$ satisfies $(\mathrm{I})_g$. In fact, since $g^\alpha(x,u) = b_1(x)\,|u|^{q_1-2}u - b_2(x)\,|u|^{q_2-2}u$, we get

$$\begin{aligned} s^{-\frac{p}{p-1}} & g^\alpha(x, s^{\frac{1}{p-1}} v) \\ &= s^{-\frac{p}{p-1}} \left\{ b_1(x)\, s^{\frac{q_1-1}{p-1}}\, |v|^{q_1-2} v - b_2(x)\, s^{\frac{q_2-1}{p-1}}\, |v|^{q_2-2} v \right\} \\ &= s^{\frac{q_2-p-1}{p-1}}\, |v|^{q_2-2} \left\{ b_1(x)\, s^{\frac{q_1-q_2}{p-1}}\, |v|^{q_1-q_2} v - b_2(x)\, v \right\}. \end{aligned}$$

Hence we can let

$$a(s) = s^{\frac{q_2-p-1}{p-1}}, \ \rho = q_2 - 2 > 0,\ h(x,s,v) = b_1(x)\, s^{\frac{q_1-q_2}{p-1}}\, |v|^{q_1-q_2} v - b_2(x)\, v.$$

Then it is easy to see that

$$h(x,s,v),\ h'_v(x,s,v) \in C(\overline{\Omega} \times [0,1] \times \mathbb{R}^1),\ h(x,s,0) = 0,$$
$$|a|_{L^1} = \int_0^1 s^{\frac{q_2-p-1}{p-1}}\, ds = \frac{p-1}{q_2-2}, \ |sa^2|_{L^1} = \int_0^1 s\, s^{\frac{2q_2-2p-2}{p-1}}\, ds = \frac{p-1}{2q_2-4},$$
$$|\,v\,|^\rho |\,h(x,s,v)\,| \le C\,|\,v\,|^{q_2-p-1} = C\,|\,v\,|^{1+\delta}, \quad (x,s,v) \in \omega \times [0,1] \times [0,1],$$
$$\delta = q_2 - 2 > 0.$$

EXAMPLE 5.2. Let $g(x,u)$, $g'_u(x,u)$, $g''_u(x,u) \in C(\overline{\omega} \times \mathbb{R}^1)$, $g'_u(x,0) = -\alpha < 0$, $g(x,0) = 0$. Then $g(\cdot,u)$ satisfies $(\mathrm{I})_g$.

In fact, we first note that

$$g(x,u) = g(x,0) + g'_u(x,0)\,u + \int_0^u (u-t)\,g''_u\,dt, \quad g^\alpha(x,u) = \int_0^u (u-t)\,g''_u\,dt.$$

Then we get

$$\begin{aligned} s^{-\frac{p}{p-1}}\,g^\alpha(x, s^{\frac{1}{p-1}} v) &= s^{-\frac{p}{p-1}} \int_0^{s^{\frac{1}{p-1}} v} (s^{\frac{1}{p-1}}\,v - t)\,g''_v(x,t)dt \\ &= s^{-\frac{p}{p-1}} \int_0^v (s^{\frac{1}{p-1}}\,v - s^{\frac{1}{p-1}}\,t)\,g''_v(x, s^{\frac{1}{p-1}} t)\,s^{\frac{1}{p-1}}\,dt \\ &= s^{\frac{2-p}{p-1}} \int_0^v (v-t)\,g''_v(x, s^{\frac{1}{p-1}}\,t)\,dt, \end{aligned}$$

and we let

$$\rho = 0, \quad h(x,s,v) = \int_0^v (v-t)\,g''_v(x, s^{\frac{1}{p-1}}\,t)\,dt, \quad a(s) = s^{\frac{2-p}{p-1}}.$$

Hence we obtain

$$\begin{aligned} &h(x,s,v),\ h'_v(x,s,v) = \int_0^v g''(x, s^{\frac{1}{p-1}}\,t)\,dt \in C(\overline{\omega}\times[0,1]\times\mathbb{R}^1),\ h(x,s,0)=0, \\ &|a|_{L^1} = \int_0^1 s^{\frac{2-p}{p-1}}\,ds = p-1, \quad |sa^2|_{L^1} = \int_0^1 s\,s^{\frac{4-2p}{p-1}}\,ds = \frac{p-1}{2}, \\ &|v|^\rho |h(x,s,v)| = \left| \int_0^v (v-t)\,g''_v(x, s^{\frac{1}{p-1}}\,t)\,dt \right| \le \int_0^{|v|} \left|1 - \frac{t}{v}\right| \left|g''_v(x, s^{\frac{1}{p-1}} t)\right| dt\,|v| \\ &\qquad \le \max_{(x,s)\in\overline{\omega}\times[0,1]} |g''(x,s)|\,|v|^2, \quad (x,s,v)\in\omega\times[0,1]\times[0,1], \quad \delta = 1. \end{aligned}$$

In order to establish the estimate from below for the $\varepsilon$-Kolmogorov entropy of our global attractor $\mathcal{A}$, we rely on the following fact.

LEMMA 5.1. *Let $\mathcal{K}^-$ be the set of all bounded solutions of* (E1)–(E2) *on $\mathbb{R}^1_-$, i.e.,* $\mathcal{K}^- = \{\, \{u(t)\}_{t\in\mathbb{R}^1_-}\,;\ u(t)$ *satisfies* (E1)–(E2) *on* $\mathbb{R}^1_-,\ \sup_{t\in\mathbb{R}^1_-} \|u(t)\|_{L^1} < +\infty \,\}$, *where $\mathbb{R}^1_- := (-\infty, 0]$ and let $\mathcal{K}^-(t)$ be the section of $\mathcal{K}^-$ at $t = t \in \mathbb{R}^1_-$, i.e., $\mathcal{K}^-(t) = \{\, u(t)\,;\ \{u(t)\}_{t\in\mathbb{R}^1_-} \in \mathcal{K}^- \,\}$. Then $\mathcal{K}^-(0) \subset \mathcal{A}$ holds true.*

PROOF. Let $\mathcal{B} = \bigcup_{t\in\mathbb{R}^1_-} \mathcal{K}^-(t)$. Then, since $\mathcal{B}$ is bounded in $L^1(\Omega)$, for arbitrary $\eta > 0$, there exists $T > 0$ such that $\mathrm{dist}_{L^1(\Omega)}(S(T)\,\mathcal{B}, \mathcal{A}) < \eta$. For any $a_0 \in \mathcal{K}^-(0)$, there exists $a_T \in \mathcal{K}^-(-T) \subset \mathcal{B}$ such that $S(T)\,a_T = a_0$. Hence we get

$$\mathrm{dist}_{L^1(\Omega)}(a_0, \mathcal{A}) = \mathrm{dist}_{L^1(\Omega)}(S(T)\,a_T, \mathcal{A}) \le \mathrm{dist}_{L^1(\Omega)}(S(T)\,\mathcal{B}, \mathcal{A}) < \eta, \quad \forall \eta > 0,$$

which implies $\mathrm{dist}_{L^1(\Omega)}(a_0, \mathcal{A}) = 0$, i.e., $a_0 \in \bar{\mathcal{A}} = \mathcal{A}\ \forall a_0 \in \mathcal{K}^-(0)$. Thus $\mathcal{K}^-(0) \subset \mathcal{A}$ is derived. □

Before we proceed to the proof of Theorem 5.3, we prepare a couple of results on the following auxiliary equation:

$$(5.8) \quad (\mathrm{E})^t_p \begin{cases} w_t = p_\alpha\left(\Delta\left(|w|^{p-1}w\right) - a(t)\,|\,w\,|^\rho\,h(x,t,w)\right), \text{ in } \omega\times(0,1), \\ w|_{\partial\omega} = 0, \quad t\in[0,1], \quad w(x,0) = w_0(x), \quad x\in\omega, \end{cases}$$

where $p_\alpha = \dfrac{1}{\alpha(p-1)}$. As for the solvability of this equation, the following result holds.

LEMMA 5.2. *Let* $(\mathrm{I})_g$ *be satisfied. Then for every* $w_0 \in L^\infty(\Omega)$, *there exists* $T_0 = T_0(\|u_0\|_{L^\infty}) > 0$ *such that* $(\mathrm{E})_p^t$ *admits a unique solution* $w$ *on* $[0, T_0]$ *satisfying*

$$(5.9) \qquad \begin{aligned} &w \in C([0,T_0]; L^1(\omega)) \cap C((0,T_0]; W_0^{1,p}(\omega)) \cap L^\infty(\omega \times (0,T_0)), \\ &|u|^{p-1}u \in L^2([t,T], H_0^1(\omega)). \end{aligned}$$

*Furthermore, there exists a (sufficiently small)* $\varepsilon_0 > 0$ *such that if* $\| w_0 \|_{L^\infty} \le \varepsilon_0$, *then the solution* $w$ *of* $(\mathrm{E})_p^t$ *given above can be continued up to* $[0,1]$ *and satisfies* $\sup_{t\in[0,1]} \| w(t) \|_{L^\infty} \le 1$.

Next we prepare a comparison result which enables us to compare solutions of $(\mathrm{E})_p^t$ with solutions of simplified equations.

LEMMA 5.3. *Let* $w$ *be a solution of* $(\mathrm{E})_p^t$ *on* $[0,1]$ *satisfying* $\| w(x,t) \|_{L^\infty(\omega\times[0,1])} \le 1$ *and* (5.9) *with* $T_0 = 1$, *and let* $w^\pm$ *satisfy* (5.9) *with* $T_0 = 1$ *and*

$$\begin{aligned} &w_t^- \le p_\alpha \left(\Delta \left(|w^-|^{p-1}w^-\right) - C\, a(t)\, w^-\right), \quad t \in (0,1), \\ &p_\alpha(\Delta \left(|w^+|^{p-1}w^+\right) + C\, a(t)\, w^+) \le w_t^+, \quad t \in (0,1), \\ &w^-(x,0) \le w(x,0) \le w^+(x,0). \end{aligned}$$

*Then it holds that* $w^-(x,t) \le w(x,t) \le w^+(x,t)$ *for a.e.* $x \in \omega$, $\forall\, t \in [0,1]$.

PROOF. Since $\| w(x,t) \|_{L^\infty(\omega\times[0,1])} \le 1$, by $(\mathrm{I})_2$, it is easy to see that $w$ satisfies

$$p_\alpha(\Delta \left(|w|^{p-1}w\right) - C\, a(t)\, w\,) \le w_t \le p_\alpha(\Delta \left(|w|^{p-1}w\right) + C\, a(t)\, w\,) \quad \forall\, t \in (0,1).$$

Hence we get

$$(\, w(t) - w^+(t)\,)_t \le \Delta \left(|w|^{p-1}w\right)(t) - \Delta \left(|w^+|^{p-1}w^+\right)(t) + C\, a(t)\, (\, w(t) - w^+(t)\,).$$

Multiplying this by $\operatorname{sgn}[\, w - w^+\,]^+$ and using the Kato inequality, we have

$$\frac{1}{2}\frac{d}{dt} \| [w - w^+]^+(t) \|_{L^1} \le C\, a(t) \| [w - w^+]^+(t) \|_{L^1} \quad \text{a.e. } t \in (0,1).$$

Then, integrating this on $[\delta, t]$ with $\delta > 0$ and applying the Gronwall inequality, we obtain

$$(5.10) \qquad \| [w - w^+]^+(t) \|_{L^1} \le \| [w - w^+]^+(\delta) \|_{L^1} e^{2C \int_0^t a(\xi)\, d\xi}.$$

Since $\| [w - w^+]^+(+0) \|_{L^1} = 0$, letting $\delta \to 0$ in (5.10), we conclude that $w(x,t) \le w^+(x,t)$ for a.e. $x \in \omega$, $\forall\, t \in [0,1]$. The assertion that $w^-(x,t) \le w(x,t)$ can be verified by much the same arguments as above. □

Now we are ready to prove Theorem 5.3.

PROOF OF THEOREM 5.3. We introduce a new time scale $s^\pm = s^\pm(t)$ by

$$(5.11) \qquad (\mathrm{S})^\pm \begin{cases} \dfrac{d}{dt} s^\pm(t) = e_p^\pm(t) := p_\alpha \left(e^{\mp \int_0^t p_\alpha C\, a(\xi)\, d\xi}\right)^{1-p}, & t \in (0,1), \\ s^\pm(0) = 0. \end{cases}$$

Since $e_p^\pm(t)$ is strictly positive and bounded on $[0,1]$, there exist unique solutions $s^\pm(t)$ of $(\mathrm{S})^\pm$, which are strictly increasing on $[0,1]$. Define

$$T_1^\pm := s^\pm(1) = \int_0^1 e_p^\pm(\xi)\,d\xi.$$

Then $0 < T_1^- < 1 < T_1^+$ and $s^\pm(t) \in [0, T_1^\pm]\ \forall\, t \in [0,1]$.

Consider

$$(\mathrm{P}) \quad \frac{\partial}{\partial s} w(x,s) = \Delta\left(|w|^{p-1}w\right)(x,s), \quad (x,s) \in \mathbb{R}^N \times (0,+\infty).$$

Then the following facts are well known:

(P1) $L^r$-norms are Lyapunov functions for (P); i.e., every solution $w(s)$ of (P) satisfies $\|\, w(s)\,\|_{L^r} \le \|\, w(0)\,\|_{L^r}\ \forall\, s \in [0,+\infty) \quad \forall\, r \in [1,\infty]$.

(P2) (P) admits the following Barenblatt type solutions:

$$w_{\gamma,\beta}(x,s) = s^{-\beta N}\left(\gamma - \frac{(p-1)^\beta}{2p}|x|^2 s^{-2\beta}\right)_+^{\frac{1}{p-1}}, \tag{5.12}$$

where $\gamma$ is determined from the conservation of mass and $\beta = \frac{1}{N(p-1)+2}$.

(P3) $\operatorname{supp} w(x,s)$ is monotone increasing in $s$.

Property (P1) is well addressed in [80], [81], [82], property (P2) is a classical result (see, for example, [57]), and property (P3) can be deduced by simple calculations from the explicit form of the Barenblatt type solutions given in (5.12).

Let $K$ be the unit ball in $\mathbb{R}^N$ centered at the origin. We can choose a solution $w_1(x,s)$ of (P) so that

$$\begin{aligned} &0 \le w_1(x,s) \le 1 \quad \forall\,(x,s) \in \mathbb{R}^N \times [0,T_1^+], \\ &\operatorname{supp}\, w_1(\cdot,s) \subset K \quad \forall\, t \in [0,T_1^+], \\ &\|\, w_1(\cdot,T_1^+)\,\|_{L^r} \ge \delta_0, \quad \forall\, r \in [1,\infty], \end{aligned} \tag{5.13}$$

where $\delta_0$ is a positive constant independent of $r$. (Since we can assume that $\omega$ is small enough to satisfy $|\omega| \le 1$ without loss of generality, if we take $\|w_1(\cdot,T_1^+)\|_{L^1} \ge \delta_0$. Then (5.13) is satisfied for all $r \in [1,\infty]$.) Furthermore, it is easy to see that $w_\varepsilon(x,s) := \varepsilon\, w_1(\varepsilon^{\frac{1-p}{2}}x, s)$ is a solution of (P) and satisfies

$$\begin{aligned} &0 \le w_\varepsilon(x,s) \le \varepsilon \quad \forall\,(x,s) \in \mathbb{R}^N \times [0,T_1^+], \\ &\operatorname{supp}\, w_\varepsilon(\cdot,s) \subset K_\varepsilon := \varepsilon^{\frac{p-1}{2}}\, K \quad \forall\, t \in [0,T_1^+]. \end{aligned} \tag{5.14}$$

Moreover, $w_\varepsilon(x - x_i, s)$ also gives a solution of (P) and for sufficiently small $\varepsilon$ it follows from (5.14) and (P2) that there exists a finite set $R_\varepsilon := \{x_i\} \subset \omega$ such that

(R1) $(x_i + K_\varepsilon) \cap (x_j + K_\varepsilon) = \emptyset, \quad \forall x_i,\ x_j \in R_\varepsilon,\ i \ne j$,

(R2) $\displaystyle \#R_\varepsilon \ge C_\omega \left(\frac{1}{\varepsilon}\right)^{\frac{N(p-1)}{2}}$,

(R3) $\displaystyle \bigcup_{i=1}^{\#R_\varepsilon} (x_i + K_\varepsilon) \subset\subset \omega$.

Consequently, for every $\vec{m} \in \{0,1\}^{\#R_\varepsilon} := \{ \vec{m} = (m_1, m_2, \ldots, m_{\#R_\varepsilon});\ m_j = 0 \text{ or } 1,\ j = 1, 2, \ldots, m_{\#R_\varepsilon} \}$, the function

$$w_{\vec{m},\varepsilon}(x,s) = \sum_{i=1}^{\#R_\varepsilon} m_i\, w_\varepsilon(x - x_i, s) \tag{5.15}$$

solves (P) and is supported in $\omega$. On the other hand, for $\vec{m}^1 \neq \vec{m}^2$, we obviously have

$$\| w_{\vec{m}^1,\varepsilon}(x,s) - w_{\vec{m}^2,\varepsilon}(x,s) \|_{L^\infty} = \varepsilon \| w_1(x,s) \|_{L^\infty} \quad \forall\, s \in [0, T_1^+]. \tag{5.16}$$

Thus we find $2^{\#R_\varepsilon}$ different solutions of (P) supported in $\omega$ having the form (5.15). Furthermore, as for the measurement in the topology of $L^r$ $(1 \le r < \infty)$, instead of (5.16), we get

$$\| w_{\vec{m}^1,\varepsilon}(x,s) - w_{\vec{m}^2,\varepsilon}(x,s) \|_{L^r} \ge \varepsilon^{\frac{2r+N(p-1)}{2r}} \| w_1(x,s) \|_{L^r} \quad \forall\, s \in [0, T_1^+].$$

Here we define the new functions $V^\pm_{\vec{m},\varepsilon}(x,t)$ via the new time scales $s^\pm(t)$ defined by (5.11) as follows:

$$V^\pm_{\vec{m},\varepsilon}(x,t) := w_{\vec{m},\varepsilon}(x, s^\pm(t)), \quad t \in [0,1]. \tag{5.17}$$

Then, by (5.11), we easily find

$$\begin{cases} \frac{\partial}{\partial t} V^\pm_{\vec{m},\varepsilon}(x,t) = \frac{d}{dt} s^\pm(t)\, \frac{\partial}{\partial s^\pm} w_{\vec{m},\varepsilon}(x, s^\pm) = e_p^\pm(t)\, \Delta \left( |V^\pm_{\vec{m},\varepsilon}|^{p-1} V^\pm_{\vec{m},\varepsilon} \right)(x,t), \\ V^\pm_{\vec{m},\varepsilon}(x,0) = w_{\vec{m},\varepsilon}(x,0), \\ \operatorname{supp}\, w_{\vec{m},\varepsilon}(\cdot,0) \subset \operatorname{supp}\, V^\pm_{\vec{m},\varepsilon}(\cdot,t) \subset \operatorname{supp}\, w_{\vec{m},\varepsilon}(\cdot, T_1^\pm) \quad \forall\, t \in [0,1]. \end{cases} \tag{5.18}$$

We further introduce new functions $\widetilde{w}^\pm_{\vec{m},\varepsilon}$ by

$$\widetilde{w}^\pm_{\vec{m},\varepsilon}(x,t) := e^\pm(t)\, V^\pm_{\vec{m},\varepsilon}(x,t), \quad e^\pm(t) := e^{\pm \int_0^t p_\alpha C a(\xi)\, d\xi}, \quad t \in [0,1]. \tag{5.19}$$

Then, by (5.19), it is easy to see that

$$\begin{aligned} \frac{\partial}{\partial t} \widetilde{w}^\pm_{\vec{m},\varepsilon}(x,t) &= e^\pm(t)\, \frac{\partial}{\partial t} V^\pm_{\vec{m},\varepsilon}(x,t) \pm p_\alpha C a(t)\, e^\pm(t)\, V^\pm_{\vec{m},\varepsilon}(x,t) \\ &= p_\alpha \left(e^\pm(t)\right)^{p-1} \Delta \left( |V^\pm_{\vec{m},\varepsilon}|^{p-1} V^\pm_{\vec{m},\varepsilon} \right)(x,t) \pm p_\alpha C a(t)\, \widetilde{w}^\pm_{\vec{m},\varepsilon}(x,t). \end{aligned}$$

Thus we find

$$\begin{cases} \frac{\partial}{\partial t} \widetilde{w}^\pm_{\vec{m},\varepsilon}(x,t) = p_\alpha \left( \Delta \left( |\widetilde{w}^\pm_{\vec{m},\varepsilon}|^{p-1} \widetilde{w}^\pm_{\vec{m},\varepsilon} \right)(x,t) \pm C a(t) \widetilde{w}^\pm_{\vec{m},\varepsilon}(x,t) \right), \ t \in [0,1], \\ \widetilde{w}^\pm_{\vec{m},\varepsilon}(x,0) = w_{\vec{m},\varepsilon}(x,0), \\ \operatorname{supp}\, w_{\vec{m},\varepsilon}(\cdot,0) \subset \operatorname{supp}\, \widetilde{w}^\pm_{\vec{m},\varepsilon}(\cdot,t) \subset \operatorname{supp}\, w_{\vec{m},\varepsilon}(\cdot, T_1^\pm) \quad \forall\, t \in [0,1]. \end{cases}$$

Let $\widetilde{w}_{\vec{m},\varepsilon}(x,t)$ be the unique solution of $(\mathrm{E})_p^t$ on $[0,1]$ with $\widetilde{w}_{\vec{m},\varepsilon}(x,0) = w_{\vec{m},\varepsilon}(x,0)$, whose existence is ensured by Lemma 5.2 for all $\varepsilon \in (0, \varepsilon_0]$. Then, by the comparison theorem given in Lemma 5.3 (note that $\widetilde{w}^\pm_{\vec{m},\varepsilon}(x,t)$ satisfies the regularity required there), we observe

(5.20)
$$\begin{cases} \frac{\partial}{\partial t}\,\widetilde{w}_{\vec{m},\varepsilon}(x,t) = p_\alpha\left(\Delta\left(|\widetilde{w}_{\vec{m},\varepsilon}|^{p-1}\widetilde{w}_{\vec{m},\varepsilon}\right)(x,t) - a(t)\,|\,\widetilde{w}_{\vec{m},\varepsilon}\,|^{\rho}\,h(x,t,\widetilde{w}_{\vec{m},\varepsilon})\right), \\ \widetilde{w}_{\vec{m},\varepsilon}(x,0) = w_{\vec{m},\varepsilon}(x,0), \\ \widetilde{w}^-_{\vec{m},\varepsilon}(x,t) \le \widetilde{w}_{\vec{m},\varepsilon}(x,t) \le \widetilde{w}^+_{\vec{m},\varepsilon}(x,t) \quad \text{a.e. } \omega,\ \forall t\in[0,1], \\ \text{supp } w_{\vec{m},\varepsilon}(\cdot,0) \subset \text{supp } \widetilde{w}_{\vec{m},\varepsilon}(\cdot,t) \subset \text{supp } w_{\vec{m},\varepsilon}(\cdot,T_1^\pm) \quad \forall t\in[0,1]. \end{cases}$$

Hence, by virtue of the choice of $R_\varepsilon$, supp $\widetilde{w}_{\vec{m},\varepsilon}(\cdot,t)$ does not touch the boundary of $\omega$ for all $t\in[0,1]$. Therefore, the zero extension of $\widetilde{w}_{\vec{m},\varepsilon}(\cdot,t)$ to $\Omega$, denoted again by $\widetilde{w}_{\vec{m},\varepsilon}(\cdot,t)$, becomes a solution of $(\mathrm{E})^t_p$ with $\omega$ replaced by $\Omega$.

We introduce another time scale $\tau=\tau(t)\in(-\infty,0]$ by

$$\tau(t) = p_\alpha\,\log t,\quad t\in(0,1] \iff t\,(\tau) = e^{\alpha(p-1)\tau},\quad \tau\in(-\infty,0],\quad \frac{dt}{d\tau} = \alpha\,(p-1)\,t$$

and define a new function $\widetilde{W}_{\vec{m},\varepsilon}(x,\tau)$ by

$$\widetilde{W}_{\vec{m},\varepsilon}(x,\tau) = \widetilde{w}_{\vec{m},\varepsilon}(x,t\,(\tau)),\quad (x,\tau)\in\Omega\times(-\infty,0].$$

Then we have

$$\begin{aligned} &\frac{\partial}{\partial\tau}\,\widetilde{W}_{\vec{m},\varepsilon}(x,\tau) \\ &\quad = \frac{dt}{d\tau}\,\frac{\partial}{\partial t}\,\widetilde{w}_{\vec{m},\varepsilon}(x,t) \\ &\quad = \alpha\,(p-1)\,t\,p_\alpha\left(\Delta\left(|\widetilde{w}_{\vec{m},\varepsilon}|^{p-1}\widetilde{w}_{\vec{m},\varepsilon}\right) - a(t)\,|\,\widetilde{w}_{\vec{m},\varepsilon}\,|^{\rho}\,h(x,t,\widetilde{w}_{\vec{m},\varepsilon}(t))\right) \\ &\quad = t\left(\Delta\left(|\widetilde{w}_{\vec{m},\varepsilon}|^{p-1}\widetilde{w}_{\vec{m},\varepsilon}\right) - t^{-\frac{p}{p-1}}\,g^\alpha(x,t^{\frac{1}{p-1}}\,\widetilde{w}_{\vec{m},\varepsilon}(t))\right). \end{aligned}$$

Therefore, we get

$$\begin{cases} \frac{\partial}{\partial\tau}\,\widetilde{W}_{\vec{m},\varepsilon} = e^{\alpha(p-1)\tau}\Delta\left(|\widetilde{W}_{\vec{m},\varepsilon}|^{p-1}\widetilde{W}_{\vec{m},\varepsilon}\right) - e^{-\alpha\tau}\,g^\alpha(x,e^{\alpha\tau}\,\widetilde{W}_{\vec{m},\varepsilon}(x,\tau)), \\ \widetilde{W}_{\vec{m},\varepsilon}(x,0) = \widetilde{w}_{\vec{m},\varepsilon}(x,1), \\ \text{supp } w_{\vec{m},\varepsilon}(\cdot,0) \subset \text{supp } \widetilde{W}_{\vec{m},\varepsilon}(\cdot,\tau) \subset \text{supp } w_{\vec{m},\varepsilon}(\cdot,T_1^\pm) \quad \forall\tau\in(-\infty,0]. \end{cases}$$

Now define

$$W_{\vec{m},\varepsilon}(x,\tau) = e^{\alpha\tau}\,\widetilde{W}_{\vec{m},\varepsilon}(x,\tau),\quad (x,\tau)\in\Omega\times(-\infty,0].$$

Then

$$\begin{aligned} \frac{\partial}{\partial\tau}\,W_{\vec{m},\varepsilon}(x,\tau) &= \alpha\,W_{\vec{m},\varepsilon}(x,\tau) + e^{\alpha\tau}\,\frac{\partial}{\partial\tau}\,\widetilde{W}_{\vec{m},\varepsilon}(x,\tau) \\ &= \alpha\,W_{\vec{m},\varepsilon}(x,\tau) + e^{\alpha p\tau}\Delta\left(|\widetilde{W}_{\vec{m},\varepsilon}|^{p-1}\widetilde{W}_{\vec{m},\varepsilon}\right)(x,\tau) - g^\alpha(x,W_{\vec{m},\varepsilon}(x,\tau)), \end{aligned}$$

whence it follows that

(5.21)
$$\begin{cases} \frac{\partial}{\partial\tau}\,W_{\vec{m},\varepsilon} = \Delta\left(|W_{\vec{m},\varepsilon}|^{p-1}W_{\vec{m},\varepsilon}\right) - g^\alpha(x,W_{\vec{m},\varepsilon}),\ (x,\tau) \text{ in } \Omega\times(-\infty,0], \\ W_{\vec{m},\varepsilon}(x,0) = \widetilde{w}_{\vec{m},\varepsilon}(x,1), \\ \text{supp } w_{\vec{m},\varepsilon}(\cdot,0) \subset \text{supp } W_{\vec{m},\varepsilon}(\cdot,\tau) \subset \text{supp } w_{\vec{m},\varepsilon}(\cdot,T_1^\pm) \quad \forall\tau\in(-\infty,0]. \end{cases}$$

Thus we observe that

(5.22) $$\{W_{\vec{m},\varepsilon}(x,\tau)\}_{\tau\in\mathbb{R}^1_-}\in\mathcal{K}^- \quad\text{and}\quad W_{\vec{m},\varepsilon}(x,0) = \widetilde{w}_{\vec{m},\varepsilon}(x,1)\in\mathcal{K}^-(0).$$

Next we let $\overrightarrow{e_i} = (0, \ldots, 1, \ldots, 0)$, i.e., the $j$-th component of $\overrightarrow{e_i} = \delta_{ij}$, and define

$$\widetilde{w}_\varepsilon(x - x_i, t) = \widetilde{w}_{\overrightarrow{e_i},\varepsilon}(x, t), \quad t \in [0, 1].$$

Then, by virtue of (5.14), (R1), and (5.20), we see that

$$\operatorname{supp} \widetilde{w}_\varepsilon(x - x_i, t) \cap \operatorname{supp} \widetilde{w}_\varepsilon(x - x_j, t) = \emptyset, \quad i \neq j, \ \forall\, t \in [0, 1].$$

Hence we can write

$$\widetilde{w}_{\overrightarrow{m},\varepsilon}(x, t) = \sum_{i=1}^{\# R_\varepsilon} m_i \, \widetilde{w}_\varepsilon(x - x_i, t), \quad t \in [0, 1].$$

Therefore, if $\overrightarrow{m_1}$ and $\overrightarrow{m_2}$ differ at the $i$-th component, we have by (5.22)

$$\begin{aligned} \| W_{\overrightarrow{m_1},\varepsilon}(x, 0) - W_{\overrightarrow{m_2},\varepsilon}(x, 0) \|_{L^r} &= \| \widetilde{w}_{\overrightarrow{m_1},\varepsilon}(x, 1) - \widetilde{w}_{\overrightarrow{m_2},\varepsilon}(x, 1) \|_{L^r} \\ &\geq \| \widetilde{w}_\varepsilon(x - x_i, 1) \|_{L^r}. \end{aligned} \tag{5.23}$$

Recalling (5.20), (5.17), and (5.19), we get

$$\widetilde{w}_\varepsilon(x - x_i, 1) \geq \widetilde{w}_\varepsilon^-(x - x_i, 1) = e^-(1)\, w_\varepsilon(x - x_i, T_1^-).$$

Hence, since $0 < T_1^- < T_1^+$, from (P1) and (5.13), we deduce that

$$\begin{aligned} \| \widetilde{w}_\varepsilon(x - x_i, 1) \|_{L^r} &\geq e^-(1) \, \| w_\varepsilon(x - x_i, T_1^-) \|_{L^r} \\ &\geq e^-(1) \, \| w_\varepsilon(x - x_i, T_1^+) \|_{L^r} \\ &= e^-(1) \, \varepsilon^{\frac{2r+N(p-1)}{2r}} \, \| w_1(x, T_1^+) \|_{L^r} \\ &\geq e^-(1) \, \delta_0 \, \varepsilon^{\frac{2r+N(p-1)}{2r}}. \end{aligned} \tag{5.24}$$

Then, combining (5.23) with (5.24), we obtain

$$\| W_{\overrightarrow{m_1},\varepsilon}(x, 0) - W_{\overrightarrow{m_2},\varepsilon}(x, 0) \|_{L^r} \geq e^-(1) \, \delta_0 \, \varepsilon^{\frac{2r+N(p-1)}{2r}},$$

which can be rewritten as

$$\begin{aligned} &\| W_{\overrightarrow{m_1},(k\,\varepsilon)^\beta}(x, 0) - W_{\overrightarrow{m_2},(k\,\varepsilon)^\beta}(x, 0) \|_{L^r} \geq 2\,\varepsilon, \quad \forall\, \varepsilon \in (0, (\varepsilon_0/k)^{1/\beta}), \\ &\beta = \frac{2\,r}{2\,r + N\,(p-1)}, \quad k = \frac{2}{e^-(1)\,\delta_0}. \end{aligned}$$

This estimate implies that balls in $L^r$ with radius $\varepsilon > 0$ can contain at most one element belonging to $\{\, W_{\overrightarrow{m},(k\varepsilon)^\beta}(x, 0) \,;\, \overrightarrow{m} \in \{0, 1\}^{\# R_{(k\,\varepsilon)^\beta}} \,\} \subset \mathcal{K}^-(0)$ and its cardinality can be estimated, by (R2) and (5.21), as

$$\begin{aligned} &\#\{\, W_{\overrightarrow{m},(k\varepsilon)^\beta}(x, 0) \,;\, \overrightarrow{m} \in \{0, 1\}^{\# R_{(k\,\varepsilon)^\beta}} \,\} \geq 2^{\# R_{(k\,\varepsilon)^\beta}}, \\ &\# R_{(k\,\varepsilon)^\beta} \geq C_\omega \left( \frac{1}{(k\,\varepsilon)^\beta} \right)^{\frac{N\,(p-1)}{2}} = C_\omega \left( \frac{1}{k\,\varepsilon} \right)^{\frac{N\,r\,(p-1)}{2\,r + N\,(p-1)}}. \end{aligned}$$

Hence, by the definition of $\mathcal{N}_\varepsilon$, $\mathcal{H}_\varepsilon$ and Lemma 5.1, we get

$$\begin{aligned} &\mathcal{N}_\varepsilon(\mathcal{K}^-(0), L^r) \geq 2^{\# R_{(k\,\varepsilon)^\beta}}, \\ &\mathcal{H}_\varepsilon(\mathcal{A}, L^r) \geq C_\omega \left( \frac{1}{k\,\varepsilon} \right)^{\frac{N\,r\,(p-1)}{2\,r + N\,(p-1)}}, \quad r \in [1, \infty]. \end{aligned}$$

Moreover, since Kolmogorov $\varepsilon$-entropy of a bounded set of $C^1(\bar{\Omega})$ in the topology of $L^\infty$ is estimated from above by $C\left(\frac{1}{\varepsilon}\right)^N$ (see [61]), we obtain

$$C_\omega \Big( \frac{1}{k\,\varepsilon} \Big)^{\frac{N\,(p-1)}{2 + N\,(p-1)}} \leq C_\omega \Big( \frac{1}{k\,\varepsilon} \Big)^{\frac{N\,r\,(p-1)}{2\,r + N\,(p-1)}} \leq \mathcal{H}_\varepsilon(\mathcal{A}, L^r) \leq C \Big( \frac{1}{\varepsilon} \Big)^N \ \forall\, r \in [1, \infty]. \tag{5.25}$$

To complete the proof of Theorem 5.3, it remains to recall that $\dim_F(\mathcal{A}, L^r)$, the fractal dimension of $\mathcal{A}$, can be expressed in terms of Kolmogorov $\varepsilon$-entropy via $\dim_F(\mathcal{A}, L^r) = \limsup_{\varepsilon\to 0} \dfrac{\mathcal{H}_\varepsilon(\mathcal{A}, L^r)}{\log_2 \frac{1}{\varepsilon}}$. Letting $\varepsilon \to 0$ in (5.25), we conclude that

$$\dim_F(\mathcal{A}, L^r) = \infty \quad \forall r \in [1, \infty].$$

□

CHAPTER 6

# Long-time dynamics of $p$-Laplacian equations: Homogeneous media

The main goal of the present chapter is to give a detailed study of the attractors for the following class of parabolic equations with $p$-Laplacian, that is

$$(\mathrm{E})_p \begin{cases} \frac{\partial u}{\partial t} \in \Delta_p u(x,t) - g(u(x,t)) + h(x), \qquad (x,t) \in \Omega \times [0,\infty), & \text{(E1)} \\ u(x,t) = 0, \qquad (x,t) \in \partial\Omega \times [0,\infty), & \text{(E2)} \\ u(x,0) = u_0(x), \qquad x \in \Omega, & \text{(E3)} \end{cases}$$

where $\Delta_p u = \operatorname{div}(|\nabla u|^{p-2}\nabla u)$, $p \in [\frac{2N}{N+2}, \infty)$, and $\Omega$ is a bounded domain in $\mathbf{R}^N$ with smooth boundary $\partial\Omega$.

More precisely, we present some new features, to the best of our knowledge, related to the attractor of such equations which one cannot observe in nondegenerate cases, namely,

(a) the infinite dimensionality of the attractor,
(b) the polynomial asymptotics of its $\varepsilon$-Kolmogorov entropy,
(c) the difference in the asymptotics of the $\varepsilon$-Kolmogorov entropy depending on the choice of the underlying phase spaces.

We especially emphasize that, although we obtain similar results (a)–(c) for $p$-Laplacian equation, as for the porous media equation studied in Chapter 4, the analysis is essentially different. It is also worth noting that in contrast to the porous medium equation (which has degeneracy at the point $u$=0), the $p$-Laplacian equation has one-dimensional degeneracy $u = \text{const}$.

Chapter 6 is organized as follows: in Section 6.1, we fix some basic results on the existence, uniqueness, and regularity of solutions. In Section 6.2, we prove the existence of the global attractor for the semigroup generated by (E1)–(E3) as well as the infinite dimensionality of the attractor and the asymptotics of its Kolmogorov entropy. In particular, we show that its Kolmogorov entropy admits polynomial asymptotics, which sheds light on the completely new phenomena (a)–(c) mentioned above.

## 6.1. Existence of global solutions and a priori estimates

Consider the initial boundary value problem $(\mathrm{E})_p$. We assume that, $g(\xi) = g_0(\xi) + g_1(\xi)$, where

(a) $g_0(\cdot)$ is a maximal monotone graph in $\mathbf{R}^2$ such that $0 \in g_0(0)$, $D(g_0) = \{\xi \in \mathbf{R}^1; |g_0(\xi)| < +\infty\} = \mathbf{R}^1$.

(b) $g_1(\cdot)$ is a globally Lipschitz function from $\mathbf{R}^1$ into $\mathbf{R}^1$ and $g_1(0) = 0$.

Note that condition (a) implies that there exists a lower semicontinous convex function $j(\cdot)$ from $\mathbf{R}^1$ into $[0,+\infty]$ such that $g_0(u) = \partial j(u)$. Here $\partial j$ denotes the subdifferential of $j$.

Let $D(\varphi) = \{u; u \in W_0^{1,p}(\Omega) \text{ and } j(u) \in L^1(\Omega)\}$ and define $\varphi$ by

$$\varphi(u) = \begin{cases} \dfrac{1}{p}\displaystyle\int_\Omega |\nabla u|^p\, dx + \int_\Omega j(u(x))\, dx & \text{if } u \in D(\varphi), \\ +\infty & \text{otherwise.} \end{cases}$$

Then $\varphi$ becomes a lower semicontinuous convex functional from $L^2(\Omega)$ into $[0,\infty]$. Furthermore, since for the Yosida approximation $(g_0)_\lambda$ of $g_0$ we obtain

$$(-\Delta_p u, (g_0)_\lambda(u))_{L^2(\Omega)} = \int_\Omega |\nabla u|^p (g_0)'_\lambda(u)\, dx \geq 0 \quad \forall \lambda > 0,$$

by virtue of Proposition 2.17 in [12] and parts (i) and (ii) of Theorem 4.4 in [12], we find that $u \mapsto -\Delta_p u + g_0(u)$ is maximal monotone in $L^2(\Omega)$. Hence we obtain

$$\partial\varphi(u) = -\Delta_p u + g_0(u) \qquad \forall u \in D(\partial\varphi),$$

$$D(\partial\varphi) = \left\{u \in D(\varphi); \Delta_p u \in L^2(\Omega),\ \exists b \in L^2(\Omega) \text{ s.t. } b(x) \in g_0(u(x)) \text{ a.e. } x \in \Omega\right\}.$$

Then, by letting $B(u(x)) = g_1(u(x)) - h(x)$, we see that $(\mathrm{E})_p$ can be reduced to the following abstract Cauchy problem in $L^2(\Omega)$:

$$\frac{du(t)}{dt} + \partial\varphi(u(t)) + B(u(t)) \ni 0, \quad u(0) = u_0.$$

By virtue of (b), $B(u)$ becomes a Lipschitz function from $L^2(\Omega)$ into itself, if $h(\cdot) \in L^2(\Omega)$. Thus, Proposition 3.12 in [12] ensures the existence of unique global solutions of $(\mathrm{E})_p$.

THEOREM 6.1. *Assume that* (a) *and* (b) *are satisfied. Then, for any* $u_0(\cdot)$, $h(\cdot) \in L^2(\Omega)$, *there exists a unique solution* $u \in C([0,\infty); L^2(\Omega)) \cap W^{1,2}_{loc}((0,\infty); L^2(\Omega)) \cap L^p_{loc}([0,\infty); W_0^{1,p}(\Omega))$ *of* $(\mathrm{E})_p$ *which depends continuously on the initial data in the strong topology of* $L^2(\Omega)$.

REMARK 6.1. If $g \in C^1(\mathbf{R}^1)$ satisfies $g(0) = 0$ and $g'(\tau) \geq -K$ for all $\tau \in \mathbf{R}^1$, then by letting $g_0(\tau) = g(\tau) + K\tau$ and $g_1(\tau) = -K\tau$, we find that $g(\cdot)$ falls within our class. In particular, $g(\tau) = C_1|\tau|^{q_1-2}\tau - C_2|\tau|^{q_2-2}\tau$, where $2 \leq q_2 < q_1 < +\infty$ with a suitable decomposition satisfies (a) and (b).

As to the a priori bounds for solutions of (E1)–(E3), we obtain the following result.

THEOREM 6.2. *Assume that* (a) *and* (b) *are satisfied,* $\frac{2N}{N+2} < p < \infty$, *and* $h \in L^\infty(\Omega)$. *For the case* $\frac{2N}{N+2} < p \leq 2$, *we further assume*

$$|\overset{\circ}{g}_0(s)| \geq k_0|s|^{1+\theta} - k_1 \quad (\theta, k_0, k_1 > 0), \tag{6.1}$$

*where* $\overset{\circ}{g}_0(s)$ *denotes the minimal section of* $g_0(s)$, *i.e., the unique nearest point of* $g_0(s)$ *from the origin. Then there exist constants* $C_1, C_2$, *and* $C_3$, *independent of the initial data* $u_0 \in L^2(\Omega)$, *such that every solution of* (E1)–(E3) *satisfies*

$$\|u(t)\|_{L^2} \leq C_1 \qquad \text{for all } t \in [1,+\infty), \tag{6.2}$$

$$\|u(t)\|_{L^\infty} \leq C_2 \qquad \text{for all } t \in [2,\infty), \tag{6.3}$$

$$\|u(t)\|_{C^{1,\alpha}(\bar{\Omega})} \leq C_3 \qquad \text{for all } t \in [3,\infty). \tag{6.4}$$

PROOF. Multiply (E1) by $u$. Then integration by parts and using (a)–(b) gives

$$\frac{1}{2}\frac{d}{dt}\|u(t)\|_{L^2} + \|\nabla u\|_{L^p}^p \leq C_4\|u(t)\|_{L^2}^2 + \|h\|_{L^2}\cdot\|u\|_{L^2}. \tag{6.5}$$

Then, by the Poincaré inequality, we get for the case $p > 2$

$$\frac{1}{2}\frac{d}{dt}\|u(t)\|_{L^2}^2 + \gamma_1\|u\|_{L^2}^p \leq \gamma_2, \tag{6.6}$$

for some $\gamma_1, \gamma_2 > 0$. Hence a Ghidaglia type estimate (see [93]) ensures (6.2).

As for the case $\frac{2N}{N+2} \leq p \leq 2$, the lower bound (6.1) yields the same estimate as (6.6) with $p = 2+\theta$, from which (6.2) follows.

In order to derive the $L^\infty$-estimate, we need the following lemma. □

LEMMA 6.1. *Let $v^\pm$ be the unique solution of*

$$v_t^\pm = \Delta_p v^\pm - \widetilde{k}_0|v^\pm|^\theta v^\pm - g_1(v^\pm) \pm (|h| + \widetilde{k}_1), \quad v^\pm|_{\partial\Omega} = 0, \tag{6.7}$$

*with the initial condition $v^\pm(x,0) = \pm|u_0(x)|$, respectively, where $\widetilde{k}_i = 0$ for $p > 2$ and $\widetilde{k}_i = k_i$ for $p \leq 2$ $(i = 0,1)$. Then the solution $u$ of* $(\mathrm{E})_p$ *satisfies*

$$v^-(x,t) \leq u(x,t) \leq v^+(x,t) \qquad \text{for a.e. } (x,t) \in \Omega\times[0,\infty). \tag{6.8}$$

PROOF. Let $u^+(x,t)$ be the unique solution of $(\mathrm{E})_p$ with $h$ and $u_0$ replaced by $|h|$ and $|u_0|$, respectively. Then it is easy to see that $u^+ \geq 0$ and $(u^+ - v^+)$ satisfies

$$(u^+ - v^+)_t \in \Delta_p u^+ - \Delta_p v^+ - g_0(u^+) + \widetilde{k}_0|v^+|^\theta v^+ - \widetilde{k}_1 - g_1(u^+) + g_1(v^+). \tag{6.9}$$

Multiply (6.9) by $[u^+ - v^+]^+ = \max(u^+(x,t) - v^+(x,t), 0)$. Then, noting that

$$(\Delta_p u^+ - \Delta_p v^+, [u^+ - v^+]^+) = -|\nabla([u^+ - v^+]^+)|_{L^p}^p \leq 0,$$

$$(-g_0(u^+) + \widetilde{k}_0|v^+|^\theta v^+ - \widetilde{k}_1, [u^+ - v^+]^+) \leq -\widetilde{k}_0\left(|u^+|^\theta u^+ - |v^+|^\theta v^+, [u^+ - v^+]^+\right) \leq 0,$$

and

$$(-g_1(u^+) + g_1(v^+), [u^+ - v^+]^+) \leq L|[u^+ - v^+]^+|_{L^2}^2,$$

we get

$$|[u^+ - v^+]^+(t)|_{L^2}^2 \leq |[u^+ - v^+]^+(0)|_{L^2}^2 e^{2tL},$$

whence it follows that $0 \leq u^+ \leq v^+$. By much the same argument as above, we also find that $u \leq u^+$. Thus, repeating this procedure for $u^-$ (the solution of $(\mathrm{E})_p$ with $h$ and $u_0$ replaced by $-|h|$ and $-|u_0|$) and $v^-$, we obtain (6.8). □

PROOF OF THEOREM 6.2 (CONTINUED). Equation (6.7) has a simple form, and $L^\infty$-estimates for $v^+$ and $v^-$ have been fully investigated by many authors. For instance, Theorem 3.2 of Chapter 5 of [27] ensures the estimate

$$\sup_{x\in\Omega}|v^\pm(x,t)| \leq \max\left(1, C\left(t^{\frac{p}{N}} + \frac{1}{t}\right)^{\frac{1}{q-\delta}}\left(\int_0^t\int_\Omega |v^\pm|^\delta\, dx d\tau\right)^{\frac{p}{N(q-\delta)}}\right) \tag{6.10}$$

where $\delta = 2+\theta$ for $p \leq 2$, $\delta = p$ for $p > 2$, and $q = p(N+2)/N$. Here [27] requires the condition $p \leq \delta < p\frac{N+2}{N}$, which is obviously satisfied for $p > 2$. As for the case where $\frac{2N}{N+2} < p \leq 2$, since $\frac{2N}{N+2} < p$ implies $2 < p\frac{N+2}{N}$, we can choose a sufficiently small $\theta_0$ such that $p \leq 2+\theta_0 < p\frac{N+2}{N}$. If $\theta_0 < \theta$, it is clear that (6.1) is satisfied with $\theta$ and $k_1$ replaced by $\theta_0$ and $k_0+k_1$, respectively. Thus the estimate (6.10) is derived.

On the other hand, by the same argument used in the first half of the proof of Theorem 6.2, we easily see that the estimate similar to (6.6) ensures

$\sup_{t\geq 1}|v^{\pm}(t)|_{L^{\delta}(t,t+1;L^{\delta}(\Omega))}\leq C_1$. Thus (6.3) can be derived from (6.8) and (6.10).

Now we can rewrite (E1) as $u_t(x,t)=\Delta_p u(x,t)+\tilde{h}(x,t)$, where $\tilde{h}(x,t)=-h_0(x,t)-g_1(u(x,t))+h(x)$ and $h_0(x,t)$ is the section of $g_0(u(x,t))$ satisfying $u_t(x,t)=\Delta_p u(x,t)-h_0(x,t)-g_1(u(x,t))+h(x)$. Note that (6.3) ensures $\tilde{h}\in L^{\infty}(\Omega\times[2,+\infty))$. Consequently, by virtue of Theorem 1.2 in Chapter 10 of [27], we can derive the $C^{1,\alpha}(\bar{\Omega})$-bound for $u$ on $[3,+\infty)$. □

## 6.2. The global attractor and its Kolmogorov $\varepsilon$-entropy

The next result states the existence of the global attractor for the problem (E1)–(E3) in the phase space $\Phi:=L^2(\Omega)$.

THEOREM 6.3. *Let all the assumptions of Theorem* 6.2 *be satisfied. Then semigroup* $S(t)$ *associated with* (E1)–(E3) *possesses a global attractor* $\mathcal{A}$ *in the phase space* $L^2(\Omega)$ *which is globally bounded in* $C^{1,\alpha}(\bar{\Omega})$ *(for some* $\alpha$*) and has the following structure:* $\mathcal{A}=\mathcal{K}\big|_{t=0}$ *where* $\mathcal{K}$ *is a set of all bounded solutions of* (E1) *defined for all* $t\in\mathbb{R}$ *. Moreover, this set satisfies* $\|\mathcal{K}\|_{C^{1,\alpha}(\mathbb{R}\times\Omega)}\leq Q(\|h\|_{L^{\infty}(\Omega)})$ *for some monotone increasing function* $Q$.

A proof of the existence of a global attractor $\mathcal{A}$ is an immediate consequence of Theorem 6.2 and the abstract existence result due to Chapter 2. The characterization of $\mathcal{A}$ in terms of $\mathcal{K}$ is derived from Chapter 2.

Next we present lower bounds for the Kolmogorov $\varepsilon$-entropy of attractor $\mathcal{A}$. In compliance with Definition 1.1 we again denote by $N_{\varepsilon}(\mathcal{A},\Phi)$ the minimal number of $\varepsilon$-balls in $\Phi$ that cover $\mathcal{A}$ (recall that $\mathcal{A}$ is a compact in $\Phi$).

From now on we assume that $p>2$, $h\equiv 0$, and $g(u)=\tilde{g}(u)-u$, where $\tilde{g}\equiv 0$ in some small neighborhood of $u=0$, say in $\{u;|u|\leq\rho_0\}$. The special form of $g(u)$ is made for simplicity of presentation. Then we can show that $\dim_F(\mathcal{A},L^2(\Omega))$, the fractal dimension of $\mathcal{A}$ in $L^2(\Omega)$, is infinite by the following arguments. Let $\mathcal{M}^+_{\rho_0}=\{u_0\in L^{\infty}(\Omega);\ \exists u(t,u_0)\in\mathcal{K}^-\text{ s.t. }u(0,u_0)=u_0,\ \|u(t,u_0)\|_{L^{\infty}}\leq\rho_0\,\forall t\in(-\infty,0]\}$, where $u\in\mathcal{K}^-$ implies that $u$ is a bounded solution of (E1)–(E3) on $(-\infty,0]$. Since $\mathcal{M}^+_{\rho_0}\subset\mathcal{A}$, it is enough to prove that $\dim_F\mathcal{M}^+_{\rho_0}=+\infty$.

We consider the following:

(6.11)
$$\begin{cases}\dfrac{\partial u}{\partial t}=\operatorname{div}\left(|\nabla u|^{p-2}\nabla u\right)+u, & (x,t)\in\Omega\times(-\infty,0],\\ u|_{\partial\Omega}=0,\quad t\in(-\infty,0],\quad u|_{t=0}=u_0\in L^{\infty}(\Omega), & x\in\Omega.\end{cases}$$

Note that any solution $u(t,x)$ with small $L^{\infty}$-norm (less than $\rho_0$) of (6.11) is also a solution of (E1)–(E2) on $(-\infty,0]$. Let $u$ be a solution of (6.11) and let $u(t,x)=e^t v(t,x)$. Then $v(t,x)$ satisfies

(6.12)
$$\begin{cases}\dfrac{\partial v}{\partial t}=e^{(p-2)t}\operatorname{div}\left(|\nabla_x v|^{p-2}\nabla_x v\right), & (x,t)\in\Omega\times(-\infty,0],\\ v|_{\partial\Omega}=0,\quad t\in(-\infty,0],\quad v|_{t=0}=u_0\in L^{\infty}(\Omega), & x\in\Omega.\end{cases}$$

Let $s=e^{(p-2)t}$ and $\tilde{v}(s,x)=v(t,x)$. Then we get

$$\begin{cases}\dfrac{\partial\tilde{v}}{\partial s}=\dfrac{1}{p-2}\operatorname{div}\left(|\nabla\tilde{v}|^{p-2}\nabla\tilde{v}\right), & (x,s)\in\Omega\times(0,1],\\ \tilde{v}|_{\partial\Omega}=0,\quad s\in(0,1],\quad \tilde{v}(1,x)=u_0(x), & x\in\Omega.\end{cases}\tag{6.13}$$

We now consider the following initial boundary value problem:

$$\begin{cases} \partial_t w = \dfrac{1}{p-2}\operatorname{div}\left(|\nabla w|^{p-2}\nabla w\right), \\ w|_{\partial\Omega} = 0, \quad t \in [0,1], \quad w|_{t=0} = w_0 \in L^\infty(\Omega), \quad x \in \Omega. \end{cases} \tag{6.14}$$

For any (sufficiently small) compact set $K \subset \Omega$, we fix the initial data $w_0$ such that the corresponding solution $w(x,t)$ of (6.14) satisfies $\|w(\cdot,1)\|_{L^\infty} = 1$, $\sup_{0\le t\le 1}\|w(\cdot,t)\|_{L^\infty} < \rho_1$, and supp $w(x,t) \subset K$ $\forall t \in [0,1]$. This is always possible since the $L^\infty$-norm is a Lyapunov function for (6.14) (cf. [80]) which also has the finite-speed propagation property. (Such a solution can be given explicitly.) Furthermore, it is easy to see that $w_\varepsilon(x,t) = \varepsilon w(\varepsilon^{\frac{2-p}{p}}x,t)$ becomes a solution of (6.14) and $\sup_{0\le t\le 1}\|w_\varepsilon\|_{L^\infty} < \varepsilon\rho_1$, supp $w_\varepsilon \subset \varepsilon^{\frac{p-2}{p}}K$ for all $t \in [0,1]$. Hence, in view of the transformations given above, we find that $w_\varepsilon(x,1) \in \mathcal{M}^+_{\rho_0}$ for sufficiently small $\varepsilon$. Moreover, $w_\varepsilon(x-x_i,t)$ also gives a solution of (6.14), as long as $x_i + K \subset \Omega$. Therefore, for sufficiently small $\varepsilon$, there exists a finite set $R_\varepsilon := \{x_i\} \subset \Omega$ such that

(1) $(x_i + K_\varepsilon)\cap(x_j + K_\varepsilon) = \emptyset$, $\forall x_i, x_j \in R_\varepsilon$, $i \ne j$, $K_\varepsilon := \varepsilon^{\frac{p-2}{p}}K$,
(2) $\#R_\varepsilon \ge C\left(\frac{1}{\varepsilon}\right)^{\frac{N(p-2)}{p}}$,
(3) $x_i + K_\varepsilon \subset\subset \Omega$, $\forall x_i \in R_\varepsilon$.

Consequently, for every $\vec{m} \in \{0,1\}^{R_\varepsilon}$ the function

$$w_{\vec{m},\varepsilon}(x,t) = \sum_{i=1}^{\#R_\varepsilon} m_i w_\varepsilon(x - x_i, t) \tag{6.15}$$

solves (6.14) in $\Omega$. On the other hand, obviously for $\vec{m}^1 \ne \vec{m}^2$

$$\|w_{\vec{m}^1,\varepsilon}(x,1) - w_{\vec{m}^2,\varepsilon}(x,1)\|_{L^\infty} = \varepsilon\|w(x,1)\|_{L^\infty} = \varepsilon. \tag{6.16}$$

Thus we find $2^{\#R_\varepsilon}$ different functions of the form (6.15) in $\mathcal{M}^+_{\rho_0}$. Hence

$$H_\varepsilon(\mathcal{A}, L^\infty) \ge H_\varepsilon(\mathcal{M}^+_{\rho_0}, L^\infty) \ge \#R_\varepsilon \ge C\left(\frac{1}{\varepsilon}\right)^{\frac{N(p-2)}{p}}. \tag{6.17}$$

As for the measurement in the topology of $L^r$ $(1 \le r < \infty)$, instead of (6.16), we get

$$\|w_{\vec{m}^1,\varepsilon}(x,1) - w_{\vec{m}^2,\varepsilon}(x,1)\|_{L^r} \ge \varepsilon^{\frac{pr+(p-2)N}{pr}}\|w(x,1)\|_{L^r}.$$

Hence it is easy to derive $H_\varepsilon(\mathcal{A}, L^r) \ge \left(\frac{1}{\varepsilon}\right)^{\frac{Nr(p-2)}{pr+N(p-2)}}$. Moreover, since the Kolmogorov $\varepsilon$-entropy of a bounded set of $C^1(\overline{\Omega})$ in the topology of $L^\infty$ is estimated above by $C\left(\frac{1}{\varepsilon}\right)^N$ (see [61]), we obtain

$$\left(\frac{1}{\varepsilon}\right)^{\frac{Nr(p-2)}{pr+N(p-2)}} \le H_\varepsilon(\mathcal{A}, L^r) \le C\left(\frac{1}{\varepsilon}\right)^N \qquad \forall r \in [1,\infty].$$

It remains to recall that $\dim_F(\mathcal{A}, L^r)$, the fractal dimension of $\mathcal{A}$, can be expressed in terms of Kolmogorov $\varepsilon$-entropy via $\dim_F(\mathcal{A}, L^r) = \limsup_{\varepsilon\to 0}\frac{\mathcal{H}_\varepsilon(\mathcal{A},L^r)}{\log_2\frac{1}{\varepsilon}}$. Letting $\varepsilon \to 0$ in (20), we find that

$$\dim_F(\mathcal{A}, L^r) = \infty \qquad \forall r \in [1,\infty].$$

CHAPTER 7

# Long-time dynamics of $p$-Laplacian equations: Heterogeneous media

The main purpose of Chapter 7 is to extend the results of Chapter 6 on the global attractor to a much wider class of $p$-Laplacian equations in a heterogeneous medium:

$$(\mathrm{E})_p \begin{cases} \dfrac{\partial u}{\partial t} = \Delta_p u(x,t) - g(x, u(x,t)) + h(x), & (x,t) \in \Omega \times [0,\infty), \quad (\mathrm{E1}) \\ u(x,t) = 0, & (x,t) \in \partial\Omega \times [0,\infty), \quad (\mathrm{E2}) \\ u(x,0) = u_0(x), & x \in \Omega, \quad (\mathrm{E3}) \end{cases}$$

where $\Delta_p u = \operatorname{div}\left(|\nabla u|^{p-2}\nabla u\right),\ p \in \left[\frac{2N}{N+2}, \infty\right)$, and $\Omega$ is a bounded domain in $\mathbf{R}^N$ with smooth boundary $\partial\Omega$.

We especially emphasize that one cannot apply the direct approach developed in Chapter 6 for homogeneous media, so achieving our goal requires new ideas. To this end, we use some new comparison results and special scale transformations which will play a very important role in our arguments.

The present chapter is composed as follows. In Section 7.1, we first prepare some basic results on the existence, uniqueness, and regularity of solutions of (E1)–(E3) and under these preparations we then present in Section 7.2 the existence of the global attractor for the semigroup generated by (E1)–(E3). The infinite dimensionality of the global attractors and the asymptotics of their Kolmogorov entropy are also given in Section 7.2. In particular, we show that their Kolmogorov entropy admits polynomial asymptotics which shed light on completely new phenomena [39].

## 7.1. Existence of global solutions and a priori estimates

In this section, we investigate the solvability of the initial boundary value problem $(\mathrm{E})_p$. To this end, we assume that $g(x,\xi)$ can be decomposed into two parts, the monotone part $g_0(x,\xi)$ and the nonmonotone part $g_1(x,\xi)$, i.e., $g(x,\xi) = g_0(x,\xi)+g_1(x,\xi)$, and we further assume that $g_0, g_1$ satisfy the following conditions:

(a) $g_0(x,0) = 0,\ g_0(x,\xi) \in C(\overline{\Omega} \times \mathbb{R}^1)$, and $g_0(x,\xi)$ is monotone increasing with respect to $\xi$ for all $x \in \Omega$.

(b) $g_1(x,\xi) \in C(\overline{\Omega} \times \mathbb{R}^1)$ and $g_1(x,\xi)$ is a globally Lipschitz function with respect to $\xi$, i.e., there exists a constant $L > 0$ such that

$$\sup_{x\in\Omega} |g_1(x,\xi) - g_1(x,\eta)| \le L|\xi - \eta| \quad \forall \xi, \eta \in \mathbb{R}^1. \tag{7.1}$$

Then $(\mathrm{E})_p$ admits a unique global solution in the following sense.

THEOREM 7.1. *Assume that* (a) *and* (b) *are satisfied. Then, for any* $u_0$, $h \in L^2(\Omega)$, *there exists a unique solution* $u$ *of* $(\mathrm{E})_p$ *satisfying*

$$u \in C([0,T];L^2(\Omega)),\ \sqrt{t}\,u_t \in L^2(0,T;L^2(\Omega)),\ \Delta_p u \in L^2_{loc}((0,T];L^2(\Omega)),$$
$$u \in L^p(0,T;W_0^{1,p}(\Omega)),\ \mathcal{G}_0(u) \in L^1(0,T),$$
$$t\,\|\nabla u\|^p_{L^p(\Omega)},\ t\,\mathcal{G}_0(u) \in L^\infty(0,T), \quad \forall T \in (0,\infty),$$

*where* $\mathcal{G}_0(u) = \int_\Omega G_0(x,u(x,t))\,dx$, $\quad G_0(x,\xi) = \int_0^\xi g_0(x,s)\,ds$. *Furthermore* $S(t):\ u_0(\cdot) \mapsto u(\cdot,t)$ *is continuous in the strong topology of* $L^2(\Omega)$.

PROOF. By assumption (a), it is immediate to see that $u \mapsto -\Delta_p u + g_0(\cdot,u)$ is monotone in $L^2(\Omega)$. However, from (a) alone it is impossible to say whether it is maximal monotone in $L^2(\Omega)$. Hence, to show the existence of solutions for $(\mathrm{E})_p$, we cannot rely on the solvability of abstract evolution equations governed by maximal monotone operators with Lipschitz perturbations. To cope with this difficulty, we introduce some approximation proceduress and make use of the $L^\infty$-energy method and the smoothing effect of the $p$-Laplacian in $L^\infty$-space.

**Step 1.** Let $u_0$, $h \in L^\infty(\Omega)$ and consider

$$(7.2) \qquad (\mathrm{E})_p^M \begin{cases} \frac{\partial u}{\partial t} - \Delta_p u + \partial I_M(u) + g(x,u) \ni h, & (x,t) \in \Omega \times [0,\infty), \\ u(x,t) = 0, & (x,t) \in \partial\Omega \times [0,\infty), \\ u(x,0) = u_0(x), & x \in \Omega, \end{cases}$$

where $I_M(\cdot)$ is the indicator function of the closed convex set $\{u \in L^2(\Omega); |u| \leq M \text{ a.e. } x \in \Omega\}$ :

$$I_M(u) = \begin{cases} 0 & \text{if } |u(x)| \leq M \text{ a.e. } x \in \Omega, \\ +\infty & \text{otherwise,} \end{cases}$$

and $\partial I_M(\cdot)$ denotes its subdifferential operator given by

$$(7.3) \qquad \partial I_M(u) = \begin{cases} 0 & \text{if } |u(x)| < M, \\ [0,+\infty] & \text{if } u(x) \geq M, \\ [-\infty,0] & \text{if } u(x) \leq -M. \end{cases}$$

Here we define $\varphi_M(\cdot)$ by

$$\varphi_M(u) = \begin{cases} \frac{1}{p}\|\nabla u\|^p_{L^p} + I_M(u) & \text{if } |u(x)| \leq M \text{ a.e. } x \in \Omega,\ u \in W_0^{1,p}(\Omega), \\ +\infty & \text{otherwise.} \end{cases}$$

Then $\varphi_M(\cdot)$ is a lower semicontinous convex functional from $L^2(\Omega)$ into $[0,+\infty]$ and its subdifferential $\partial\varphi_M(\cdot)$ satisfies

$$\partial\varphi_M(u) = -\Delta_p u + \partial I_M(u),$$
$$D(\varphi_M) = \{u \in W_0^{1,p}(\Omega); |u(x)| \leq M \text{ a.e. } x \in \Omega\},$$
$$D(\partial\varphi_M) = \{u \in D(\varphi_M); \Delta_p u \in L^2(\Omega)\}.$$

By letting $B(u) = g(\cdot,u(\cdot))$, we can reduce our approximate equation $(\mathrm{E})_p^M$ to the following abstract equation:

$$(\mathrm{AE})_p^M \quad \frac{du}{dt}(t) + \partial\varphi_M(u(t)) + B(u(t)) \ni h, \quad u(0) = u_0.$$

In order to ensure the existence of global solutions of $(\mathrm{AE})_p^M$, we apply Theorem III and Corollary IV of [82] by taking $H = L^2(\Omega)$, $\varphi^t(\cdot) = \varphi_M(\cdot)$, $B(t,\cdot) = B(\cdot) = g(\cdot, u(\cdot))$. To this end, we need to check compactness condition (A1), demiclosedness condition (A2), and boundedness conditions (A5) and (A6) given in [82]. In fact, for any $L > 0$, the level set $\{u; \varphi_M(u) + \|u\|_H^2 \le L\}$ is compact in $L^2(\Omega)$ by virtue of the Rellich compact embedding theorem, which ensures (A1). The demiclosedness of the operator $B : u \mapsto B(u) = g(\cdot, u(\cdot))$ in $L^2(\Omega) \times L^2(\Omega)$ is easily derived from (a) and (b), whence (A2) follows. As for the boundedness condition, since $u \in D(\partial\varphi_M)$ implies that $|u(x)| \le M$, we easily get

$$|B(u)|_{L^2(\Omega)} \le C_M \quad \forall u \in D(\partial\varphi_M),$$

whence follow (A5) and (A6). Thus we see that for any $u_0 \in \overline{D(\partial\varphi_M)} = \{u; |u(x)| \le M\}$ and $h \in L^2(\Omega)$, $(\mathrm{AE})_p^M$ admits a global solution $u$ satisfying $u \in C([0,T]; L^2(\Omega))$, $\sqrt{t}\, u_t, \sqrt{t}\, \Delta_p u \in L^2(0,T; L^2(\Omega)) \quad \forall T > 0$. Furthermore, uniqueness follows easily from the monotonicity of $u \mapsto -\Delta_p u + g_0(\cdot, u(\cdot))$ in $L^2(\Omega)$ and the Lipschitz continuity of $u \mapsto g_1(\cdot, u(\cdot))$ in $L^2(\Omega)$.

**A priori estimate 1:** We multiply $(\mathrm{E})_p^M$ by $|u|^{r-2}u$ and integrate over $\Omega$ to get

$$\begin{aligned}\| u(t) \|_{L^r}^{r-1} \frac{d}{dt} \| u(t) \|_{L^r} &+ (r-1) \int_\Omega |u|^{r-2} |\nabla u|^p \, dx + \int_\Omega \partial I_M(u) |u|^{r-2} u \, dx \\ &+ \int_\Omega g_0(x,u)\, |u|^{r-2} u \, dx \\ &\le \int_\Omega \{ |g_1(x,u)| + |h(x)| \} \, |u|^{r-2} u \, dx \\ &\le L \| u \|_{L^r}^r + C(1 + \| h \|_{L^\infty}) \| u \|_{L^r}^{r-1}.\end{aligned}$$

Here we used (7.1) and the facts that $\partial I_M(u)|u|^{r-2}u \ge 0$, $g_0(x,u)\ |u|^{r-2}u \ge 0$, which are ensured by (7.3) and (a). Hence, by the Gronwall inequality, we have

$$\sup_{0 \le t \le T} \|u(t)\|_{L^r} \le (\, \|u_0\|_{L^r} + C\,(1 + \|h\|_{L^\infty})\, T\,)\, e^{LT}.$$

Here letting $r \to \infty$, we obtain (see [80], [81])

$$\sup_{0 \le t \le T} \|u(t)\|_{L^\infty} \le (\, \|u_0\|_{L^\infty} + C\,(1 + \|h\|_{L^\infty})\, T\,)\, e^{LT} =: C_T. \tag{7.4}$$

For each $u_0 \in L^\infty(\Omega)$ and $T > 0$, fix $M$ so that $C_T < M$. Then by virtue of (7.3) and (7.4), we find that $\partial I_M(u(t)) = 0\ \forall\, t \in [0,T]$, which implies that $u(t)$ gives a solution of $(\mathrm{E})_p$ on [0,T].

**Step 2.** Let $u_0^n,\ h^n \in L^\infty(\Omega)$, $u_0^n \to u_0$, $h^n \to h$ in $L^2(\Omega)$ as $n \to \infty$, and let $u^n$ be the global solution of $(\mathrm{E})_p$ with $u_0 = u_0^n$, $h = h^n$, i.e., $(\mathrm{E})_{p,n}$: $u_t^n - \Delta_p u^n + g_0(x,u^n) = -g_1(x,u^n) + h^n$, $u_0^n(0) = u_0^n$.

**A priori estimate 2:** We first note that the monotonicity of $g_0(\cdot,\xi)$ and the definition of $G_0$ imply

$$G_0(x,\xi) = \int_0^\xi g_0(x,s)\, ds \le g_0(x,\xi)\, \xi.$$

Hence, multiplying $(\mathrm{E})_{p,n}$ by $u^n$, we get

$$\begin{aligned}\frac{1}{2}\frac{d}{dt}\|u^n(t)\|_{L^2}^2 + \|\nabla u^n(t)\|_{L^p}^p + \mathcal{G}_0(u^n(t)) \\ \leq (\|g_1(\cdot,u^n)\|_{L^2} + \|h^n\|_{L^2})\,\|u^n(t)\|_{L^2} \\ \leq L\,\|u^n(t)\|_{L^2}^2 + (\,C + \|h^n\|_{L^2}\,)\,\|u^n(t)\|_{L^2},\end{aligned} \tag{7.5}$$

whence it follows that

$$\sup_{0\leq t\leq T}\|u^n(t)\|_{L^2} + \int_0^T (\,\|\nabla u^n(t)\|_{L^p}^p + \mathcal{G}_0(u^n(t))\,)\,dt \leq C_0(\|u_0\|_{L^2}, \|h\|_{L^2}). \tag{7.6}$$

Next multiply $(\mathrm{E})_{p,n}$ by $t\,(u^n)_t$ and integrate over $[0,T]$ with respect to $t$. Then we have

$$\begin{aligned} t\,\|(u^n)_t\|_{L^2}^2 + t\,\frac{d}{dt}\left(\frac{1}{p}\,\|\nabla u^n(t)\|_{L^p}^p + \mathcal{G}_0(u^n(t))\right) \\ \leq (\,\|g_1(u^n(t))\|_{L^2} + \|h^n\|_{L^2}\,)\,\|(u^n)_t\|_{L^2} \\ \leq \frac{1}{2}\,\|(u^n)_t\|_{L^2}^2 + C_0,\end{aligned}$$

$$\sup_{0\leq t\leq T} t\left(\frac{1}{p}\,\|\nabla u^n(t)\|_{L^p}^p + \mathcal{G}_0(u^n(t))\right) + \int_0^T t\,\|(u^n)_t\|_{L^2}^2\,dt \leq C_0. \tag{7.7}$$

Hence, from equation $(\mathrm{E})_{p,n}$, we derive the a priori bound

$$\|\sqrt{t}\,(\,-\Delta_p\,u^n + g_0(x,u^n)\,)\|_{L^2(Q)} \leq C_0, \quad Q = \Omega\times(0,T). \tag{7.8}$$

However, this does not ensure the boundedness of $\|\sqrt{t}\,\Delta_p\,u^n\|_{L^2(Q)}$. Nevertheless, by applying a comparison theorem, which will be given in Lemma 7.1, between $u^n$ and $v^\pm$, solutions of (7.17) satisfying (7.20), we can derive some a priori estimate for $u^n$ in $L^\infty(\Omega)$ as follows. We first note that $v^\pm$ also satisfies (7.15), so by the same reasoning used for (7.12), we can obtain the a priori bound for $\|v^\pm\|_{L^2(\Omega)}$. Then integrating (7.15) on $[0,T]$, we derive, by the Poincaré inequality, the a priori bound for $\|v^\pm\|_{L^p(Q)}$ depending only on $\|u_0\|_{L^2(\Omega)}, \|h\|_{L^2(\Omega)}$. Hence the right-hand side of (7.20) is bounded by some constant independent of $n$ on $[\delta,T]$ for each $\delta>0$, which together with (7.18) yields the estimate

$$\forall\,\delta>0,\ \exists\,C_\delta = C(\delta,C_0)\ \text{s.t.}\ \sup_{\delta\leq t\leq T}\|u^n(\cdot,t)\|_{L^\infty(\Omega)} \leq C_\delta,$$

whence, by (7.8), it follows that

$$\sup_{\delta\leq t\leq T}\|g_0(\cdot,u^n(t))\|_{L^\infty(\Omega)} \leq C_\delta, \quad \|\Delta_p\,u^n\|_{L^2(\delta,T;L^2(\Omega))} \leq C_\delta. \tag{7.9}$$

**Convergence:** $w = u^m - u^n$ satisfies

$$\begin{aligned} w_t - (\,\Delta_p\,u^m - \Delta_p\,u^n\,) + (\,g_0(x,u^m) - g_0(x,u^n)\,) \\ = -(\,g_1(x,u^m) - g_1(x,u^n)\,).\end{aligned} \tag{7.10}$$

Multiplying (7.10) by $w$ and using the monotonicity of $u \mapsto -\Delta_p\,u + g_0(\cdot,u)$ in $L^2(\Omega)$ and condition (b), we get

$$\frac{1}{2}\frac{d}{dt}\|w(t)\|_{L^2}^2 \leq L\,\|w(t)\|_{L^2}^2,$$

whence it follows that

$$\|w(t)\|_{L^2} \leq \|u_0^m - u_0^n\|_{L^2}\, e^{Lt},$$

which implies that $\{u^n\}$ forms a Cauchy sequence in $L^2(\Omega)$. Thus, in view of (7.6), (7.7), and (7.9), we find that there exists a subsequence of $\{u^n\}$ denoted again by $\{u^n\}$ such that

$$\begin{aligned}
u^n &\to u \quad \text{strongly in } C([0,T];L^2(\Omega)), \text{ and a.e. } (x,t)\in Q,\\
g_1(x,u^n) &\to g_1(x,u) \quad \text{strongly in } \ C([0,T];L^2(\Omega)),\\
\nabla u^n &\to \nabla u \quad \text{weakly in } \ L^p([0,T];L^p(\Omega)),\\
\sqrt{t}\, u_t^n &\to \sqrt{t}\, u_t \quad \text{weakly in } L^2(Q),\\
g_0(x,u^n) &\to g_0(x,u) \quad \text{weakly-}*\text{ in } \ L^\infty(Q_\delta),\\
\Delta_p u^n &\to \Delta_p u \quad \text{weakly in } L^2(Q_\delta), \quad Q_\delta = \Omega\times[\delta,T], \quad \forall\delta>0.
\end{aligned}$$

Here we used the semiclosedness of $g_0(\cdot,u)$, $-\Delta_p u$. Furthermore, by virtue of the lower-semicontinuity of $\mathcal{G}_0(u), \|\nabla u\|_{L^p}^p$, we easily see that $\|\mathcal{G}_0(u)\|_{L^1(0,T)}$, $\|t\,|\nabla u|^p\,\|_{L^\infty(0,T;L^1(\Omega))}$, $\|t\,\mathcal{G}_0(u)\|_{L^\infty(0,T)}$ are all bounded. □

REMARK 7.1. If $g(x,u)\in C(\overline{\Omega}\times\mathbb{R}^1)$ satisfies $g(x,0)=0$, $g'_\tau(x,\tau)\geq -K\ \forall(x,\tau)\in\Omega\times\mathbb{R}^1$, then by letting $g_0(x,\tau)=g(x,\tau)+K\tau$ $((g_0)'_\tau(x,\tau)\geq 0)$, $g_1(x,\tau)=-K\tau$, we find that $g(\cdot,u)$ falls within our class. In particular, $g(x,\tau)=C_1(x)\,|\tau|^{q_1-2}\,\tau - C_2(x)|\tau|^{q_2-2}\tau$ $(2\leq q_2<q_1<+\infty)$ satisfies (a) and (b), provided that $C_1(x)$, $C_2(x)\in L^\infty(\Omega)$ and $0<c_1\leq C_1(x)\quad \forall x\in\Omega$, since

$$g'_\tau(x,\tau)=|\tau|^{q_2-1}(C_1(x)\,(q_1-1)|\tau|^{q_1-q_2}-C_2(x)\,(q_2-1))\geq -K\ \forall\,(x,\tau)\in\Omega\times\mathbb{R}^1.$$

As for the a priori bounds for solutions of (E1)–(E3), we obtain the following result.

THEOREM 7.2. *Assume that* (a) *and* (b) *are satisfied and let* $\frac{2N}{N+2}<p<\infty$, $u_0\in L^2(\Omega)$, *and* $h\in L^\infty(\Omega)$. *For the case* $\frac{2N}{N+2}<p\leq 2$, *we further assume*

$$|\,g_0(s)|\geq k_0\,|s|^{1+\theta}-k_1 \quad (\theta,\ k_0,\ k_1>0\,). \tag{7.11}$$

*Then, for any* $0<\delta_1<\delta_2<\delta_3\leq 1$, *there exist constants* $C_1,C_2,C_3$ *depending on* $\delta_1,\delta_2,\delta_3$ *but not on the initial data* $u_0\in L^2(\Omega)$ *such that every solution of* (E1)–(E3) *satisfies*

$$\|u(t)\|_{L^2(\Omega)}\leq C_1 \qquad \forall\,t\in[\,\delta_1,+\infty), \tag{7.12}$$

$$\|u(t)\|_{L^\infty(\Omega)}\leq C_2 \qquad \forall\,t\in[\,\delta_2,+\infty), \tag{7.13}$$

$$\|u(t)\|_{C^{1,\alpha}(\overline{\Omega})}\leq C_3 \qquad \forall\,t\in[\,\delta_3,+\infty). \tag{7.14}$$

PROOF. Multiply (E1) by $u$. Then the same argument for (7.5) gives

$$\frac{1}{2}\frac{d}{dt}\|u(t)\|_{L^2}^2+\|\nabla u\|_{L^p}^p\leq L\,\|u(t)\|_{L^2}^2+(\,C+\|h\|_{L^2})\,\|u\|_{L^2}. \tag{7.15}$$

Then by the Poincaré inequality, we get for the case $p>2$

$$\frac{1}{2}\frac{d}{dt}\|u(t)\|_{L^2}^2+\gamma_1\|u\|_{L^2}^p\leq\gamma_2, \tag{7.16}$$

for some $\gamma_1,\gamma_2>0$. Hence the Ghidaglia type estimate (see [93]) ensures (7.12). For the case $\frac{2N}{N+2}\leq p\leq 2$, (7.11) yields the same estimate as (7.15) with $p=2+\theta$, whence follows (7.12).

In order to derive the $L^\infty$-estimate, we need the following lemma.

LEMMA 7.1. *Let $v^\pm$ be the unique solution of*

$$v^\pm_t = \Delta_p\, v^\pm - \widetilde{k}_0\, |v^\pm|^\theta v^\pm - g_1(v^\pm) \pm (\,|h| + \widetilde{k}_1), \quad v^\pm|_{\partial\Omega} = 0, \tag{7.17}$$

*with the initial condition* $v^\pm(x,0) = \pm\,|u_0(x)|$, *respectively, where* $\widetilde{k}_i = 0$ *for* $p > 2$ *and* $\widetilde{k}_i = k_i$ *for* $p \le 2$ $(i = 0,1)$. *Then the solution $u$ of* $(\mathrm{E})_p$ *satisfies*

$$v^-(x,t) \le u(x,t) \le v^+(x,t) \qquad \textit{for a.e. } (x,t) \in \Omega \times [\,0,+\infty). \tag{7.18}$$

PROOF OF LEMMA 7.1. Let $u^+(x,t)$ be the unique solution of $(\mathrm{E})_p$ with $h$ and $u_0$ replaced by $|h|$ and $|u_0|$, respectively. Then it is easy to see that $u^+ \ge 0$ and $(u^+ - v^+)$ satisfies

$$(u^+ - v^+)_t = \Delta_p\, u^+ - \Delta_p\, v^+ - g_0(u^+) + \widetilde{k}_0 |v^+|^\theta v^+ - \widetilde{k}_1 - g_1(u^+) + g_1(v^+). \tag{7.19}$$

Multiply (7.19) by $[\,u^+ - v^+]^+ = \max\,(\,u^+(x,t) - v^+(x,t),\, 0\,)$. Then noting that

$$\begin{aligned}
&(\,\Delta_p\, u^+ - \Delta_p\, v^+,\, [\,u^+ - v^+]^+\,) = -\,\|\,\nabla(\,[\,u^+ - v^+]^+\,)\,\|^p_{L^p} \le 0,\\
&(\,-g_0(u^+) + \widetilde{k}_0\,|\,v^+|^\theta v^+ - \widetilde{k}_1,\, [\,u^+ - v^+]^+\,)\\
&\qquad \le -\,\widetilde{k}_0\,(\,|\,u^+|^\theta u^+ - |\,v^+|^\theta v^+,\, [\,u^+ - v^+]^+\,) \le 0,\\
&\qquad (\,-g_1(u^+) + g_1(v^+),\, [\,u^+ - v^+]^+\,) \le L\,\|\,[\,u^+ - v^+]^+\,\|^2_{L^2},
\end{aligned}$$

we get

$$\|\,[\,u^+ - v^+]^+(t)\,\|^2_{L^2} \le \|\,[\,u^+ - v^+]^+(0)\,\|^2_{L^2}\, e^{2Lt},$$

whence follows $0 \le u^+ \le v^+$. By much the same argument as above, we also find that $u \le u^+$. Thus repeating this procedure for $u^-$ (the solution of $(\mathrm{E})_p$ with $h$ and $u_0$ replaced by $-|h|$ and $-|u_0|$) and $v^-$, we obtain (7.18). □

PROOF OF THEOREM 7.2 (CONTINUED). Equation (7.17) has the simple form, and the $L^\infty$-estimates for $v^\pm$ have been fully investigated by many authors. For instance, Theorem 3.2 of Chapter 5 of [27] gives the estimate

$$\sup_{x\in\Omega} |v^\pm(x,t)| \le \max\left(1,\, C\left(t^{\frac{p}{N}} + \frac{1}{t}\right)^{\frac{1}{q-\delta}} \left(\int_0^t \int_\Omega |v^\pm|^\delta\, dx d\tau\right)^{\frac{p}{N(q-\delta)}}\right) \tag{7.20}$$

where $\delta = 2 + \theta$ for $p \le 2$, $\delta = p$ for $p > 2$, and $q = p(N+2)/N$. Here [27] requires the condition $p \le \delta < p\,\frac{N+2}{N}$, which is obviously satisfied for $p > 2$. As for the case where $\frac{2N}{N+2} < p \le 2$, since $\frac{2N}{N+2} < p$ implies $2 < p\,\frac{N+2}{N}$, we can choose a sufficiently small $\theta_0$ such that $p \le 2 + \theta_0 < p\,\frac{N+2}{N}$. If $\theta_0 < \theta$, it is clear that (7.11) is satisfied with $\theta$ and $k_1$ replaced by $\theta_0$ and $k_0 + k_1$, respectively. As is seen in the proof of Theorem 7.1, a priori estimate 2, $v^\pm$ satisfies (7.16) with $p = p$ or $p = 2 + \theta$. Hence, integrating this on $[t, t+1]$, we get by (7.12)

$$\sup_{t\ge 1}\, \|\, v^\pm\,\|_{L^\delta(t,t+1;L^\delta(\Omega))} \le C_1. \tag{7.21}$$

Thus the estimate (7.13) is derived from (7.18), (7.20), and (7.21). Now we can rewrite (E1) as $u_t(x,t) = \Delta_p\, u(x,t) + \widetilde{h}(x,t)$, where $\widetilde{h}(x,t) = -g_0(x, u(x,t)) - g_1(x, u(x,t)) + h(x)$. Note that (7.13) ensures $\widetilde{h} \in L^\infty(\Omega \times [\delta_2, +\infty))$. Consequently, by virtue of Theorem 1.2 in Chapter 10 of [27], we can derive the $C^{1,\alpha}(\overline{\Omega})$-bound for $u$ on $[\,\delta_3, +\infty)$. □

## 7.2. The global attractor and Kolmogorov $\varepsilon$-entropy

In this section we prove the main result of the chapter, that is, Theorem 7.4 (see below) on the infinite dimensionality of the global attractor of a semigroup generated by $(\mathrm{E})_p$. Our existence result for global attractors of (E1)–(E3) in the phase space $\Phi = L^2(\Omega)$ can be stated as follows.

THEOREM 7.3. *Let all the assumptions in Theorem 7.2 be satisfied. Then the semigroup $S(t)$ associated with* (E1)–(E3) *possesses a global attractor $\mathcal{A}$ in the phase space $L^2(\Omega)$ which is globally bounded in $C^{1,\alpha}(\overline{\Omega})$ with $\alpha \in (0,1]$ and has the following structure : $\mathcal{A} := \mathcal{K}_0 := \{\, u(0)\ ;\ \{u(t)\}_{t\in\mathbb{R}^1} \in \mathcal{K}\,\}$, where $\mathcal{K}$ is the set of all bounded solutions of* (E1)–(E2) *defined on $\mathbb{R}^1$, i.e.,*

$$\mathcal{K} = \{\{u(t)\}_{t\in\mathbb{R}^1}\ ;\ u(t) \textit{ is a solution of } (\mathrm{E1})\text{–}(\mathrm{E2}) \textit{ on } \mathbb{R}^1,\ \sup_{t\in\mathbb{R}^1} \|u(t)\|_{L^2} < +\infty\,\}.$$

PROOF. In order to prove the existence of the global attractor $\mathcal{A}$ for (E1)–(E3), it suffices to show that the semigroup $S(t)$ associated with (E1)–(E3) is continuous in the topology of $L^2(\Omega)$ for each $t > 0$ and that there exists a pre-compact absorbing set $\mathcal{B}$ in $L^2(\Omega)$ such that for every $x \in L^2(\Omega)$, there exists $T = T(x) > 0$ such that $S(t)\,x \in \mathcal{B}\ \forall\, t \in [\,T, +\infty)$. (See Theorem 1.1 of [93].) For our case, the first property is ensured by Theorem 7.1 and the second by Theorem 7.2. The characterization of $\mathcal{A}$ in terms of $\mathcal{K}$ is derived by standard arguments. $\square$

Next we present lower bounds for Kolmogorov $\varepsilon$-entropy of the attractor $\mathcal{A}$ in $\Phi = L^p(\Omega)$ $(1 \le p \le \infty)$. In compliance with Definition 1.1 we denote by $\mathcal{N}_\varepsilon(\mathcal{A}, \Phi)$ the minimal number of $\varepsilon$-balls in $\Phi$ that cover $\mathcal{A}$ (recall that $\mathcal{A}$ is a compact set in $\Phi$). From now on we assume that $p > 2$, $h \equiv 0$, and $g(\cdot, u)$ satisfies the following assumption:

$(\mathrm{I})_g$ There exists an open bounded subset $\omega$ of $\Omega$ and $\alpha > 0$ such that $g^\alpha(x,u) = g(x,u) + \alpha\, u$ satisfies the following:

$(\mathrm{I})_1$ There exist $a(s)$, $\rho \ge 0$, and $h(x,s,v)$ satisfying

$$s^{\frac{-p+1}{p-2}}\, g^\alpha(x, s^{\frac{1}{p-2}} v) = a(s)\,|v|^\rho\, h(x,s,v), \quad (x,s,v) \in \omega \times [0,1] \times \mathbb{R}^1,$$

$$h(x,s,v),\ h'_v(x,s,v) \in C(\overline{\omega} \times [0,1] \times \mathbb{R}^1), \quad h(x,s,0) = 0,$$

$$a \ge 0,\ a \in L^1(0,1),\ s\,a^2 \in L^1(0,1).$$

$(\mathrm{I})_2$ There exist $C > 0$ and $\delta > 0$ such that

$$|\,v\,|^\rho\,|\,h(x,s,v)\,| \le C\,|\,v\,|^{1+\delta}, \quad (x,s,v) \in \omega \times [0,1] \times [0,1].$$

Then our result on the infinite dimensionality of global attractors for (E1)–(E3) is as follows.

THEOREM 7.4. *Let $(\mathrm{I})_g$ be satisfied and assume that* (E1)–(E3) *possess a global attractor $\mathcal{A}$ in the topology of $L^2(\Omega)$. Then the fractal dimension of $\mathcal{A}$ is infinite.*

The presentation of condition $(\mathrm{I})_g$ might seem to be somewhat obscure. In order to clarify the meaning of this condition, we will show below that it covers a very large class of nonlinearity.

EXAMPLE 7.1. Let $g(x,u) = -\alpha\, u + b_1(x)\,|u|^{q_1-2}u - b_2(x)\,|u|^{q_2-2}u$, $\alpha > 0$, $2 < q_2 < q_1$, $b_1, b_2 \in C(\overline{\omega})$. Then $g(\cdot,u)$ satisfies $(\mathrm{I})_g$. In fact, since $g^\alpha(x,u) =$

$b_1(x)\,|u|^{q_1-2}u - b_2(x)\,|u|^{q_2-2}u$, we get

$$\begin{aligned} s^{\frac{-p+1}{p-2}} g^\alpha(x, s^{\frac{1}{p-2}} v) &= s^{\frac{-p+1}{p-2}} \left\{ b_1(x)\, s^{\frac{q_1-1}{p-2}}\, |v|^{q_1-2} v - b_2(x)\, s^{\frac{q_2-1}{p-2}}\, |v|^{q_2-2} v \right\} \\ &= s^{\frac{q_2-p}{p-2}}\, |v|^{q_2-2} \left\{ b_1(x)\, s^{\frac{q_1-q_2}{p-2}}\, |v|^{q_1-q_2} v - b_2(x)\, v \right\}. \end{aligned}$$

Hence we can let

$$a(s) = s^{\frac{q_2-p}{p-2}},\ \rho = q_2 - 2 > 0,\ h(x,s,v) = b_1(x)\, s^{\frac{q_1-q_2}{p-2}}\, |v|^{q_1-q_2} v - b_2(x)\, v.$$

Then it is easy to see that

$$\begin{aligned} & h(x,s,v),\ h'_v(x,s,v) \in C(\overline{\Omega} \times [0,1] \times \mathbb{R}^1),\ h(x,s,0) = 0, \\ & |a|_{L^1} = \int_0^1 s^{\frac{q_2-p}{p-2}}\, ds = \frac{p-2}{q_2-2},\ |sa^2|_{L^1} = \int_0^1 s\, s^{\frac{2q_2-2p}{p-2}}\, ds = \frac{p-2}{2q_2-4}, \\ & |\,v\,|^\rho |\,h(x,s,v)\,| \le C\,|\,v\,|^{q_2-1} = C\,|\,v\,|^{1+\delta}, \quad (x,s,v) \in \omega \times [0,1] \times [0,1], \\ & \quad \delta = q_2 - 2 > 0. \end{aligned}$$

EXAMPLE 7.2. Let $g(x,u),\ g'_u(x,u),\ g''_u(x,u) \in C(\overline{\omega}\times\mathbb{R}^1),\ g'_u(x,0) = -\alpha < 0$, $g(x,0) = 0$. Then $g(\cdot,u)$ satisfies $(\mathrm{I})_g$. In fact, we first note that

$$g(x,u) = g(x,0) + g'_u(x,0)\,u + \int_0^u (u-t)\, g''_u(x,t)\, dt,\ g^\alpha(x,u) = \int_0^u (u-t)\, g''_u(x,t)\, dt.$$

Then we get

$$\begin{aligned} s^{\frac{-p+1}{p-2}}\, g^\alpha(x, s^{\frac{1}{p-2}} v) &= s^{\frac{-p+1}{p-2}} \int_0^{s^{\frac{1}{p-2}} v} (s^{\frac{1}{p-2}}\, v - t)\, g''_v(x,t) dt \\ &= s^{\frac{-p+1}{p-2}} \int_0^v (s^{\frac{1}{p-2}}\, v - s^{\frac{1}{p-2}}\, t)\, g''_v(x, s^{\frac{1}{p-2}} t)\, s^{\frac{1}{p-2}}\, dt \\ &= s^{\frac{3-p}{p-2}} \int_0^v (v-t)\, g''_v(x, s^{\frac{1}{p-2}}\, t)\, dt, \end{aligned}$$

and we let

$$\rho = 0, \quad h(x,s,v) = \int_0^v (v-t)\, g''_v(x, s^{\frac{1}{p-2}}\, t)\, dt, \quad a(s) = s^{\frac{3-p}{p-2}}.$$

Hence we obtain

$$\begin{aligned} & h(x,s,v),\ h'_v(x,s,v) = \int_0^v g''(x, s^{\frac{1}{p-2}}\, t)\, dt \in C(\overline{\omega} \times [0,1] \times \mathbb{R}^1),\ h(x,s,0) = 0, \\ & |a|_{L^1} = \int_0^1 s^{\frac{3-p}{p-2}}\, ds = (p-2), \quad |sa^2|_{L^1} = \int_0^1 s\, s^{\frac{6-2p}{p-2}}\, ds = \frac{p-2}{2}, \\ & |v|^\rho |h(x,s,v)| = |\int_0^v (v-t)\, g''_v(x, s^{\frac{1}{p-2}}\, t)\, dt| \le \int_0^{|v|} |1 - \frac{t}{v}|\, |g''_v(x, s^{\frac{1}{p-2}} t)|\, dt\, |v| \\ & \quad \le \max_{(x,s)\in\overline{\omega}\times[0,1]} |g''(x,s)|\, |v|^2, \quad (x,s,v) \in \omega \times [0,1] \times [0,1], \quad \delta = 1. \end{aligned}$$

In order to establish the estimate from below for the fractal dimension of our global attractor $\mathcal{A}$, we rely on the following fact.

LEMMA 7.2. *Let $\mathcal{K}^-$ be the set of all bounded solutions of* (E1)–(E2) *on* $\mathbb{R}^1_-$, *i.e.*, $\mathcal{K}^- = \{\{u(t)\}_{t\in\mathbb{R}^1_-}\,;\, u(t)$ *satisfies* (E1)–(E2) *on* $\mathbb{R}^1_-$, $\sup_{t\in\mathbb{R}^1_-}\|u(t)\|_{L^2} < +\infty\}$, *where* $\mathbb{R}^1_- := (-\infty, 0]$ *and let* $\mathcal{K}^-(t)$ *be the section of* $\mathcal{K}^-$ *at* $t = t \in \mathbb{R}^1_-$, *i.e.*, $\mathcal{K}^-(t) = \{\, u(t)\,;\, \{u(t)\}_{t\in\mathbb{R}^1_-} \in \mathcal{K}^-\,\}$. *Then* $\mathcal{K}^-(0) \subset \mathcal{A}$ *holds true.*

PROOF. The proof repeats literally the proof of Lemma 5.1 but for $L^1$ replaced by $L^2$. Let $\mathcal{B} = \bigcup_{t\in\mathbb{R}^1_-}\mathcal{K}^-(t)$. Then, since $\mathcal{B}$ is bounded in $L^2(\Omega)$, for arbitrary $\eta > 0$, there exists $T > 0$ such that $\mathrm{dist}_{L^2(\Omega)}(S(T)\,\mathcal{B}, \mathcal{A}) < \eta$. For any $a_0 \in \mathcal{K}^-(0)$, there exists $a_T \in \mathcal{K}^-(-T) \subset \mathcal{B}$ such that $S(T)\,a_T = a_0$. Hence we get

$$\mathrm{dist}_{L^2(\Omega)}(a_0, \mathcal{A}) = \mathrm{dist}_{L^2(\Omega)}(S(T)\,a_T, \mathcal{A}) \le \mathrm{dist}_{L^2(\Omega)}(S(T)\,\mathcal{B}, \mathcal{A}) < \eta, \quad \forall\,\eta > 0,$$

which implies $\mathrm{dist}_{L^2(\Omega)}(a_0, \mathcal{A}) = 0$, i.e., $a_0 \in \bar{\mathcal{A}} = \mathcal{A}\ \forall a_0 \in \mathcal{K}^-(0)$. Thus $\mathcal{K}^-(0) \subset \mathcal{A}$ is derived. □

Before we proceed to the proof of Theorem 7.4, we prepare a couple of results on the following auxiliary equation:

$$\text{(7.22)}\qquad (\mathrm{E})^t_p \begin{cases} w_t = p_\alpha\left(\Delta_p w - a(t)\,|\,w\,|^\rho\, h(x,t,w)\right), & (x,t) \in \omega \times (0,1), \\ w|_{\partial\omega} = 0, \quad t \in [0,1], \quad w(x,0) = w_0(x), & x \in \omega, \end{cases}$$

where $p_\alpha = \dfrac{1}{\alpha(p-2)}$. As for the solvability of this equation, the following result holds.

LEMMA 7.3. *Let* $(\mathrm{I})_g$ *be satisfied. Then for every* $w_0 \in L^\infty(\Omega)$, *there exists* $T_0 = T_0(\|u_0\|_{L^\infty}) > 0$ *such that* $(\mathrm{E})^t_p$ *admits a unique solution* $w$ *on* $[0, T_0]$ *satisfying*

$$\text{(7.23)}\qquad \begin{aligned} &w \in C([0,T_0]; L^2(\omega)) \cap C((0,T_0]; W^{1,p}_0(\omega)) \cap L^\infty(\omega \times (0,T_0)), \\ &\sqrt{t}\, w_t,\ \sqrt{t}\,\Delta_p w \in L^2(\omega \times (0,T_0)). \end{aligned}$$

*Furthermore, there exists a (sufficiently small)* $\varepsilon_0 > 0$ *such that if* $\|\, w_0\,\|_{L^\infty} \le \varepsilon_0$, *then the solution* $w$ *of* $(\mathrm{E})^t_p$ *given above can be continued up to* $[0,1]$ *and satisfies* $\sup_{t\in[0,1]}\|\,w(t)\,\|_{L^\infty} \le 1$.

PROOF. We again apply the $L^\infty$-energy method as in the proof of Theorem 7.1 (Step 1). Take $0 \le a_n(t) \in C([0,1])$ so that $a_n(t) \to a(t)$ in $L^1(0,T)$ and $\sqrt{t}\,a_n(t) \to \sqrt{t}\,a(t)$ in $L^2(0,T)$ as $n \to \infty$. Let $M = \|\,w_0\,\|_{L^\infty} + 1$ and consider the following equations:

$$\text{(7.24)}\qquad (\mathrm{E})^{t,M}_{p,n} \begin{cases} w^n_t \in p_\alpha\left(\Delta_p w^n - \partial I_M(w^n) - a_n(t)\,|\,w^n\,|^\rho\, h(x,t,w^n)\right) \text{ in } \omega \times (0,1), \\ w^n|_{\partial\omega} = 0, \quad t \in [0,1], \quad w^n(x,0) = w_0(x), \quad x \in \omega, \end{cases}$$

where $\partial I_M(\cdot)$ is the subdifferential of $I_M(\cdot)$, given by (7.3). As in the proof of Theorem 7.1, we can reduce $(\mathrm{E})^{t,M}_{p,n}$ to some abstract Cauchy problem such as $(\mathrm{AE})^M_p$ and apply Theorem III and Corollary IV of [82] by taking $H = L^2(\Omega)$, $\varphi^t(\cdot) =$

$p_\alpha\, \varphi_M(\cdot)$, $B(t,w) = p_\alpha\, a_n(t)\,|\,w\,|^\rho\, h(x,t,w)$. Here we note that $w_0 \in \overline{D(\partial\varphi_M)} = \{u; |u(x)| \le M = \|\, w_0\,\|_{L^\infty} + 1\}$. Thus the existence of solutions of (7.24) is verified.

**A priori estimates:** We first note that

$$|\, h(x,s,v)\,| = |\int_0^v h'_\tau(x,s,\tau)\, d\tau| \le \ell_0(|\,v\,|)\,|\,v\,|,$$
$$\ell_0(r) = \max\{\,|\,h'_v(x,s,v)\,|\,;\, x \in \overline{\omega},\ s \in [0,1],\ |v| \le r\,\}.$$

Here $\ell_0(r)$ is monotone increasing in $r \in [0,+\infty)$. Then, as in the proof of Theorem 7.1 (a priori estimate 1), by multiplying $(\mathrm{E})^{t,M}_{p,n}$ by $|u|^{r-2}u$, we get

$$\frac{1}{r}\frac{d}{dt}\|\, w^n(t)\,\|^r_{L^2} \le p_\alpha\, a_n(t)\,\|\, w^n(t)\,\|^\rho_{L^\infty}\, \ell_0(\|\, w^n(t)\,\|_{L^\infty})\,\|\, w^n(t)\,\|^r_{L^r},$$
$$\|\, w^n(t)\,\|_{L^r} \le \|\, w_0\,\|_{L^r} + \int_0^t a_n(s)\,\ell_1(\|\, w^n(s)\,\|_{L^\infty})\,\|\, w^n(s)\,\|_{L^r}\, ds,$$
$$\text{with } \ell_1(r) = p_\alpha\, r^\rho\, \ell_0(r).$$

Then letting $r \to \infty$, we obtain (see [80], [81])

$$\|\, w^n(t)\,\|_{L^\infty} \le \|\, w_0\,\|_{L^\infty} + \int_0^t a_n(s)\,\ell(\|\, w^n(s)\,\|_{L^\infty})\, ds, \text{ with } \ell(r) = \ell_1(r)\, r. \tag{7.25}$$

Here define a positive number $\delta$ and choose $T_0$ such that

$$\delta = \frac{1}{\ell(M)+1}, \qquad \int_0^{T_0} a(s)\, ds < \frac{\delta}{2}. \tag{7.26}$$

Then, since $a_n \to a$ in $L^1(0,1)$, there exists $N$ such that

$$\int_0^{T_0} a_n(s)\, ds < \delta, \quad \forall n \ge N. \tag{7.27}$$

We claim that

$$\|\, w^n(t)\,\|_{L^\infty} \le \|\, w_0\,\|_{L^\infty} + 1 = M \quad \forall t \in [0,T_0],\ \forall n \ge N. \tag{7.28}$$

To see this, we let

$$z^n(t) = \|\, w_0\,\|_{L^\infty} + \int_0^t a_n(s)\,\ell(\|\, w^n(s)\,\|_{L^\infty})\, ds.$$

Then it is clear that $z^n(t) \in C([0,1])$ and $\|\, w^n(t)\,\|_{L^\infty} \le z^n(t) \quad \forall t \in [0,1]$. Hence, since $\ell(\cdot)$ is monotone increasing, $z^n(\cdot)$ satisfies

$$z^n(t) \le \|\, w_0\,\|_{L^\infty} + \int_0^t a_n(s)\,\ell(z^n(s))\, ds \quad \forall t \in [0,1], \tag{7.29}$$

and hence

$$z^n(t) \le \|\, w_0\,\|_{L^\infty} + 1 = M \quad \forall t \in [0,T_0],\ \forall n \ge N. \tag{7.30}$$

In fact, if (7.30) does not hold, there exists $t_0 \in (0,T_0)$ such that $z^n(t_0) > M$. Then since $z^n(t)$ is continuous on $[0,1]$ and $z^n(0) = \|\, w_0\,\|_{L^\infty} < M$, there exists $t_1 \in (0,t_0)$ such that

$$z^n(t_1) = M, \quad z^n(t) < M \quad \forall t \in [0,t_1).$$

Hence, by (7.26), (7.27), and (7.29), we get

$$M = z^n(t_1) = \| w_0 \|_{L^\infty} + \int_0^{t_1} a_n(s)\, \ell(z^n(s))\, ds \le \| w_0 \|_{L^\infty} + \ell(M) \int_0^{t_1} a_n(s)\, ds$$
$$\le \| w_0 \|_{L^\infty} + \ell(M)/\{\ell(M)+1\} < \| w_0 \|_{L^\infty} + 1 = M,$$

which leads to a contradiction. This yields (7.30), whence follows (7.28). Hence $\partial I_M(w^n(t)) = 0 \quad \forall\, t \in [0, T_0],\ \forall\, n \ge N$, so $w^n(t)$ satisfies

$$w_t^n \in p_\alpha \left(\Delta_p w^n - a_n(t) \,|\, w^n \,|^\rho\, h(x,t,w^n)\right),\ (x,t) \in \omega \times (0,T_0),\ n \ge N. \tag{7.31}$$

By virtue of (7.28), multiplication of (7.31) by $w^n$ gives

$$\frac{1}{2} \frac{d}{dt} \| w^n(t) \|_{L^2}^2 + p_\alpha \| \nabla w^n(t) \|_{L^p}^p \le C_0\, a_n(t), \quad t \in (0,T_0), \quad n \ge N,$$

where $C_0$ denotes a general constant depending only on $\| w_0 \|_{L^\infty}$. Hence we get

$$\sup_{0\le t\le T_0} \| w^n(t) \|_{L^2}^2 + \int_0^{T_0} \| \nabla w^n(t) \|_{L^p}^p\, dt \le C_0 \quad \forall\, n \ge N. \tag{7.32}$$

Furthermore, the multiplication of (7.31) by $t\, w_t^n$ gives

$$t \,\| w_t^n(t) \|_{L^2}^2 + \frac{p_\alpha t}{p} \frac{d}{dt} \| \nabla w^n(t) \|_{L^p}^p \le C_0\, a_n(t)\, t\, \| w_t^n(t) \|_{L^2}$$
$$\le \frac{1}{2}\, t\, \| w_t^n(t) \|_{L^2}^2 + \frac{1}{2}\, C_0^2\, t\, |a_n(t)|^2.$$

Then integrating this on $[0, T_0]$, by (7.32) we get

$$\| \sqrt{t}\, w_t^n \|_{L^2(\omega\times(0,T_0))} + \| \sqrt{t}\, \Delta_p w^n \|_{L^2(\omega\times(0,T_0))} + \sup_{0\le t\le T_0} t \,\| \nabla w^n(t) \|_{L^p}^p \le C_0. \tag{7.33}$$

**Convergence:** Since $-\Delta_p$ is monotone in $L^2$, we easily see that $U(t) = w^n(t) - w^m(t)$, satisfies

$$\frac{1}{2} \frac{d}{dt} \| U(t) \|_{L^2}^2 \le p_\alpha \left( a_n(t)\, \widetilde{h}(x,t,w^n(t)) - a_m(t)\, \widetilde{h}(x,t,w^m(t)), U(t) \right)_{L^2},$$

where $\widetilde{h}(x,s,v) = |v|^\rho\, h(x,s,v)$ satisfies

$$\widetilde{h}'_v(x,s,v) = \rho\, |v|^{\rho-2} v\, h(x,s,v) + |v|^\rho\, h'_v(x,s,v)$$
$$= |v|^\rho \left( \rho \int_0^1 h'_v(x,s,\tau\, v)\, d\tau + h'_u(x,s,v) \right) \in C(\overline{\omega} \times [0,1] \times \mathbb{R}^1).$$

Then, in view of (7.28), we have, for all $n, m \ge N$

$$|\widetilde{h}(x,t,w^n(x,t)) - \widetilde{h}(x,t,w^m(x,t))\,| \le C_0 |w^n(x,t) - w^m(x,t)|, \quad (x,t) \in \omega \times [0,T_0].$$

Hence, again by (7.28), we get

$$\frac{1}{2} \frac{d}{dt} \| U(t) \|_{L^2}^2 \le |\, a_n(t) - a_m(t)\,|\, \| \widetilde{h}(\cdot,t,w^n(\cdot,t)) \|_{L^2} \| U(t) \|_{L^2}$$
$$+ C_0\, a_m(t)\, \| U(t) \|_{L^2}^2,$$

$$\frac{d}{dt} \| U(t) \|_{L^2} \le C_0\, |\, a_n(t) - a_m(t)\,| + C_0\, a_m(t)\, \| U(t) \|_{L^2}, \quad n, m \ge N.$$

Then we obtain

$$\sup_{0\le t\le T_0} \| U(t) \|_{L^2} \le C_0 \left( \int_0^{T_0} | a_n(t) - a_m(t) | \, dt \right) e^{\int_0^{T_0} a_m(t)\,dt}, \qquad n, m \ge N, \tag{7.34}$$

which implies that $\{w^n(\cdot,t)\}_{n\ge N}$ forms a Cauchy sequence in $C([0,T_0]; L^2(\omega))$. Therefore, by virtue of (7.33), we can extract a subsequence of $\{w^n(t)\}$ denoted again by $\{w^n(t)\}$ such that

$$\begin{aligned} w^n &\to w \quad \text{strongly in } C([0,T_0]; L^2(\omega)), \text{ and a.e. } (x,t) \in \omega \times [0,T_0], \\ \nabla w^n &\to \nabla w \quad \text{weakly in } L^p([0,T_0]; L^p(\omega)), \\ \sqrt{t}\, w_t^n &\to \sqrt{t}\, w_t \quad \text{weakly in } L^2(\omega \times [0,T_0]), \\ \sqrt{t}\, \Delta_p w^n &\to \sqrt{t}\, \Delta_p w \quad \text{weakly in } L^2(\omega \times [0,T_0]), \end{aligned}$$
$$\sqrt{t}\, a_n(t) \, |w^n|^\rho \, h(x,t,w^n(x,t)) \to \sqrt{t}\, a(t) \, |w|^\rho \, h(x,t,w(x,t)) \quad \text{weakly in } L^2(\omega \times [0,T_0]).$$

Furthermore, the fact that $t \, | \nabla w(t) |^p \in L^\infty(0,T_0; L^1(\omega))$ ensures the continuity of $w(t)$ on $(0,T_0]$ in the weak topology of $W_0^{1,p}(\omega)$ and the fact that $\sqrt{t}\, w_t$, $\sqrt{t}\, \Delta_p w \in L^2(\omega \times [0,T_0])$ ensures the absolute continuity of $\| \nabla w(t) \|_{L^p(\omega)}$ on $(0,T_0]$. Hence, by virtue of the uniform convexity of $W_0^{1,p}$, we find that $w(t) \in C((0,T_0]; W_0^{1,p}(\omega))$. The uniqueness of the solution is derived by exactly the same arguments used for (7.34) with $a_n(\cdot) = a_m(\cdot) = a(\cdot)$.

**Global existence:** We first note that by assumption $(\mathrm{I})_2$ and the same verification as for (7.25) with $a_n(t), w_n(t)$ replaced by $a(t), w(t)$, we can obtain

$$\| w(t) \|_{L^\infty} \le \| w_0 \|_{L^\infty} + \int_0^t a(s) \, p_\alpha \, C \, \| w(s) \|_{L^\infty}^{1+\delta} \, ds, \tag{7.35}$$

as long as $\sup_{0\le s\le t} \| w(s) \|_{L^\infty} \le 1$ holds true. Here we define $\varepsilon_0 \in (0, 1/2)$ by

$$\varepsilon_0 = \frac{1}{2} \left( \frac{1}{2^{1+\delta} \, C \, p_\alpha \, |a|_{L^1(0,1)} + 1} \right)^{1/\delta} \tag{7.36}$$

and claim that if $\| w_0 \|_{L^\infty} \le \varepsilon_0$, then

$$\sup_{0\le t\le 1} \| w(t) \|_{L^\infty} \le 2\, \varepsilon_0 < 1, \tag{7.37}$$

which ensures the existence of global solutions. To see this, we let

$$z(t) = \| w_0 \|_{L^\infty} + \int_0^t a(s) \, p_\alpha \, C \, \| w(s) \|_{L^\infty}^{1+\delta} \, ds.$$

Then $z(t)$ is continuous and satisfies $\| w(t) \|_{L^\infty} \le z(t)$ and

$$z(t) \le \| w_0 \|_{L^\infty} + \int_0^t a(s) \, p_\alpha \, C \, z(s)^{1+\delta} \, ds. \tag{7.38}$$

In order to prove (7.37), it suffices to show that $\sup_{0\le t\le 1} z(t) \le 2\, \varepsilon_0$, provided that $\| w_0 \|_{L^\infty} \le \varepsilon_0$. Suppose that this does not hold. Then there exists $t_1 \in (0,1)$ such that

$$z(t_1) = 2\, \varepsilon_0, \quad z(t) < 2\, \varepsilon_0 \quad \forall t \in [0, t_1).$$

Hence, by (7.36) and (7.38), we get

$$\begin{aligned}2\,\varepsilon_0 = z(t_1) &\le \|\, w_0 \,\|_{L^\infty} + \int_0^{t_1} a(s)\, p_\alpha\, C\, z(s)^{1+\delta}\, ds \\ &\le \varepsilon_0 + |a|_{L^1(0,t_1)}\, p_\alpha\, C\, 2^{1+\delta}\, \varepsilon_0^\delta\, \varepsilon_0 < 2\,\varepsilon_0,\end{aligned}$$

which leads to a contradiction. □

Here we prepare a comparison result which enables us to compare solutions of $(\mathrm{E})_p^t$ with solutions of simplified equations.

LEMMA 7.4. *Let $w$ be a solution of $(\mathrm{E})_p^t$ on $[0,1]$ satisfying $\|\, w(x,t) \,\|_{L^\infty(\omega\times[0,1])} \le 1$ and (7.23) with $T_0 = 1$, and let $w^\pm$ satisfy (7.23) with $T_0 = 1$ and*

$$\begin{aligned}&w_t^- \le p_\alpha\,(\,\Delta_p\, w^- - C\, a(t)\, w^-), \quad p_\alpha\,(\,\Delta_p\, w^+ + C\, a(t)\, w^+) \le w_t^+, \quad t\in(0,1),\\ &w^-(x,0) \le w(x,0) \le w^+(x,0).\end{aligned}$$

*Then it follows that $w^-(x,t) \le w(x,t) \le w^+(x,t)$ for a.e. $x \in \omega$, $\forall\, t \in [0,1]$.*

PROOF. Since $\|\, w(x,t) \,\|_{L^\infty(\omega\times[0,1])} \le 1$, by $(\mathrm{I})_2$, it is easy to see that $w$ satisfies

$$p_\alpha\,(\,\Delta_p\, w - C\, a(t)\, w\,) \le w_t \le p_\alpha\,(\,\Delta_p\, w + C\, a(t)\, w\,) \quad \forall\, t \in (0,1).$$

Hence we get

$$(\, w(t) - w^+(t)\,)_t \le \Delta_p\, w(t) - \Delta_p\, w^+(t) + C\, a(t)\,(\, w(t) - w^+(t)\,).$$

Multiplying this by $[\, w - w^+\,]^+(t) = \max(w(t) - w^+(t), 0)$, we have

$$\frac{1}{2}\frac{d}{dt}\,\|\, [w - w^+]^+(t)\,\|_{L^2}^2 \le C\, a(t)\,\|\, [w - w^+]^+(t)\,\|_{L^2}^2 \quad \text{a.e. } t \in (0,1).$$

Then, integrating this on $[\delta, t]$ with $\delta > 0$ and applying the Gronwall inequality, we obtain

$$\|\, [w - w^+]^+(t)\,\|_{L^2}^2 \le \|\, [w - w^+]^+(\delta)\,\|_{L^2} e^{2C\int_0^t a(\xi)\, d\xi}. \tag{7.39}$$

Since $\|\, [w - w^+]^+(+0)\,\|_{L^2} = 0$, letting $\delta \to 0$ in (7.39), we conclude that $w(x,t) \le w^+(x,t)$ for a.e. $x \in \omega$, $\forall\, t \in [0,1]$. The assertion $w^-(x,t) \le w(x,t)$ can be verified by much the same arguments as above. □

Now we are ready to prove Theorem 7.4.

PROOF OF THEOREM 7.4. We introduce a new time scale $s^\pm = s^\pm(t)$ by

$$(\mathrm{S})^\pm \quad \begin{cases} \dfrac{d}{dt}\, s^\pm(t) = e_p^\pm(t) := p_\alpha \big(e^{\mp\int_0^t p_\alpha C a(\xi)\, d\xi}\big)^{2-p}, & t \in (0,1),\\ s^\pm(0) = 0. \end{cases} \tag{7.40}$$

Since $e_p^\pm(t)$ is strictly positive and bounded on $[0,1]$, there exist unique solutions $s^\pm(t)$ of $(\mathrm{S})^\pm$, which are strictly increasing on $[0,1]$. Define

$$T_1^\pm := s^\pm(1) = \int_0^1 e_p^\pm(\xi)\, d\xi.$$

Then $0 < T_1^- < 1 < T_1^+$ and $s^\pm(t) \in [0, T_1^\pm]\ \forall\, t \in [0,1]$. Consider

$$(\mathrm{P}) \quad \frac{\partial}{\partial s}\, w(x,s) = \Delta_p\, w(x,s), \quad (x,s) \in \mathbb{R}^N \times (0,+\infty).$$

Then the following facts are well known:

(P1) $L^r$-norms are Lyapunov functions for (P); i.e., every solution $w(s)$ of (P) satisfies $\| w(s) \|_{L^r} \le \| w(0) \|_{L^r}\ \forall s \in [0, +\infty) \quad \forall r \in [1, \infty]$.

(P2) (P) admits the following Barenblatt type solutions:

$$w_{\delta,\gamma}(x,t) = (t+\delta)^{-k}\Big[\gamma - q\Big(\frac{|x|}{(t+\delta)^{k/N}}\Big)^{\frac{p}{p-1}}\Big]_+^{\frac{p-1}{p-2}}, \quad [r]_+ = \max(r,0),$$

$$k = \Big(p - 2 + \frac{p}{N}\Big)^{-1}, \quad q = \frac{p-2}{p}\Big(\frac{k}{N}\Big)^{\frac{1}{p-1}}, \quad \delta,\ \gamma > 0.$$

(P3) $\operatorname{supp} w_{\delta,\gamma}(x,s)$ is monotone increasing in $s$.

(P4) $(w_{\delta,\gamma}(x,s))_s,\ \Delta_p w_{\delta,\gamma}(x,s)$ belong to $C(\mathbb{R}^n \times [0,+\infty))$.

Let $K$ be the unit ball in $\mathbb{R}^N$ centered at the origin. Then by (P2), fixing the parameters $\delta, \gamma$ suitably, we can choose a solution $w_1(x,s)$ of (P) so that

$$\begin{aligned} &0 \le w_1(x,s) \le 1 \quad \forall (x,s) \in \mathbb{R}^n \times [0, T_1^+], \\ &\operatorname{supp}\, w_1(\cdot,s) \subset K \quad \forall t \in [0, T_1^+], \\ &\| w_1(\cdot, T_1^+) \|_{L^r} \ge \delta_0, \quad \forall r \in [1,\infty], \end{aligned} \tag{7.41}$$

where $\delta_0$ is a positive constant independent of $r$. (Since we can assume that $\omega$ is small enough to satisfy $|\omega| \le 1$ without loss of generality, if we take $\|w_1(\cdot,T_1^+)\|_{L^1} \ge \delta_0$, then (7.41) is satisfied for all $r \in [1,\infty]$.) Furthermore, it is easy to see that $w_\varepsilon(x,s) := \varepsilon\, w(\varepsilon^{\frac{2-p}{p}} x, s)$ is a solution of (P) and satisfies

$$\begin{aligned} &0 \le w_\varepsilon(x,s) \le \varepsilon \quad \forall (x,s) \in \mathbb{R}^n \times [0, T_1^+], \\ &\operatorname{supp}\, w_\varepsilon(\cdot,s) \subset K_\varepsilon := \varepsilon^{\frac{p-2}{p}} K \quad \forall t \in [0, T_1^+]. \end{aligned} \tag{7.42}$$

Moreover, $w_\varepsilon(x - x_i, s)$ also gives a solution of (P) and for sufficiently small $\varepsilon$ it follows from (7.42) and (P2) that there exists a finite set $R_\varepsilon := \{x_i\} \subset \omega$ such that

(R1) $(x_i + K_\varepsilon) \cap (x_j + K_\varepsilon) = \emptyset, \quad \forall x_i,\ x_j \in R_\varepsilon,\ i \neq j,$

(R2) $\# R_\varepsilon \ge C_\omega \Big(\dfrac{1}{\varepsilon}\Big)^{\frac{N(p-2)}{p}},$

(R3) $\displaystyle\bigcup_{i=1}^{\# R_\varepsilon} (x_i + K_\varepsilon) \subset\subset \omega.$

Consequently, for every $\vec{m} \in \{0,1\}^{\# R_\varepsilon} := \{\, \vec{m} = (m_1, m_2, \ldots, m_{\# R_\varepsilon})\ ;\ m_j = 0 \text{ or } 1,\ j = 1,2,\ldots, m_{\# R_\varepsilon}\,\}$, the function

$$w_{\vec{m},\varepsilon}(x,s) = \sum_{i=1}^{\# R_\varepsilon} m_i\, w_\varepsilon(x - x_i, s) \tag{7.43}$$

solves (P) and is supported in $\omega$. On the other hand, for $\vec{m}^1 \neq \vec{m}^2$, we obviously have

$$\| w_{\vec{m}^1,\varepsilon}(x,s) - w_{\vec{m}^2,\varepsilon}(x,s) \|_{L^\infty} = \varepsilon \| w_1(x,s) \|_{L^\infty} \quad \forall s \in [0, T_1^+]. \tag{7.44}$$

Thus we find $2^{\# R_\varepsilon}$ different solutions of (P) supported in $\omega$ having the form (7.43). Furthermore, as for the measurement in the topology of $L^r$ $(1 \le r < \infty)$, instead of (7.44), we get

$$\| w_{\vec{m}^1,\varepsilon}(x,s) - w_{\vec{m}^2,\varepsilon}(x,s) \|_{L^r} \ge \varepsilon^{\frac{pr + N(p-2)}{pr}} \| w_1(x,s) \|_{L^r} \quad \forall s \in [0, T_1^+].$$

Here we define the new functions $V^{\pm}_{\vec{m},\varepsilon}(x,t)$ via the new time scales $s^{\pm}(t)$ defined by (7.40) as follows:

$$V^{\pm}_{\vec{m},\varepsilon}(x,t) := w_{\vec{m},\varepsilon}(x, s^{\pm}(t)), \quad t \in [0,1]. \tag{7.45}$$

Then, by (7.40), we easily find

$$\begin{cases} \frac{\partial}{\partial t} V^{\pm}_{\vec{m},\varepsilon}(x,t) = \frac{d}{dt} s^{\pm}(t) \frac{\partial}{\partial s^{\pm}} w_{\vec{m},\varepsilon}(x,s^{\pm}) = e^{\pm}_{p}(t)\, \Delta_p V^{\pm}_{\vec{m},\varepsilon}(x,t), \\ V^{\pm}_{\vec{m},\varepsilon}(x,0) = w_{\vec{m},\varepsilon}(x,0), \\ \text{supp } w_{\vec{m},\varepsilon}(\cdot,0) \subset \text{supp } V^{\pm}_{\vec{m},\varepsilon}(\cdot,t) \subset \text{supp } w_{\vec{m},\varepsilon}(\cdot,T^{\pm}) \quad \forall t \in [0,1]. \end{cases} \tag{7.46}$$

We further introduce the new functions $\widetilde{w}^{\pm}_{\vec{m},\varepsilon}$ by

$$\widetilde{w}^{\pm}_{\vec{m},\varepsilon}(x,t) := e^{\pm}(t)\, V^{\pm}_{\vec{m},\varepsilon}(x,t), \quad e^{\pm}(t) := e^{\pm \int_0^t p_\alpha C a(\xi)\, d\xi}, \quad t \in [0,1]. \tag{7.47}$$

Then, by (7.46), it is easy to see that

$$\begin{aligned} \frac{\partial}{\partial t} \widetilde{w}^{\pm}_{\vec{m},\varepsilon}(x,t) &= e^{\pm}(t) \frac{\partial}{\partial t} V^{\pm}_{\vec{m},\varepsilon}(x,t) \pm p_\alpha\, C\, a(t)\, e^{\pm}(t)\, V^{\pm}_{\vec{m},\varepsilon}(x,t) \\ &= p_\alpha \left(e^{\pm}(t)\right)^{p-1} \Delta_p V^{\pm}_{\vec{m},\varepsilon}(x,t) \pm p_\alpha\, C\, a(t)\, \widetilde{w}^{\pm}_{\vec{m},\varepsilon}(x,t). \end{aligned}$$

Thus we find

$$\begin{cases} \frac{\partial}{\partial t} \widetilde{w}^{\pm}_{\vec{m},\varepsilon}(x,t) = p_\alpha \left( \Delta_p \widetilde{w}^{\pm}_{\vec{m},\varepsilon}(x,t) \pm C\, a(t)\, \widetilde{w}^{\pm}_{\vec{m},\varepsilon}(x,t) \right), \quad t \in [0,1], \\ \widetilde{w}^{\pm}_{\vec{m},\varepsilon}(x,0) = w_{\vec{m},\varepsilon}(x,0), \\ \text{supp } w_{\vec{m},\varepsilon}(\cdot,0) \subset \text{supp } \widetilde{w}^{\pm}_{\vec{m},\varepsilon}(\cdot,t) \subset \text{supp } w_{\vec{m},\varepsilon}(\cdot,T^{\pm}) \quad \forall t \in [0,1]. \end{cases}$$

Let $\widetilde{w}_{\vec{m},\varepsilon}(x,t)$ be the unique solution of $(\mathrm{E})^t_p$ on $[0,1]$ with $\widetilde{w}_{\vec{m},\varepsilon}(x,0) = w_{\vec{m},\varepsilon}(x,0)$, whose existence is ensured by Lemma 7.3 for all $\varepsilon \in (0,\varepsilon_0]$. Then, by the comparison theorem given in Lemma 7.4 (note that $\widetilde{w}^{\pm}_{\vec{m},\varepsilon}(x,t)$ satisfies the regularity required there), we observe

$$\begin{cases} \frac{\partial}{\partial t} \widetilde{w}_{\vec{m},\varepsilon}(x,t) = p_\alpha \left( \Delta_p \widetilde{w}_{\vec{m},\varepsilon}(x,t) - a(t)\, |\, \widetilde{w}_{\vec{m},\varepsilon}\,|^\rho\, h(x,t,\widetilde{w}_{\vec{m},\varepsilon}) \right), \\ \widetilde{w}_{\vec{m},\varepsilon}(x,0) = w_{\vec{m},\varepsilon}(x,0), \\ \widetilde{w}^{-}_{\vec{m},\varepsilon}(x,t) \le \widetilde{w}_{\vec{m},\varepsilon}(x,t) \le \widetilde{w}^{+}_{\vec{m},\varepsilon}(x,t) \quad \text{a.e. } \omega, \ \forall t \in [0,1], \\ \text{supp } w_{\vec{m},\varepsilon}(\cdot,0) \subset \text{supp } \widetilde{w}_{\vec{m},\varepsilon}(\cdot,t) \subset \text{supp } w_{\vec{m},\varepsilon}(\cdot,T^{\pm}) \quad \forall t \in [0,1]. \end{cases} \tag{7.48}$$

Hence, by virtue of the choice of $R_\varepsilon$, supp $\widetilde{w}_{\vec{m},\varepsilon}(\cdot,t)$ does not touch the boundary of $\omega$ for all $t \in [0,1]$. Therefore, the zero extension of $\widetilde{w}_{\vec{m},\varepsilon}(\cdot,t)$ to $\Omega$, denoted again by $\widetilde{w}_{\vec{m},\varepsilon}(\cdot,t)$, becomes a solution of $(\mathrm{E})^t_p$ with $\omega$ replaced by $\Omega$. We introduce another time scale $\tau = \tau(t) \in (-\infty, 0]$ by

$$\tau(t) = p_\alpha \log t, \quad t \in (0,1] \iff t(\tau) = e^{\alpha(p-2)\tau}, \ \tau \in (-\infty,0], \quad \frac{dt}{d\tau} = \alpha\,(p-2)\,t$$

and define a new function $\widetilde{W}_{\vec{m},\varepsilon}(x,\tau)$ by

$$\widetilde{W}_{\vec{m},\varepsilon}(x,\tau) = \widetilde{w}_{\vec{m},\varepsilon}(x, t(\tau)), \quad (x,\tau) \in \Omega \times (-\infty, 0].$$

Then we have

$$\begin{aligned}\frac{\partial}{\partial\tau}\widetilde{W}_{\vec{m},\varepsilon}(x,\tau) &= \frac{dt}{d\tau}\,\frac{\partial}{\partial t}\,\widetilde{w}_{\vec{m},\varepsilon}(x,t)\\ &= \alpha\,(p-2)\,t\,p_\alpha\big(\Delta_p\,\widetilde{w}_{\vec{m},\varepsilon} - a(t)\,|\,\widetilde{w}_{\vec{m},\varepsilon}\,|^\rho\,h(x,t,\widetilde{w}_{\vec{m},\varepsilon}(t))\big)\\ &= t\,\Big(\Delta_p\,\widetilde{w}_{\vec{m},\varepsilon} - t^{\frac{-p+1}{p-2}}\,g^\alpha(x,t^{\frac{1}{p-2}}\,\widetilde{w}_{\vec{m},\varepsilon}(t))\Big)\,.\end{aligned}$$

Therefore, we get

$$\begin{cases}\frac{\partial}{\partial\tau}\widetilde{W}_{\vec{m},\varepsilon}(x,\tau) = e^{\alpha(p-2)\tau}\Delta_p\,\widetilde{W}_{\vec{m},\varepsilon}(x,\tau) - e^{-\alpha\tau}\,g^\alpha(x,e^{\alpha\tau}\,\widetilde{W}_{\vec{m},\varepsilon}(x,\tau)),\\ \widetilde{W}_{\vec{m},\varepsilon}(x,0) = \widetilde{w}_{\vec{m},\varepsilon}(x,1),\\ \operatorname{supp}\,w_{\vec{m},\varepsilon}(\cdot,0)\subset\operatorname{supp}\,\widetilde{W}_{\vec{m},\varepsilon}(\cdot,\tau)\subset\operatorname{supp}\,w_{\vec{m},\varepsilon}(\cdot,T^\pm)\quad\forall\tau\in(-\infty,0].\end{cases}$$

Now define

$$W_{\vec{m},\varepsilon}(x,\tau) = e^{\alpha\tau}\,\widetilde{W}_{\vec{m},\varepsilon}(x,\tau),\qquad (x,\tau)\in\Omega\times(-\infty,0].$$

Then

$$\begin{aligned}\frac{\partial}{\partial\tau}\,W_{\vec{m},\varepsilon}(x,\tau) &= \alpha\,W_{\vec{m},\varepsilon}(x,\tau) + e^{\alpha\tau}\,\frac{\partial}{\partial\tau}\,\widetilde{W}_{\vec{m},\varepsilon}(x,\tau)\\ &= \alpha\,W_{\vec{m},\varepsilon}(x,\tau) + e^{\alpha(p-1)\tau}\Delta_p\,\widetilde{W}_{\vec{m},\varepsilon}(x,\tau) - g^\alpha(x,W_{\vec{m},\varepsilon}(x,\tau)),\end{aligned}$$

whence it follows that

$$\begin{cases}\frac{\partial}{\partial\tau}\,W_{\vec{m},\varepsilon}(x,\tau) = \Delta_p\,W_{\vec{m},\varepsilon}(x,\tau) - g^\alpha(x,W_{\vec{m},\varepsilon}),\ (x,\tau)\in\Omega\times(-\infty,0],\\ W_{\vec{m},\varepsilon}(x,0) = \widetilde{w}_{\vec{m},\varepsilon}(x,1),\\ \operatorname{supp}\,w_{\vec{m},\varepsilon}(\cdot,0)\subset\operatorname{supp}\,W_{\vec{m},\varepsilon}(\cdot,\tau)\subset\operatorname{supp}\,w_{\vec{m},\varepsilon}(\cdot,T^\pm)\quad\forall\tau\in(-\infty,0].\end{cases}\tag{7.49}$$

Thus we observe that

$$\{W_{\vec{m},\varepsilon}(x,\tau)\}_{\tau\in\mathbb{R}^1_-}\in\mathcal{K}^-\quad\text{and}\quad W_{\vec{m},\varepsilon}(x,0) = \widetilde{w}_{\vec{m},\varepsilon}(x,1)\in\mathcal{K}^-(0).\tag{7.50}$$

Next we let $\vec{e_i} = (0,\dots,1,\dots,0)$, i.e., the $j$-th component of $\vec{e_i} = \delta_{ij}$, and define

$$\widetilde{w}_\varepsilon(x-x_i,t) = \widetilde{w}_{\vec{e_i},\varepsilon}(x,t),\quad t\in[0,1].$$

Then, by virtue of (7.42), (R1), and (7.48), we see that

$$\operatorname{supp}\,\widetilde{w}_\varepsilon(x-x_i,t)\cap\operatorname{supp}\,\widetilde{w}_\varepsilon(x-x_j,t) = \emptyset,\quad i\neq j,\ \forall\,t\in[0,1].$$

Hence we can write

$$\widetilde{w}_{\vec{m},\varepsilon}(x,t) = \sum_{i=1}^{\#R_\varepsilon} m_i\,\widetilde{w}_\varepsilon(x-x_i,t),\quad t\in[0,1].$$

Therefore, if $\vec{m_1}$ and $\vec{m_2}$ differ at the $i$-th component, we have by (7.50)

$$\begin{aligned}\|\,W_{\vec{m_1},\varepsilon}(x,0) - W_{\vec{m_2},\varepsilon}(x,0)\,\|_{L^r} &= \|\,\widetilde{w}_{\vec{m_1},\varepsilon}(x,1) - \widetilde{w}_{\vec{m_2},\varepsilon}(x,1)\,\|_{L^r}\\ &\geq \|\,\widetilde{w}_\varepsilon(x-x_i,1)\,\|_{L^r}.\end{aligned}\tag{7.51}$$

Recalling (7.48), (7.45), and (7.47), we get

$$\widetilde{w}_\varepsilon(x-x_i,1)\geq\widetilde{w}_\varepsilon^-(x-x_i,1) = e^-(1)\,w_\varepsilon(x-x_i,T_1^-).$$

Hence, since $0 < T_1^- < T_1^+$, from (P1) and (7.41), we deduce that

$$\begin{aligned} \| \widetilde{w}_\varepsilon(x - x_i, 1) \|_{L^r} &\geq e^-(1) \| w_\varepsilon(x - x_i, T_1^-) \|_{L^r} \\ &\geq e^-(1) \| w_\varepsilon(x - x_i, T_1^+) \|_{L^r} \\ &= e^-(1)\, \varepsilon^{\frac{pr+N(p-2)}{pr}} \| w_1(x, T_1^-) \|_{L^r} \\ &\geq e^-(1)\, \delta_0\, \varepsilon^{\frac{pr+N(p-2)}{pr}}. \end{aligned} \tag{7.52}$$

Then, combining (7.51) with (7.52), we obtain

$$| W_{\overrightarrow{m1},\varepsilon}(x, 0) - W_{\overrightarrow{m2},\varepsilon}(x, 0) \|_{L^r} \geq e^-(1)\, \delta_0\, \varepsilon^{\frac{pr+N(p-2)}{pr}},$$

which can be rewritten as

$$\| W_{\overrightarrow{m1},(k\varepsilon)^\beta}(x, 0) - W_{\overrightarrow{m2},(k\varepsilon)^\beta}(x, 0) \|_{L^r} \geq 2\varepsilon, \quad \forall \varepsilon \in (0, (\varepsilon_0/k)^{1/\beta}),$$
$$\beta = \frac{p\,r}{p\,r + N\,(p-2)}, \quad k = \frac{2}{e^-(1)\,\delta_0}.$$

This estimate implies that balls in $L^r$ with radius $\varepsilon > 0$ can contain at most one element belonging to $\{ W_{\vec{m},(k\varepsilon)^\beta}(x, 0) \,;\, \vec{m} \in \{0, 1\}^{\# R_{(k\varepsilon)^\beta}} \} \subset \mathcal{K}^-(0)$ and its cardinality can be estimated, by (R2) and (7.49), as

$$\#\{ W_{\vec{m},(k\varepsilon)^\beta}(x, 0) \,;\, \vec{m} \in \{0, 1\}^{\# R_{(k\varepsilon)^\beta}} \} \geq 2^{\# R_{(k\varepsilon)^\beta}},$$
$$\# R_{(k\varepsilon)^\beta} \geq C_\omega \left( \frac{1}{(k\,\varepsilon)^\beta} \right)^{\frac{N\,(p-2)}{p}} = C_\omega \left( \frac{1}{k\,\varepsilon} \right)^{\frac{N\,r\,(p-2)}{p\,r + N\,(p-2)}}.$$

Hence, by the definition of $\mathcal{N}_\varepsilon$, $\mathcal{H}_\varepsilon$, and Lemma 7.2, we get

$$\mathcal{N}_\varepsilon(\mathcal{K}^-(0), L^r) \geq 2^{\# R_{(k\varepsilon)^\beta}},$$
$$\mathcal{H}_\varepsilon(\mathcal{A}, L^r) \geq C_\omega \left( \frac{1}{k\,\varepsilon} \right)^{\frac{N\,r\,(p-2)}{p\,r + N\,(p-2)}}, \quad r \in [1, \infty].$$

Moreover, since Kolmogorov $\varepsilon$-entropy of a bounded set of $C^1(\bar{\Omega})$ in the topology of $L^\infty$ is estimated from above by $C\left(\frac{1}{\varepsilon}\right)^N$ (see [61]), we obtain

$$C_\omega \Big( \frac{1}{k\,\varepsilon} \Big)^{\frac{N\,(p-2)}{p + N\,(p-2)}} \leq C_\omega \Big( \frac{1}{k\,\varepsilon} \Big)^{\frac{N\,r\,(p-2)}{p\,r + N\,(p-2)}} \leq \mathcal{H}_\varepsilon(\mathcal{A}, L^r) \leq C \Big( \frac{1}{\varepsilon} \Big)^N \;\; \forall\, r \in [1, \infty]. \tag{7.53}$$

To complete the proof of Theorem 7.4, it remains to recall that $\dim_F(\mathcal{A}, L^r)$, the fractal dimension of $\mathcal{A}$, can be expressed in terms of Kolmogorov $\varepsilon$-entropy via $\dim_F(\mathcal{A}, L^r) = \limsup_{\varepsilon \to 0} \dfrac{\mathcal{H}_\varepsilon(\mathcal{A}, L^r)}{\log_2 \frac{1}{\varepsilon}}$. Letting $\varepsilon \to 0$ in (7.53), we conclude that

$$\dim_F(\mathcal{A}, L^r) = \infty \quad \forall\, r \in [1, \infty]. \qquad \square$$

**Concluding Remarks.** (1) For the sake of the simplicity of the presentation, the monotone part $g_0(x, \xi)$ of $g(x, \xi)$ is here assumed to be single-valued and continuous with respect to $\xi$. However, Theorems 7.1, 7.2, and 7.3 hold true with obvious modifications for any $g_0(x, \xi)$, which is a (possibly multivalued) maximal monotone graph in $\mathbb{R}^2$ for a.e. $x \in \Omega$ such that

$$0 \in g_0(x, 0), \quad \sup\{ \,|\, z \,|\,;\, z \in g_0(x, \xi),\; x \in \Omega,\; |\,\xi\,| \leq M \} \leq C_M \quad \forall M > 0.$$

(2) It is clear that Theorem 7.4 holds true for unbounded domains $\Omega$, since the arguments in the proof are always localized in a bounded domain $\omega$.

CHAPTER 8

# Doubly nonlinear degenerate parabolic equations

It is well known that many problems of the modern material sciences can be reduced to the following abstract doubly nonlinear equations in the appropriate functional spaces:

$$A\left(\frac{d}{dt}u\right) = B(u) + f, \tag{8.1}$$

$$\frac{d}{dt}A(u) = B(u) + f \tag{8.2}$$

(where $A$ and $B$ are some (nonlinear, unbounded) operators and $f$ is external force) which have been intensively studied by many authors; see [1], [24], [25], [30], [49], [51], [71], [93] and the references therein. The standard approach to equations (8.1) and (8.2) uses the assumption that the operators $A$ and $B$ are maximal monotone in the proper functional spaces and is based on the general theory of monotone operators; see [49].

It is, however, worth noting that, being a very effective technical tool for establishing the existence of solutions for such equations, the monotone operators approach has essential drawbacks. Indeed, this method usually gives only *weak* energy solutions of (8.1) or (8.2) which are not regular enough to be unique, and the existence of more regular solutions is a much more delicate problem which requires principally different methods. Thus, even after proving the existence of weak energy solutions via the monotonicity methods, the analytic structure of the problem considered can remain completely unclear especially in the case where one or two of operators $A$ and $B$ are singular or degenerate.

We intend to give a detailed study of the following relatively simple (but still nontrivial) model example of a doubly nonlinear equation of the form (8.2):

$$f(x, \partial_t u) = \Delta_x u - g(x, u), \quad u\big|_{\partial\Omega} = 0, \quad u\big|_{t=0} = u_0 \tag{8.3}$$

in a smooth bounded domain $x \in \Omega \subset \mathbb{R}^3$ avoiding the usage of monotone operator theory. In particular, doubly nonlinear equations of the form

$$b(u, \partial_t u)\partial_t u = \Delta_x u - g(u) + h(x) \tag{8.4}$$

were introduced by M. Gurtin in order to generalize the classical Allen-Cahn model; see [51]. In the case where $b$ is independent of $u$, (8.4) has the form of (8.3) with $f(\partial_t u) = b(\partial_t u)\partial_t u$.

Moreover, we allow the function $f$ to have polynomial degeneration with respect to $\partial_t u$. To be more precise, $f$ can be represented in the form

$$f(x, \partial_t u) = a(x)\partial_t u + \phi(x, \partial_t u) \tag{8.5}$$

where $a(x) \geq 0$ and $\phi$ satisfies

$$C(1 + |v|^p) \geq \phi'_v(x, v) \geq \alpha|v|^p, \quad C, \alpha > 0, \tag{8.6}$$

for some $p > 0$. Thus, the function $f(x, v)$ degenerates as $v|v|^p$ at all points $x$ where $a(x) = 0$. Concerning the second nonlinearity $g$, we assume the standard dissipativity assumptions to be satisfied; see Section 8.1 for their precise formulation.

We also note that equation (8.3) is a fully nonlinear degenerate second-order parabolic problem, so the highly developed classical theory of *quasilinear* parabolic equations (see, e.g., [68]) is not formally applicable to it. Moreover, the structure of equation (8.3) does not fit the assumptions of the general fully nonlinear theory of Krylov and Safonov; see, e.g., [64]. However, as we will see below, equation (8.3) possesses very good regularity properties and, in contrast to the quasilinear case, even with *degenerate* nonlinearity $f$, it has classical solutions. The key idea of our method is to differentiate equation (8.3) by $t$ and study the formally quasilinear equation obtained with respect to $v = \partial_t u$ using the classical methods; see Section 8.1.

On the other hand, it is worth emphasizing that the standard energy phase space

$$\Phi_{en} := \{u_0 \in W_0^{1,2}(\Omega), \ \ G(x, u_0(x)) \in L^1(\Omega)\}, \ \ G(x, w) := \int_0^w g(x, v)\, dv, \tag{8.7}$$

surprisingly occurs to be irrelevant for problem (8.3) (even in the nondegenerate case) due to the existence of "pathological" singular weak energy solutions, e.g., of the form

$$u(t, x) = \frac{v(t, x)}{|x|^\beta}, \quad v(t, 0) \neq 0, \tag{8.8}$$

with regular $v$ and positive $\beta$; see Example 8.1 below. Instead of $\Phi_{en}$, one should take the slightly more regular phase space

$$\Phi := W_0^{1,2}(\Omega) \cap L^\infty(\Omega). \tag{8.9}$$

Then, as we will show, the "pathological" solutions cannot exist any more and we will only have the solutions with the usual and reasonable properties. We believe that this phenomenon has a general nature, and we somehow clarify the difficulties related to finding the stronger solutions of more general doubly nonlinear equations of the form (8.1) or (8.2) via the energy method.

In spite of this, the uniqueness of such solutions in the degenerate case is a more delicate problem. Indeed, as the simplest ODE example

$$y'(t)|y'(t)|^p = y(t), \quad y(0) = 0, \ p > 0, \tag{8.10}$$

shows, we cannot have uniqueness under assumptions (8.6) on the nonlinearity $f$ (if $f$ is allowed to be degenerate), so, in this case, the additional restrictions are necessary. Moreover, this example suggests that, in order to exclude the non-uniqueness example (8.10), the right-hand side $\Delta_x u - g(x, u)$ of equation (8.3) should be monotonic with respect to $u$ at all points $x \in \Omega$ where the left-hand side $f(x, \partial_t u)$ is degenerate. In order to avoid technicalities, we prove uniqueness under the additional assumption in slightly stronger form:

$$K_R[a(x)]^{1/2} + \inf_{|u| \leq R} g'_u(x, u) \geq 0, \quad x \in \Omega, \tag{8.11}$$

for every $R > 0$ and appropriate $K_R > 0$ depending on $R$; see Theorem 8.4 (and Remark 8.6 for weaker assumptions). Here, we only note that (8.11) is automatically satisfied in the nondegenerate case $a(x) \geq a_0 > 0$, so no additional restrictions are required if $f$ is nondegenerate.

We also study the long-time behavior of solutions of problem (8.3) in terms of the associated global and exponential attractors. In particular, we establish the finite dimensionality of the global attractor of (8.3) under the uniqueness assumption (8.11) and assumptions (8.5) and (8.6) on the nonlinearity $f$ which thus can be degenerate.

It is worth recalling here that the degeneracy of the equation considered can change drastically the long-time behavior and the structure of the associated global attractor. In particular, the global attractor of the degenerate analogue of the Chafee-Infante equation

$$\partial_t u = \Delta_x(u^3) + u - u^3, \quad x \in \Omega \subset\subset \mathbb{R}^n, \quad u\big|_{\partial\Omega} = 0 \tag{8.12}$$

is infinite dimensional (see Chapter 4 and [40]), although, in the nondegenerate case, the dimension of the attractors associated with dissipative systems in bounded domains is usually finite; see [7], [93] and the references therein.

Fortunately, in our case of equation (8.3), the infinite dimensionality of the global attractor is automatically excluded by the uniqueness condition (8.11), so using the proper generalization of the so-called $l$-trajectory method (see [70]), we verify the finite dimensionality for the degenerate case as well; see Theorem 8.6 below.

This chapter is organized as follows. A number of a priory estimates for the "sufficiently regular" solutions of (8.3) which are crucial for our study are given in Section 8.1. In particular, the dissipative estimate for the solutions of (8.3) in the phase space $\Phi$ and the regularity $\partial_t u(t) \in L^\infty(\Omega)$ for $t > 0$ are verified there and the examples of singular weak energy solutions which do not possess any smoothing properties are also given in this section; see Example 8.1.

Based on these estimates, we prove (in Section 8.2) the existence of a solution for (8.3) and its uniqueness under the additional assumption (8.11).

In Section 8.3, we formulate and prove the theorems on the existence of finite-dimensional global and exponential attractors for the semigroup (8.3), which can be considered as the main result of the chapter (see also [38], [41]).

Finally, some additional properties of solutions of (8.3) which are not important for the proof of our main result but (as we believe) clarify the nature of the equation considered are collected in a number of remarks throughout the chapter.

## 8.1. A priori estimates and dissipativity

In this section, we obtain a number of estimates for the solutions of the following problem in a bounded smooth domain $\Omega \subset\subset \mathbb{R}^3$:

$$\begin{cases} f(x, \partial_t u) = \Delta_x u - g(x, u), \\ u\big|_{\partial\Omega} = 0, \ u\big|_{t=0} = u_0. \end{cases} \tag{8.13}$$

Here $u = u(t, x)$ is an unknown function, $\Delta_x$ is a Laplacian with respect to the variable $x$, and $f$ and $g$ are given nonlinear functions.

We assume that the nonlinearity $f$ has the following structure:

$$f(x, v) = a(x)v + \phi(x, v) \tag{8.14}$$

where the function $a \in L^\infty(\Omega)$ is nonnegative:

$$a(x) \geq 0, \quad x \in \Omega, \tag{8.15}$$

and the function $\phi \in L^\infty(\Omega, C^2(\mathbb{R}))$ satisfies

$$\begin{cases} (1) \ \ \phi'_v(x,v) \geq \alpha |v|^p, \\ (2) \ \ \phi'_v(x,v) \leq C(|v|^p + 1) \leq C, \quad \alpha > 0, \end{cases} \tag{8.16}$$

for some $p \geq 0$. Thus, the degeneration of the form $f(x, \partial_t u) \sim \partial_t u |\partial_t u|^p$ is allowed in the subdomain $\Omega_0 \subset \Omega$ where $a(x) = 0$.

We also assume that the second nonlinearity $g \in L^\infty(\Omega, C^2(\mathbb{R}))$ satisfies the standard dissipativity assumption

$$\liminf_{|v| \to \infty} \inf_{x \in \Omega} g'_v(x,v) \geq \beta > 0. \tag{8.17}$$

We start with the standard energy estimate for sufficiently regular solutions $u(t)$ of problem (8.13) (roughly speaking, in this section "sufficiently regular" means that it can be approximated by smooth solutions of the regularized problem (8.13), which allows us to justify all a priori estimates formally deduced below; see Section 8.2 for details).

PROPOSITION 8.1. *Let the above assumptions hold and let $u(t)$ be a sufficiently regular solution of* (8.13). *Then, the following equality holds:*

$$\frac{d}{dt}[\|\nabla_x u(t)\|^2_{L^2(\Omega)} + 2(G(\cdot, u(t)), 1)] = -2(f(\cdot, \partial_t u(t)), \partial_t u(t)) \tag{8.18}$$

*where $G(x,u) := \int_0^u g(x,v)\,dv$ and $(\cdot,\cdot)$ denotes the standard inner product in $L^2(\Omega)$. In particular,*

(8.19)

$$\|\nabla_x u(t)\|^2_{L^2(\Omega)} + \|G(\cdot, u(t))\|_{L^1(\Omega)} + \int_0^t (a(\cdot)\partial_t u(s), \partial_t u(s)) + \|\partial_t u(s)\|^{p+2}_{L^{p+2}(\Omega)}\,ds$$
$$\leq C(1 + \|\nabla_x u(0)\|^2_{L^2(\Omega)} + \|G(\cdot, u(0))\|_{L^1(\Omega)})$$

*where the constant $C$ is independent of $u$ and $t$ (here and below $(\cdot,\cdot)$ denotes the standard inner product in $L^2(\Omega)$).*

PROOF. Indeed, multiplying equation (8.13) by $\partial_t u(t)$ and integrating over $x \in \Omega$, we deduce (8.18). In order to obtain estimate (8.19), it is sufficient to integrate (8.18) by $t$ and to use assumptions (8.16) and the obvious fact that $G(x,v) \geq -C$ (due to (8.17)). Proposition 8.1 is proven. □

We see that the energy estimate (8.19) gives the estimate of the $L^{p+2}$-norm of $\partial_t u$ and, consequently, due to assumptions (8.16), the $L^{\frac{p+2}{p+1}}$-norm of $f(x, \partial_t u)$. Our next task is to obtain stronger estimates for $\partial_t u$ and $f(x, \partial_t u)$. To this end, we differentiate equation (8.13) by $t$ and denote $v = \partial_t u$. Then, we get

$$\partial_t f(x,v) = \Delta_x v - g'_u(x,u)v. \tag{8.20}$$

The next theorem gives the local $L^\infty$-estimates for the quasilinear second-order parabolic equation (8.20).

THEOREM 8.1. *Let the above assumptions hold and let $u(t)$ be a sufficiently regular solution of* (8.13). *Then for every $q \in [p+2,\infty]$, the following estimate holds:*

$$\|\partial_t u(t)\|_{L^q(\Omega)} \le C_T(1+\|\partial_t u(0)\|_{L^q(\Omega)}),\quad t\in[0,T], \tag{8.21}$$

*where $C_T$ depends on $T$ but is independent of $t$ and $u$. Moreover, if, in addition, $q > 3p/2$, then the following smoothing property holds:*

$$\|\partial_t u(t)\|_{L^\infty(\Omega)} \le Q_T(1/t+\|\partial_t u(0)\|_{L^q(\Omega)}),\quad t\in(0,T], \tag{8.22}$$

*where the monotonic function $Q_T$ depends on $T$ but is independent of $u$ and $t$.*

PROOF. Indeed, for any $q>0$, multiplying equation (8.20) by $v|v|^q$, we have

$$\partial_t(F_{p+q+2}(\cdot,\partial_t u(t)),1)+\frac{4(q+1)}{(q+2)^2}\||v|^{1+q/2}\|^2_{W^{1,2}(\Omega)} \le K\|v\|^{q+2}_{L^{q+2}(\Omega)} \tag{8.23}$$

where $F_{p+q+2}(x,v) := \int_0^v f(x,s)s|s|^q\,ds$ and $K$ is independent of $u$ and $t$ (here we have used that $g'(x,u)\ge -K$ due to assumption (8.17)). Moreover, according to (8.16), one has

$$\frac{C}{q+2}(|v|^{p+q+2}+1) \ge F_{p+q+2}(x,v) \ge \frac{a(x)}{q+2}|v|^{q+2}+\frac{\alpha}{p+q+2}|v|^{p+q+2}. \tag{8.24}$$

Integrating (8.23) over $t\in[\tau,s]\subset[0,T]$, using (8.24) and the embedding $W^{1,2}\subset L^6$, we deduce that

$$\begin{aligned}&\|v(s)\|^{p+q+2}_{L^{p+q+2}(\Omega)}+\int_\tau^s \|v(t_1)\|^{q+2}_{L^{3(q+2)}(\Omega)}\,dt_1\\ &\qquad \le C(\|v(\tau)\|^2_{L^{p+q+2}(\Omega)}+1)+C(q+2)\int_\tau^s \|v(t_1)\|^{q+2}_{L^{q+2}(\Omega)}\,dt_1\end{aligned} \tag{8.25}$$

where the constant $C$ is independent of $u$, $\tau$, $s$, and $q$.

Estimate (8.21) with $q<\infty$ is now an immediate corollary of (8.25), the estimate $|v|^{q+2}\le 1+|v|^{p+q+2}$, and the Gronwall inequality. So, we now only need to verify (8.22) and (8.21) with $q=\infty$. For simplicity, we verify (8.22) for $t=1$ only (for the general case it can be obtained analogously by the appropriate time scaling). In order to do so, we are going to iterate (8.25) infinitely many times, step by step increasing the exponent $q$. To be more precise, we set

$$q_{n+1}+2+p = 3(q_n+2),\quad \text{i.e.,}\quad q_n+2=(q_0+2-\frac{p}{2})3^n+\frac{p}{2}. \tag{8.26}$$

We see that, in order to have an increasing sequence of $q_n$, we need $q_0+2-p/2>0$ or $q_0+p+2>3p/2$, which corresponds to the restriction $q>3p/2$ in the statement of (8.22). We also set

$$t_0=0,\quad t_{n+1}-t_n=\frac{\beta}{q_n},\quad \lim_{n\to\infty} t_n = 1. \tag{8.27}$$

We now assume that the estimate for

$$I_n := \|v\|^{p+q_n+2}_{L^\infty([t_n,2],L^{p+q_n+2}(\Omega))}+\|v\|^{q_n+2}_{L^{q_n+2}([t_n,2],L^{3(q_n+2)}(\Omega))} \tag{8.28}$$

is already obtained and we deduce the recurrent estimate for $I_{n+1}$. Indeed, there exists $t^* \in [t_n, t_n+1]$ such that

$$\begin{aligned} \|v(t^*)\|^{q_n+2}_{L^{p+q_{n+1}+2}(\Omega)} &= \|v(t^*)\|^{q_n+2}_{L^{3(q_n+2)}(\Omega)} \\ &\le \frac{1}{t_{n+1}-t_n}\int_{t_n}^{t_{n+1}} \|v(s)\|^{q_n+2}_{L^{3(q_n+2)}(\Omega)}\,ds \\ &\le C\beta^{-1}q_n\|v\|^{q_n+2}_{L^{q_n+2}([t_n,2],L^{3(q_n+2)}(\Omega))} \le C\beta^{-1}q_nI_n. \end{aligned} \tag{8.29}$$

Now setting $\tau = t^*$, $q = q_{n+1}$, and $s \in [t_{n+1}, 2]$ in the basic estimate (8.25) and using the last estimate, we arrive at

$$I_{n+1} \le [C_1(q_n+2)]^{\frac{p+q_{n+1}+2}{q_n+2}}\left(1 + I_n^{\frac{p+q_{n+1}+2}{q_n+2}} + I_n^{\frac{q_{n+1}+2}{p+q_n+2}}\right) \tag{8.30}$$

with the constant $C_1$ independent of $n$. Now setting

$$J_n := \max\{1, I_n^{\frac{1}{p+q_n+2}}\}, \tag{8.31}$$

we transform (8.30) as follows:

$$J_{n+1} \le [C_2(q_n+2)]^{\frac{1}{q_n+2}} J_n^{1+\frac{p}{q_n+2}}. \tag{8.32}$$

Iterating this estimate, we will have

$$J_n \le A_n J_0^{B_n} \tag{8.33}$$

with

$$B_n := \Pi_{i=0}^{n-1}(1+\frac{p}{q_i+2}), \quad A_n := \Pi_{i=0}^{n-1}[C_2(q_i+2)]^{\frac{1}{q_i+2}B_{n-i}}. \tag{8.34}$$

Since $q_n \sim 3^n$, then, obviously,

$$A_n \le A_\infty := \lim_{n\to\infty} A_n < \infty, \quad B_n \le \lim_{n\to\infty} B_n < \infty$$

and, consequently,

$$J_\infty := \limsup_{n\to\infty} J_n \le A_\infty J_0^{B_\infty}. \tag{8.35}$$

On the other hand, we have

$$\|v\|_{L^\infty([1,2]\times\Omega)} = \lim_{n\to\infty}\|v\|_{L^{p+q_n+2}([0,1]\times\Omega)} \le J_\infty.$$

Thus, we have proven that

$$\|v\|_{L^\infty([1,2]\times\Omega)} \le Q(\|v\|^{p+q_0+2}_{L^\infty([0,2],L^{p+q_0+2}(\Omega))} + \|v\|_{L^{q_0+2}([0,2],L^{3(q_0+2)}(\Omega))}).$$

Combining this estimate with estimate (8.21) and (8.25) with $q = p+q_0+2$, we finally get

$$\|v\|_{L^\infty([1,2]\times\Omega)} \le Q_1(\|v(0)\|_{L^{p+q_0+2}(\Omega)}),$$

which finishes the proof of estimate (8.22). Estimate (8.21) with $q = \infty$ can be proven analogously, but even more simply since we can now set $t_n \equiv 0$. Theorem 8.1 is proven. □

We now formulate one more simple but useful interior estimate for the derivative $\partial_t u$ of the solution of (8.13).

PROPOSITION 8.2. *Let the above assumptions hold and let $u$ be a sufficiently regular solution of* (8.13) *and let $q \in [0, \infty)$. Then, the following estimate holds:*

$$t\|\partial_t u(t)\|^{p+q+2}_{L^{p+q+2}(\Omega)} + \int_0^1 s\|\nabla_x |\partial_t u(s)|^{1+q/2}\|^2_{L^2(\Omega)}\, ds$$
$$\le C \int_0^1 \|\partial_t u(s)\|^{p+q+2}_{L^{p+q+2}(\Omega)}\, ds + C, \quad t \in [0,1], \tag{8.36}$$

*where the constant $C$ is independent of $u$ and $t$.*

Indeed, in order to obtain this estimate, it is sufficient to multiply equality (8.24) by $t$ and to integrate over $t \in [0, 1]$.

REMARK 8.1. It is worth emphasizing that the $L^\infty$-estimate for $\partial_t u$ is crucial for the theory of equation (8.13). Indeed, after obtaining this estimate, one can rewrite the term $f(x, \partial_t u)$ in the form of $l(t, x)\partial_t u$ and obtain a second-order quasilinear parabolic equation with *bounded* coefficients. Applying the De Giorgi theory to that equation, one obtains, e.g., $C^\alpha$-estimates for $u$ at least in the nondegenerate case $a(x) > a_0 > 0$. In turn, considering now equation (8.20) as a second-order quasilinear parabolic equation with respect to $v$, one obtains the standard $C^{2+\alpha}$-estimates for $v = \partial_t u$; see [68] and Section 8.2 for more details. Repeating this procedure, one can obtain as much regularity of a solution as is allowed by the smoothness of $\Omega$, $f$, and $g$.

Thus, at least in the nondegenerate case, the analytical properties of the fully nonlinear parabolic problem (8.13) will be the same as for the standard case of second-order quasilinear equations, if the $L^\infty$-estimate of $\partial_t u$ is available. Theorem 8.1 and Proposition 8.2 show that it is sufficient to be able to estimate only the integral $\int_0^1 \|\partial_t u(t)\|^q_{L^q(\Omega)}\, ds$ with $q > 3p/2$.

In particular, in the *subcritical* case $p < 4$, we have $p+2 > 3/2p$, so the boundedness of that integral follows immediately from the energy estimate of Proposition 8.1. Thus, in that case any properly defined weak energy solution of (8.13) possesses an internal estimate of the form (8.22) and becomes essentially more regular for $t > 0$.

In contrast to that, in the supercritical case $p > 4$, we have a gap between the minimal regularity obtained from the energy estimate (8.19) and the regularity required for the crucial $L^\infty$-estimate of $\partial_t u$. As we will show below, this gap is related to the existence of "pathological" weak energy solutions which are singular and do not possess any regularity increasing for an arbitrarily long time; see Example 8.1 below. Note that we do not know whether the conditions $q \ge p + 2$ and $q > 3p/2$ are optimal or not.

The next proposition gives the basic dissipative estimate for the $L^\infty$-norm of the solution $u$ of problem (8.13).

PROPOSITION 8.3. *Let the above assumptions hold and let $u(t)$ be a sufficiently regular solution of* (8.13). *Then, the following estimate holds:*

$$\|u(t)\|_{L^\infty(\Omega)} \le Q(\|u_0\|_{L^\infty(\Omega)})e^{-\gamma t} + C_* \tag{8.37}$$

*where the positive constants $C_*$ and $\gamma$ and the monotonic function $Q$ are independent of $u$ and $t$.*

PROOF. We obtain estimate (8.37) by comparing the solution $u(t)$ with the appropriate sub- and super-solutions. To this end, we note that, due to condition (8.17) there exists a constant $C_+ > 0$ such that

$$g(x,u) \geq \beta(u - C_+)/2, \quad u \geq C_+. \tag{8.38}$$

Let now the function $Y_+(t)$ solve the equation

$$\frac{\alpha}{p+1} Y_+'(t)|Y_+'(t)|^p = -\beta(Y_+(t) - C_+)/2, \quad Y_+(0) = \max\{C_+, \|u_0\|_{L^\infty(\Omega)}\} \tag{8.39}$$

where $\alpha > 0$ is the same as in (8.16). On the one hand, solving (8.39), we get

$$Y_+(t) = C_+ + \left( (\|u_0\|_{L^\infty(\Omega)} - C_+)_+^{p/(p+1)} - \delta t \right)_+^{1+1/p},$$

$$\delta := \frac{p+1}{p} \left( \frac{\beta(p+1)}{2\alpha} \right)^{1/(p+1)},$$

where $z_+ := \max\{z, 0\}$ and, consequently, $Y_+(t)$ satisfies the analogue of the dissipative estimate (8.37)

$$Y_+(t) \leq Q(\|u_0\|_{L^\infty(\Omega)} e^{-\gamma t} + C_+ \tag{8.40}$$

for the appropriate positive constant $\gamma$ and monotonic $Q$.

On the other hand, using assumptions (8.16) and (8.38) and the fact that $Y_+(t) \geq C_+ > 0$, $Y_+'(t) \leq 0$, we see that

$$f(x, \partial_t Y_+(t)) - \Delta_x Y_+(t) + g(x, Y_+(t)) \geq 0$$

and, consequently, $Y_+(t)$ is a super-solution of (8.13) and, by the comparison principle, we have

$$u(t,x) \leq Y_+(t), \quad (t,x) \in [0,T] \times \Omega,$$

which together with (8.40) gives the upper bound for $u(t,x)$ in the form of (8.37). The lower bounds can be obtained analogously by constructing the corresponding sub-solution $Y_-(t)$. Thus, estimate (8.37) is verified and Proposition 8.3 is proven. □

REMARK 8.2. As a mathematician, one needs to justify the comparison principle for the solutions of (8.13) especially in the degenerate case. It can be done in a standard way, e.g., by approximating the "sufficiently regular" solutions of (8.13) by the classical solutions of the regularized versions of equation (8.13)

$$f_\varepsilon(x, \partial_t u_\varepsilon) = \Delta_x u_\varepsilon - g_\varepsilon(x, u_\varepsilon).$$

Then, for the classical solutions of the regularized equation, the comparison principle is obvious (see, e.g., [68]), and passing to the limit $\varepsilon \to 0$, we obtain the comparison principle for the "sufficiently regular" solutions of (8.13). We will prove (in the next section) the existence of the solution $u$ by exactly such a regularization procedure (and we will not consider the solutions which cannot be obtained by this procedure). That is the reason why we have omitted the rigorous proof of the comparison principle in Proposition 8.3.

REMARK 8.3. If the nonlinearity $g$ satisfies a stronger dissipativity condition

$$g(x,u) \operatorname{sgn} u \geq -C + \beta |u|^{p+1+\delta}, \quad \delta > 0, \tag{8.41}$$

we will have a stronger equation for the super-solution $Y_+(t)$, namely

$$\frac{\alpha}{p+1}Y_+'(t)|Y_+'(t)|^p + \beta Y_+^{p+1+\delta}(t) \le C. \tag{8.42}$$

It is not difficult to see that this "superlinear" equation possesses the "dissipative" estimate in much stronger form:

$$Y_+(t) \le Q(1/t), \quad t \ge 0,$$

where the monotonic function $Q$ depends on $C$ but is independent of $Y_+(0)$. This, in turn, gives the following interior estimate for the $L^\infty$-norm of $u$:

$$\|u(t)\|_{L^\infty(\Omega)} \le Q(1/t), \tag{8.43}$$

which shows that, in that case, every sufficiently regular solution of (8.13) becomes bounded for $t > 0$. As we will see below, it is not the case if the strong dissipativity condition (8.41) is violated.

The next proposition gives the dissipative analogue of energy estimate (8.19).

PROPOSITION 8.4. *Let the above assumptions hold and let $u(t)$ be a sufficiently regular solution of* (8.13). *Then, the following estimate holds:*

$$\begin{aligned}\|u(t)\|_{L^\infty(\Omega)} + \|\nabla_x u(t)\|_{L^2(\Omega)} + \int_t^{t+1} \|\partial_t u(s)\|^{p+2}_{L^{p+2}(\Omega)}\,ds \\ \le Q(\|u_0\|_{L^\infty(\Omega)\cap W^{1,2}(\Omega)})e^{-\gamma t} + C_*\end{aligned} \tag{8.44}$$

*where the positive constants $C_*$ and $\gamma$ and the monotonic function $Q$ are independent of $t$ and $u$.*

PROOF. We note that the dissipative estimate for the $L^\infty$-norm of $u$ is already obtained in the previous proposition; therefore, due to estimate (8.19), we may assume without loss of generality that

$$\|u(t)\|_{L^\infty(\Omega)} \le 2C_* \tag{8.45}$$

for all $t \ge 0$ where $C_*$ is the same as in (8.37).

Now multiplying equation (8.13) by $2u(t)$, integrating over $\Omega$, and summing the obtained relation with (8.18), we have

$$\begin{aligned}&\frac{d}{dt}[\|\nabla_x u(t)\|^2_{L^2(\Omega)} + 2(G(\cdot, u(t)), 1)] + 2(f(\cdot, \partial_t u(t)), \partial_t u(t)) \\ &+ \|\nabla_x u(t)\|^2_{L^2(\Omega)} + 2(g(\cdot, u(t)), u(t)) \\ &\quad = -2(f(\cdot, \partial_t u(t)), u(t)).\end{aligned} \tag{8.46}$$

Now using estimate (8.45) and the obvious fact that $|f(x, \partial_t u)| \le \frac{1}{2}|f(x, \partial_t u)| \cdot |\partial_t u| + C$ (see (8.16)), we deduce that

$$\begin{aligned}&\frac{d}{dt}[\|\nabla_x u(t)\|^2_{L^2(\Omega)} + 2(G(\cdot, u(t)), 1)] + [\|\nabla_x u(t)\|^2_{L^2(\Omega)} + 2(G(\cdot, u(t)), 1)] \\ &+ \gamma\|\partial_t u(t)\|^{p+2}_{L^{p+2}(\Omega)} \\ &\le C_*'\end{aligned} \tag{8.47}$$

where the positive constants $C_*'$ and $\gamma$ are independent of $t$ and $u$. Applying the Gronwall inequality to that relation, we deduce (8.44) and finish the proof of Proposition 8.4. □

The next theorem estimates the $L^q$-norm of $\partial_t u$ in terms of the $L^\infty$-norm of $u$.

THEOREM 8.2. *Let the above assumptions hold and let $u(t)$ be a sufficiently regular solution of* (8.13). *Then, for every $q \ge p+2$, the following estimate holds:*

$$\|\partial_t u\|_{L^q([t,t+1]\times\Omega)} \le Q_T(\|u_0\|_{L^\infty(\Omega)\cap W^{1,2}(\Omega)}), \quad t \ge T > 0, \tag{8.48}$$

*where the function $Q_T$ depends on $t$ but is independent of $t$ and $u$.*

PROOF. Analogously to Theorem 8.1, we will prove estimate (8.48) by the iteration procedure. We first note that for $q = p+2$ estimate (8.48) is obtained in Proposition 8.4. We now assume that this estimate is known for $q = p+2+\kappa$ for some $\kappa \ge 0$. Then, due to Proposition 8.2, we have

$$\begin{aligned} \|\partial_t u(t)\|^{p+2+\kappa}_{L^{p+2+\kappa}(\Omega)} &+ \int_t^{t+1} (|\nabla_x \partial_t u(s)|^2, |\partial_t u(s)|^\kappa)\, ds \\ &\le Q_T(\|u_0\|_{L^\infty(\Omega)\cap W^{1,2}(\Omega)}), \quad t \ge T > 0. \end{aligned} \tag{8.49}$$

Moreover, interpreting (8.13) as an elliptic equation at every fixed $t$:

$$\Delta_x u(t) = h_u(t) := f(x, \partial_t u) + g(x,u), \tag{8.50}$$

estimating $f(x,\partial_t u)$ by (8.49) and $g(x,u)$ by (8.37), and using the classical $L^q$-regularity theorem for the Laplacian (see, e.g., [97]), we deduce

(8.51)
$$\|u(t)\|_{W^{2,\frac{p+2+\kappa}{p+1}}(\Omega)} \le C\|h_u(t)\|_{L^{\frac{p+2+\kappa}{p+1}}(\Omega)} \le Q_T(\|u_0\|_{L^\infty(\Omega)\cap W^{1,2}(\Omega)}), \quad t \ge T > 0.$$

Interpolating between $W^{2,\frac{p+2+\kappa}{p+1}}$ and $L^\infty$, we deduce from (8.51) and (8.37) that

$$\|\nabla_x u(t)\|_{L^{2\frac{p+2+\kappa}{p+1}}(\Omega)} \le Q_T(\|u_0\|_{L^\infty(\Omega)\cap W^{1,2}(\Omega)}), \quad t \ge T > 0. \tag{8.52}$$

Now multiplying equation (8.13) by $\partial_t u|\partial_t u|^{\kappa_1}$ where $\kappa_1 > \kappa$ will be determined later and integrating over $[t,t+1]\times\Omega$, we obtain the following inequality:

$$\begin{aligned} \|\partial_t u\|^{p+2+\kappa_1}_{L^{p+2+\kappa_1}([t,t+1]\times\Omega)} \le C &\int_t^{t+1} (|\nabla_x u(s)|, |\nabla_x \partial_t u(s)| \cdot |\partial_t u(s)|^{\kappa_1})\, ds \\ &+ C\int_t^{t+1} (|g(\cdot, u(s))|, |\partial_t u(s)|^{\kappa_1+1})\, ds. \end{aligned} \tag{8.53}$$

Since the $L^\infty$-estimate for $u(t)$ is already known, the last term on the right-hand side of (8.53) can be easily estimated by the Hölder inequality:

$$\begin{aligned} C&\int_t^{t+1} (|g(\cdot, u(s))|, |\partial_t u(s)|^{\kappa_1+1})\, ds \\ &\le \frac{1}{2}\|\partial_t u\|^{p+2+\kappa_1}_{L^{p+2+\kappa_1}([t,t+1]\times\Omega)} + Q(\|u_0\|_{L^\infty(\Omega)\cap W^{1,2}(\Omega)}). \end{aligned} \tag{8.54}$$

So, we only need to estimate the first integral on the right-hand side of (8.53). To this end, we use the Hölder inequality with exponents $q_1 = 2\frac{p+2+\kappa}{p+1}$, $q_2 = 1/2$, and $q_3 = 2\frac{p+2+\kappa}{1+\kappa}$:

(8.55)

$$\int_t^{t+1} (|\nabla_x u(s)|, (|\nabla_x \partial_t u(s)| \cdot |\partial_t u(s)|^{\kappa/2}) \cdot |\partial_t u(s)|^{\kappa_1-\kappa/2})\, ds$$

$$\leq \|\nabla_x u\|_{L^{q_1}([t,t+1]\times\Omega)} \left(\int_t^{t+1} (|\nabla_x \partial_t u(s)|^2, |\partial_t u(s)|^{\kappa})\, ds\right)^{1/2} \|\partial_t u\|^{\kappa_1-\kappa/2}_{L^{q_3(\kappa_1-\kappa/2)}(\Omega)}.$$

We see that the first two terms on the right-hand side of (8.55) can be estimated by (8.52) and (8.49), respectively. In order to estimate the third term, we assume that $\kappa_1 = \kappa_1(\kappa)$ is chosen in such a way that

$$(\kappa_1 - \kappa/2)q_3 = p + \kappa + 2,$$

i.e., $\kappa_1 = \kappa + 1/2$. Then, the third term is also controlled by (8.52) which, together with (8.53) and (8.54), gives

$$\|\partial_t u\|_{L^{p+2+\kappa+1/2}(\Omega)} \leq Q_T(\|u_0\|_{L^\infty(\Omega)\cap W^{1,2}(\Omega)}). \tag{8.56}$$

Thus, the above-described procedure allows us to increase the exponent $q$ in estimate (8.48) by $1/2$. Thus, starting from $q_0 = p + 2$ (for which this estimate is known) and iterating this procedure sufficiently many times, we obtain (8.48) for any finite $q > p + 2$. Theorem 8.2 is proven. □

The next corollary combines Theorems 8.1 and 8.2.

COROLLARY 8.1. *Let the above assumptions hold and let $u(t)$ be a sufficiently regular solution of* (8.13). *Then, for any $1 < q < \infty$ and every $t \geq T > 0$, the following estimate holds:*

$$\|\partial_t u(t)\|_{L^\infty(\Omega)} + \|u(t)\|_{W^{2,q}(\Omega)} \leq Q_T(\|u_0\|_{L^\infty(\Omega)\cap W^{1,2}(\Omega)})e^{-\gamma t} + C_T \tag{8.57}$$

*where the $\gamma > 0$ and the constant $C_T$ and the monotonic function $Q_T$ depend on $T$ but are independent of $u$ and $t$.*

PROOF. Indeed, due to dissipative estimate (8.44), it is sufficient to verify (8.57) for $t \leq 1$ only. In order to do this, we first estimate $\|\partial_t u\|_{L^{3p/2+\delta}([t,t+1]\times\Omega)}$, for some $\delta > 0$, via the initial data by Theorem 8.2, after estimating $\|\partial_t u\|_{L^\infty([t,t+1],L^{3p/2+\delta}(\Omega))}$ by Proposition 8.2 and we finally estimate the $L^\infty$-norm of $\partial_t u$ by Theorem 8.1. The estimate for the $W^{2,q}$-norm of $u(t)$ can be obtained after that from the elliptic equation (8.50). Corollary 8.1 is proven. □

The following example shows that interior estimate (8.57) can be violated for weak energy solutions of (8.13) with the initial data $u_0 \notin L^\infty(\Omega)$.

EXAMPLE 8.1. Let us consider the following simplest version of equation (8.13):

$$\partial_t u(t)|\partial_t u(t)|^p = \Delta_x u(t), \quad p > 1, \tag{8.58}$$

and look for its radially symmetric singular solution in the form $u_0(t,x) = \frac{\phi(t)}{|x|^{2/p}}$. Inserting this into equation (8.58), we deduce the equation for $\phi(t)$, namely,

$$\phi'(t)|\phi'(t)|^p + \frac{4(p-1)}{p^2}\phi(t) = 0.$$

Thus, (8.58) possesses a family of special solutions of the form

$$u_0(t,x) = \frac{(C-\delta t)_+^{1+1/p}}{|x|^{2/p}} \tag{8.59}$$

where $\delta = \delta(p)$ is some fixed constant and $C \geq 0$ is arbitrary.

Having this automodel solution of the degenerate equation (8.58), one can easily construct analogous singular solutions for the nondegenerate equations of the form (8.13) as well. Namely, using the obvious fact that $\partial_t u_0 = -(1+1/p)(C-t)^{-1}u_0$, one verifies that (8.59) also solves

$$\partial_t u + 2\partial_t u|\partial_t u|^p = \Delta_x u - (1+1/p)\frac{u}{(C-t)} - (1+1/p)^{p+1}\frac{u|u|^p}{(C-t)^{p+1}}. \tag{8.60}$$

Without loss of generality, we may assume that $0 \in \Omega$. Then, equation (8.60) has the form of (8.13) with nondegenerate $f(x, \partial_t u)$, nonautonomous $g = g(t,x,u)$ satisfying the dissipativity assumption (8.17) and smooth nonhomogeneous boundary conditions $u\big|_{\partial\Omega} = u_0(t,x)$ (we will consider this equation on the time interval $t \in [0, C/2]$ only). So, it is not difficult to see that all the above-verified estimates remain true (after the corresponding minor changings) for such nonautonomous equations as well.

On the other hand, this equation possesses a singular weak solution (8.59) which does not have any regularizing on the time interval $t \in [0, C/2]$. Thus, the analogue of interior estimate (8.57) clearly does not hold for such weak solutions.

We also mention that the solution (8.59) has a finite energy ($u_0(t) \in W^{1,2}(\Omega)$, $G(\cdot, u_0(t)) \in L^1(\Omega)$) if and only if $p > 4$.

REMARK 8.4. We see that, in the supercritical case $p > 4$, the energy phase space is too large to be the adequate phase space for problem (8.13) (since the nonregularizing "pathological" solutions like (8.59) are allowed). In contrast to this, the phase space $\Phi := L^\infty(\Omega) \cap W^{1,2}(\Omega)$ does not contain such solutions and, according to Corollary 8.1, gives, in a minimal sense, a reasonable phase space for that problem. That is the reason why we will consider below equation (8.13) in the phase space $\Phi$.

We also recall that, due to Remark 8.3, the above singular solutions cannot exists if $g$ satisfies the strong dissipativity assumption (8.41) with arbitrarily small positive $\delta$. As equation (8.60) shows, such singular solutions can exist if $\delta = 0$. Thus, the above regularity analysis seems to be sharp.

We conclude the section by obtaining a little more regularity of $\partial_t u$ which is, however, important for the existence of a solution in the next section.

PROPOSITION 8.5. *Let the above assumptions hold and let $u(t)$ be a sufficiently regular solution of equation* (8.13). *Then, the following estimate holds:*

$$\begin{aligned} \|\partial_t u(t)\|_{W^{1,2}(\Omega)} &+ \int_t^{t+1} (|\partial_t^2 u(s)|^2, |\partial_t u(s)|^p)\, ds + \|\partial_t u\|_{W^{\beta,2}([t,t+1]\times\Omega)} \\ &\leq Q_T(\|u_0\|_{L^\infty(\Omega)\cap W^{1,2}(\Omega)})e^{-\alpha t} + C_*, \qquad t \geq T > 0, \end{aligned} \tag{8.61}$$

*where $\alpha$, $\beta$, and $C_*$ are some positive constants and the monotonic function $Q_T$ depends on $T$ but is independent of $t$ and $u$.*

PROOF. Indeed, due to Corollary 8.1 and Proposition 8.2, it is sufficient to prove estimate (8.61) under the additional assumption that

$$\|\partial_t u(t)\|_{L^\infty(\Omega)} + \int_t^{t+1} \|\nabla_x \partial_t u(s)\|^2_{L^2(\Omega)}\, ds \le C'_*, \quad t \in \mathbb{R}_+. \tag{8.62}$$

In order to do so, let us multiply equation (8.20) by $t\partial_t v(t)$ and integrate over $x \in \Omega$. Then, using assumption (8.16), we get

$$\begin{aligned} &\alpha t(|\partial_t v(t)|^2, |v(t)|^p) + \partial_t[t\|\nabla_x v(t)\|^2_{L^2(\Omega)} + t(g'(\cdot, u(t))v(t), v(t))] \\ &\le \|\nabla_x v(t)\|^2_{L^2(\Omega)} + (g'(\cdot, u(t))v(t), v(t)) + t(g''(\cdot, u(t))v^2(t), v(t)). \end{aligned} \tag{8.63}$$

Integrating this inequality by $t$ from $\max\{0, t-1\}$ until $t+1$ and using (8.62), we deduce estimate (8.61) for the first two terms on the left-hand side. In order to obtain the last term in that estimate, we note that, from the first two terms, we infer that $|\partial_t u|^{1+p/2} \in W^{1,2}([t, t+1] \times \Omega)$ and

$$\||\partial_t u|^{1+p/2}\|_{W^{1,2}([t,t+1]\times\Omega)} \le Q_T(\|u_0\|_{L^\infty(\Omega)\cap W^{1,2}(\Omega)})e^{-\alpha t} + C_*, \ t \ge T > 0, \tag{8.64}$$

which, in turn, implies the required estimate for $\|\partial_t u\|_{W^{\beta,2}([t,t+1]\times\Omega)}$ with $\beta < \frac{2}{p+2}$ and Proposition 8.5 is proven. □

## 8.2. Existence and uniqueness of solutions

Using the a priori estimates of the previous section, we establish here the existence of the appropriate solution of problem (8.13) and, under the additional assumptions that the domains of degeneration and nonmonotonicity of equation (8.13) are well-separated, we also verify that this solution is unique. As we have already noted in Remark 8.4, we will consider equation (8.13) in the phase space $\Phi := L^\infty(\Omega) \cap W^{1,2}(\Omega)$. To be more precise, we define a solution of (8.13) as follows.

DEFINITION 8.1. A function $u = u(t, x)$ is a (bounded) solution of (8.13) if

$$u \in L^\infty([0,T] \times \Omega) \cap L^\infty([0,T], W^{1,2}(\Omega)), \quad \partial_t u \in L^{p+2}([0,T] \times \Omega), \tag{8.65}$$

and it satisfies equation (8.13) in the sense of distributions.

The next theorem gives the existence of such solutions.

THEOREM 8.3. *Let the assumptions of Section* 8.1 *hold. Then, for any* $u_0 \in \Phi$, *equation* (8.13) *possesses at least one solution* $u(t)$ *in the sense of Definition* 8.1.

PROOF. We first approximate the nonlinearities $f$ and $g$ in equation (8.13) by the smooth ones $f_\varepsilon(x, v)$ and $g_\varepsilon(x, v)$, $\varepsilon > 0$, in such a way that (8.14)–(8.17) will be satisfied uniformly with respect to $\varepsilon \to 0$ and, for every $R > 0$,

$$\sup_{|v|\le R} \|f_\varepsilon(\cdot, v) - f(\cdot, v)\|_{L^1(\Omega)} \to 0 \quad \text{as } \varepsilon \to 0$$

and analogously for $g_\varepsilon$ and $g$. Moreover, we also assume that

$$a_\varepsilon(x) > a_\varepsilon > 0 \tag{8.66}$$

and, consequently, the associated approximations

$$f_\varepsilon(x, \partial_t u_\varepsilon) = \Delta_x u_\varepsilon - g_\varepsilon(x, u_\varepsilon), \quad u\big|_{t=0} = u_0^\varepsilon \tag{8.67}$$

will be smooth and nondegenerate. Thus, all of the estimates of the previous section hold for equation (8.67) *uniformly* with respect to $\varepsilon$.

Let us first verify the existence of a solution for the approximate equation (8.67). To this end, we rewrite it with respect to the variable $v_\varepsilon := \partial_t u_\varepsilon$:

$$f_\varepsilon'(x, v_\varepsilon)\partial_t v_\varepsilon = \Delta_x v_\varepsilon - g_\varepsilon'(x, u_\varepsilon)v_\varepsilon, \quad v_\varepsilon\big|_{t=0} = v_0^\varepsilon \tag{8.68}$$

where $v_0^\varepsilon$ solves

$$f_\varepsilon(x, v_0^\varepsilon) = \Delta_x u_0^\varepsilon - g_\varepsilon(x, u_0^\varepsilon). \tag{8.69}$$

It is not difficult to verify using our assumptions on $f$ that $v_0^\varepsilon$ is uniquely defined by $u_0^\varepsilon$ and smooth if $u_0^\varepsilon$ is smooth.

Then, according to Theorem 8.1 and Proposition 8.3, we have

$$\|\partial_t u_\varepsilon\|_{L^\infty([0,T]\times\Omega)} + \|u\|_{L^\infty([0,T]\times\Omega)} \le C(u_0^\varepsilon). \tag{8.70}$$

Moreover, due to (8.66), equation (8.68) can now be interpreted as a linear nondegenerate second-order parabolic equation with the coefficients belonging to $L^\infty$. Applying the classical De Giorgi estimate to this equation, we deduce

$$\|\partial_t U_\varepsilon\|_{C^\beta([0,T]\times\Omega)} \le C(u_0^\varepsilon), \tag{8.71}$$

for some $\beta > 0$; see, e.g., [27, Chapter III, Section 10]. Returning back to equation (8.67) and using the classical $C^\beta$-estimates for parabolic equations (see, e.g., [27, Chapter IV, Theorem 5.2 and estimate (8.71)]), we have

$$\|u_\varepsilon\|_{C^{1+\beta/2,2+\beta}([0,T]\times\Omega)} \le C(u_0^\varepsilon). \tag{8.72}$$

Finally, returning again to equation (8.68) and using (8.72) and the $C^\beta$-estimates mentioned earlier, we finally deduce

$$\|\partial_t u_\varepsilon\|_{C^{1+\beta/2,2+\beta}([0,T]\times\Omega)} + \|u_\varepsilon\|_{C^{1+\beta/2,2+\beta}([0,T]\times\Omega)} \le C(u_0^\varepsilon). \tag{8.73}$$

In order to verify the existence of a solution, we introduce the integration operator

$$I(v_\varepsilon)(t) := u_0^\varepsilon + \int_0^t v_\varepsilon(s)\,ds$$

and rewrite (8.68) in the following equivalent form:

$$\partial_t v_\varepsilon = \frac{1}{f_\varepsilon'(x, v_\varepsilon)}\Delta_x v_\varepsilon - \frac{g_\varepsilon'(x, I(v_\varepsilon))}{f_\varepsilon'(x, v_\varepsilon)} v_\varepsilon. \tag{8.74}$$

The existence of a solution for such a quasilinear equation can be obtained based on a priori estimate (8.73) and the Leray-Schauder principle, exactly as in [27, Chapter VI, Section 4].

Thus, the existence of the solutions $u_\varepsilon(t)$ for the auxiliary problem (8.67) is verified for any smooth $u_0^\varepsilon$.

Let us approximate the initial data $u_0 \in \Phi$ by a sequence of smooth ones $u_0^\varepsilon$ which converges to $u_0$ as $\varepsilon \to 0$, say, in $L^q(\Omega) \cap W^{1,2}(\Omega)$, $q \gg 1$, and weakly-$*$ in $L^\infty(\Omega)$; construct the associated solutions $u_\varepsilon(t)$ of problem (8.67) and pass to the limit $\varepsilon \to 0$. We claim that this procedure gives the desired solution $u(t)$ of the initial problem (8.13)

Indeed, according to Proposition 8.4, the estimate

$$\|u(t)\|_\Phi + \int_t^{t+1} \|\partial_t u(s)\|_{L^{p+2}(\Omega)}^{p+2}\,ds \le Q(\|u_0\|_\Phi)e^{-\alpha t} + C_* \tag{8.75}$$

holds *uniformly* with respect to $\varepsilon \to 0$. Thus, the limit function $u(t)$ will also satisfy this estimate and, therefore, will belong to the class (8.65). So, we only need to verify that $u(t)$ satisfies (8.13) in the sense of distributions.

To this end, applying Corollary 8.1 and Proposition 8.5 to the auxiliary problem (8.67) we infer

$$\|\partial_t u\|_{L^\infty([t,T]\times\Omega)} + \|\partial_t u\|_{W^{\beta,2}([t,T]\times\Omega)} + \|u\|_{L^\infty([t,T]\times W^{2,q}(\Omega))} \le C_{t,T} \tag{8.76}$$

where $t > 0$ and the constant $C_{t,T}$ is independent of $\varepsilon$. Thus, without loss of generality, we may assume that, for every $t > 0$ the sequence $u_\varepsilon$ converges $*$-weakly to $u$ in the spaces involved on the right-hand side of (8.76). In particular, this weak convergence implies that

$$u_\varepsilon \to u \text{ strongly in } C([t,T]\times\Omega), \quad \partial_t u_\varepsilon \to \partial_t u \text{ strongly in } L^q([t,T]\times\Omega) \tag{8.77}$$

for every $q \ge 1$. Having this convergence, it is not difficult to check that equation (8.67) converges to (8.13) as $\varepsilon \to 0$ in the sense of distributions. In fact, the convergence of the linear term $\Delta_x u_\varepsilon$ is obvious and the convergence of the nonlinear terms $f_\varepsilon$ and $g_\varepsilon$ follows immediately from the following standard convergence:

$$\|f_\varepsilon(\cdot,\partial_t u_\varepsilon) - f(\cdot,\partial_t u)\|_{L^1([t,T]\times\Omega)} + \|g_\varepsilon(\cdot,u_\varepsilon) - g(\cdot,u)\|_{L^1([t,T]\times\Omega)} \to 0, \quad \text{as } \varepsilon \to 0. \tag{8.78}$$

Indeed, due to assumption (8.16) and the fact that the $L^\infty$-norm of $\partial_t u$ is bounded, we have

$$\begin{aligned}
&\|f_\varepsilon(\cdot,\partial_t u_\varepsilon) - f(\cdot,\partial_t u)\|_{L^1([t,T]\times\Omega)} \\
&\quad \le \|f_\varepsilon(\cdot,\partial_t u_\varepsilon) - f_\varepsilon(\cdot,\partial_t u)\|_{L^1([t,T]\times\Omega)} + \|f_\varepsilon(\cdot,\partial_t u) - f(\cdot,\partial_t u)\|_{L^1([t,T]\times\Omega)} \\
&\quad \le C_T\|\partial_t u_\varepsilon - \partial_t u\|_{L^1([t,T]\times\Omega)} + \|f_\varepsilon(\cdot,\partial_t u) - f(\cdot,\partial_t u)\|_{L^1([t,T]\times\Omega)}.
\end{aligned} \tag{8.79}$$

The first term on the right-hand side of this inequality tends to zero due to (8.77), and the second one due to (8.66) and the fact that the $L^\infty$-norm of $\partial_t u$ is bounded. The convergence of $g_\varepsilon(x,u)$ to $g(x,u)$ can be verified analogously. Theorem 8.3 is proven. □

COROLLARY 8.2. *The solution $u(t)$ of problem (8.13) constructed in Theorem 8.3 satisfies all of the estimates obtained in the previous section and, thus, gives a rigorous interpretation of the notion of a "sufficiently regular solution" used there.*

REMARK 8.5. Let us assume, in addition, that the functions $f$ and $g$ are Hölder continuous with respect to $x$. Then, a more delicate analysis of equation (8.20) based on the adaptation of the De Giorgi technique to degenerate parabolic equations (see [17], [27]) allows us to establish not only the boundedness of $\partial_t u$, but also its Hölder continuity with respect to $(t,x)$ with a sufficiently small positive Hölder exponent $\beta$ depending on $p$. Applying after that the usual $C^\beta$-regularity theorem for elliptic equations to (8.13) (and interpreting $f(x,\partial_t u)$ as the external force), we then establish that $u(t) \in C^{2+\beta}(\Omega)$. Thus, finally, we obtain that, even in the degenerate case, the solution $u$ belongs to

$$u \in C^{1+\beta,2+\beta}([t,T]\times\Omega) \tag{8.80}$$

where we can take $t = 0$ if the initial data is smooth enough.

Therefore, even in the degenerate case, equation (8.13) possesses *classical* solutions for every sufficiently smooth initial data $u_0$. This shows the principal difference between the degenerate fully nonlinear problem (8.13) and standard quasilinear degenerate/singular equations where the classical solutions usually do not exist; see [27].

Our next task is to verify the uniqueness of a solution for problem (8.13). To this end, we need the following additional assumption on the structure of the nonlinearities $f$ and $g$:

$$K_R[a(x)]^{1/2} + \inf_{|u|\le R} g'_u(x,u) \ge 0 \tag{8.81}$$

for every $R > 0$ and some $K_R > 0$ depending on $R$. The last assumption means, in particular, that $g'_v(x,v) \ge 0$ for every $x$ such that $a(x) = 0$. Thus, equation (8.13) can be nonmonotonic only in the subdomain of $\Omega$ where $f$ is nondegenerate. As we have already mentioned, this assumption is crucial for uniqueness. Indeed, the simplest ODE example,

$$y'(t)|y'(t)|^p = y(t), \tag{8.82}$$

shows that the solution $u(t)$ of (8.13) can be nonunique if (8.81) is violated. In contrast to this, the next theorem shows that the solution is indeed unique if (8.81) holds.

THEOREM 8.4. *Let the assumptions of Theorem* 8.3 *hold and let, in addition,* (8.81) *be satisfied. Then, for every two solutions* $u_1(t)$ *and* $u_2(t)$ *of* (8.13) *(in the sense of Definition* 8.1*), the following estimate holds:*

$$\|u_1(t) - u_2(t)\|_{W^{1,2}(\Omega)} \le Ce^{Kt}\|u_1(0) - u_2(0)\|_{W^{1,2}(\Omega)} \tag{8.83}$$

*where the constants $C$ and $K$ depend on the* $\Phi := L^\infty(\Omega)\cap W_0^{1,2}(\Omega)$*-norms of* $u_1(0)$ *and* $u_2(0)$ *but are independent of* $t$*. In particular, the solution* $u(t)$ *of problem* (8.13) *is uniquely determined by its initial data* $u_0 \in \Phi$.

PROOF. Let $u_1(t)$ and $u_2(t)$ be two solutions of problem (8.13) with the initial data belonging to $\Phi$ and let $v(t) := u_1(t) - u_2(t)$. Then, this function solves

$$a(x)v + [\phi(x,\partial_t u_1) - \phi(x,\partial_t u_2)] = \Delta_x v - [g(x,u_1) - g(x,u_2)]. \tag{8.84}$$

Moreover, according to Definition 8.1, we have

$$\|u_1\|_{L^\infty([0,T]\times\Omega)} + \|u_2\|_{L^\infty([0,T],\times\Omega)} \le R \tag{8.85}$$

for some finite $R$. Now multiplying equation (8.84) by $\partial_t v$ and using assumption (8.16), the obvious formula

$$\begin{aligned}&\partial_t[G(x,u_1) - G(x,u_2) - g(u_2)v]\\ &\quad= (g(x,u_1) - g(x,u_2), v) + \partial_t u_2[g(x,u_1) - g(x,u_2) - g'_u(x,u_2)v]\end{aligned} \tag{8.86}$$

with $G(x,u) := \int_0^u g(x,s)\,ds$, the fact that $g \in C^2$ with respect to $u$, and estimate (8.85), we have

$$\begin{aligned}&(a(\cdot)\partial_t v(t), \partial_t v(t)) + \partial_t[\frac{1}{2}\|\nabla_x v(t)\|^2_{L^2(\Omega)}\\ &\quad+ (G(\cdot,u_1(t)) - G(\cdot,u_2(t)) - g(\cdot,u_2(t))v(t), 1)]\\ &\qquad\le C_R(|\partial_t u_2(t), |v(t)|^2).\end{aligned} \tag{8.87}$$

We also note that

$$
\begin{aligned}
&(G(\cdot,u_1(t))-G(\cdot,u_2(t))-g(\cdot,u_2(t))v(t),1)\\
&\qquad=\left(\int_0^1\int_0^1 g'_u(x,u_2+s_1s_2v)\,ds_1\,ds_2,|v|^2\right)\\
&\qquad\geq\left(\inf_{|u|\leq 2R} g'_u(x,u),|v|^2\right).
\end{aligned}
\tag{8.88}
$$

Now taking a sum of (8.87) with the simple inequality

$$
\begin{aligned}
\partial_t(K_{2R}[a(\cdot)]^{1/2}v,v)&=2K_{2R}([a(\cdot)]^{1/2}\partial_t v,v)\\
&\leq\frac{1}{2}(a(\cdot)\partial_t v,\partial_t v)+4K_{2R}^2\|v\|^2_{L^2(\Omega)},
\end{aligned}
\tag{8.89}
$$

we get

$$
\begin{aligned}
&\partial_t\mathcal{G}(v(t))+\frac{1}{2}(a(x)\partial_t v(t),\partial_t v(t))\\
&\qquad\leq\|\partial_t u_2(t)\|_{L^{3/2}(\Omega)}\|v(t)\|^2_{W^{1,2}(\Omega)}+4K_{2R}^2\|v(t)\|^2_{L^2(\Omega)}
\end{aligned}
\tag{8.90}
$$

where

$$
\begin{aligned}
\mathcal{G}(v):=&\frac{1}{2}\|\nabla_x v\|^2_{L^2(\Omega)}\\
&+(G(\cdot,u_1(t))-G(\cdot,u_2(t))-g(\cdot,u_2(t))v(t),1)\\
&+K_{2R}([a(\cdot)]^{1/2}v(t),v(t)).
\end{aligned}
\tag{8.91}
$$

Moreover, due to assumption (8.81) and estimate (8.88), we have

$$
C_R\|\nabla_x v(t)\|^2_{L^2(\Omega)}\geq\mathcal{G}(v(t))\geq\frac{1}{2}\|\nabla_x v(t)\|^2_{L^2(\Omega)}.
\tag{8.92}
$$

Thus, (8.90) implies that

$$
\partial_t\mathcal{G}(v(t))\leq C_R(1+\|\partial_t u_2(t)\|_{L^{3/2}(\Omega)})\mathcal{G}(v(t)).
\tag{8.93}
$$

Now applying the Gronwall inequality to (8.93) and using that

$$
\int_0^T\|\partial_t u(t)\|_{L^{3/2}(\Omega)}\,dt\leq C(1+\int_0^t\|\partial_t u(t)\|^{p+2}_{L^{p+2}(\Omega)}\,dt)\leq C(T+1)
$$

(due to Proposition 8.4), we obtain estimate (8.83) and we finish the proof of Theorem 8.4. □

REMARK 8.6. The uniqueness theorem proved above shows, in particular, that any solution $u(t)$ of equation (8.13) in the sense of Definition 8.1 is "sufficiently regular" in the sense that it satisfies all of the estimates of Section 8.1.

It is also worth noting that the uniqueness condition (8.81) is automatically satisfied if equation (8.13) is nondegenerate:

$$
a(x)\geq a_0>0.
\tag{8.94}
$$

Thus, for the nondegenerate case, we have uniqueness under the assumptions of the existence Theorem 8.1.

## 8.3. Global and exponential attractors

In this section, we study the long-time behavior of solutions of problem (8.13) under the uniqueness assumption (8.81). In this case, equation (8.13) generates a dissipative semigroup $\{S(t),\, t \geq 0\}$ in the phase space $\Phi$:

$$S(t)u_0 := u(t), \quad u(t) \text{ solves (8.13) with } u(0) = u_0. \tag{8.95}$$

The following theorem establishes the existence of a global attractor for the semigroup (8.95) associated with equation (8.13).

THEOREM 8.5. *Let the assumptions of Theorem* 8.4 *hold. Then, the semigroup* $S(t)$ *associated with equation* (8.95) *possesses a compact global attractor* $\mathcal{A}$ *in* $\Phi$ *which is bounded in* $W^{2,q}(\Omega)$ *for every finite* $q$ *which can be described as follows:*

$$\mathcal{A} = \mathcal{K}\big|_{t=0} \tag{8.96}$$

*where* $\mathcal{K}$ *is a set of solutions of* (8.13) *defined for every* $t \in \mathbb{R}$ *and bounded:*

$$\mathcal{K} := \{u \in L^\infty(\mathbb{R}, \Phi), \ u \text{ solves } (8.13)\}. \tag{8.97}$$

PROOF. According to the standard attractor existence Theorem 2.1 (and see Remark 2.1), we only need to verify the following:

(1) The semigroup $S(t)$ has a closed graph in $\Phi$.

(2) This semigroup possesses a (pre)compact absorbing set $\mathcal{B}$ in the phase space $\Phi$. The latter means that for any bounded subset $B \subset \Phi$ there exists time $T = T(B)$ such that

$$S(t)B \subset \mathcal{B}, \quad t \geq T. \tag{8.98}$$

Let us verify these conditions for the semigroup (8.95). Indeed, the first condition is immediate, since, due to Theorem 8.4, the semigroup $S(t)$ is Lipschitz continuous in a weaker topology of $W^{1,2}(\Omega)$. Moreover, from Corollary 8.1 and estimate (8.57), we conclude that the set

$$\mathcal{B} = \mathcal{B}_q := \{u_0 \in W^{2,q}(\Omega), \ \ \|u_0\|_{W^{2,q}(\Omega)} \leq R\} \tag{8.99}$$

will be absorbing for the semigroup $S(t)$ if $R = R(q)$ is large enough. It remains to note that $\mathcal{B}_q$ is compactly embedded in $\Phi$ if $q$ is large enough.

Thus, the existence of the attractor $\mathcal{A}$ is proven. It is bounded in $W^{2,q}(\Omega)$ since $\mathcal{A} \subset \mathcal{B}_q$ and, finally, description (8.97) is also the standard corollary of the abstract attractor existence theorem. Theorem 8.5 is proven. □

Our next task is to verify the finite dimensionality of the global attractor and to construct the associated *exponential* attractor. The next theorem establishes the existence of an exponential attractor for problem (8.13) by the use of a modification of the so-called method of $l$-trajectories (see [70]).

THEOREM 8.6. *Let the assumptions of Theorem* 8.5 *hold. Then, semigroup* $S(t)$ *associated with equation* (8.13) *possesses an exponential attractor* $\mathcal{M}$ *in* $\Phi$ *in the sense of Definition* 3.1.

PROOF. The proof of this theorem is based on the following abstract exponential attractor existence theorem formulated as Proposition 8.6.

PROPOSITION 8.6. *Let $H, V, V_1$ be Banach spaces such that the embedding $V_1 \hookrightarrow V$ is compact. Let $B$ be a closed bounded subset of $H$, and let $\mathbb{S} : B \to B$ be a map. Assume also that there exists a uniformly Lipschitz continuous map $\mathbb{T} : B \to V_1$, i.e.,*

$$\|\mathbb{T}b_1 - \mathbb{T}b_2\|_{V_1} \le L\|b_1 - b_2\|_H, \qquad \forall b_1, b_2 \in B, \tag{8.100}$$

*for some $L \ge 0$, such that*

$$\|\mathbb{S}b_1 - \mathbb{S}b_2\|_H \le \vartheta\|b_1 - b_2\|_H + K\|\mathbb{T}b_1 - \mathbb{T}b_2\|_V, \qquad \forall b_1, b_2 \in B, \tag{8.101}$$

*for some $\vartheta < 1/2$ and $K \ge 0$. Then, there exists a (discrete) exponential attractor $\mathcal{M}_d \subset B$ of the semigroup $\{S(n) := S^n, n \in \mathbb{Z}_+\}$ with discrete time in the phase space $H$ (see Definition 3.1).*

The general proof repeats word for word the proof of Theorem3.2, where we considered the particular case $H = V_1$. We omit it here (see [35] for details).

In order to apply this abstract result to our problem (8.13), we first note that it is sufficient to verify the existence of an exponential attractor not in the whole phase space $\Phi$, but only for the restriction of $S(t)$ on some absorbing set in $\Phi$. In order to construct the proper absorbing set, we recall that, due to Proposition 8.4, the ball

$$B(R) := \{u_0 \in \Phi,\ \|u_0\|_\Phi \le R\} \tag{8.102}$$

will be absorbing for semigroup $S(t)$ if $R$ is large enough. Since we want the absorbing set to be positively invariant with respect to the semigroup, we further set

$$B_1 = [\bigcup_{t\ge 0} S(t)B(R)]_\Phi \tag{8.103}$$

where $[\cdot]$ denotes the closure in the space $\Phi$. Finally, in order to have some compactness, we set

$$B := S(1)B_1. \tag{8.104}$$

Then, on the one hand, it is not difficult to verify that $B$ is a compact positively invariant subset of the phase space $\Phi$, i.e.,

$$S(t)B \subset B, \quad t \ge 0, \tag{8.105}$$

and, on the other hand, due to Corollary 8.1, we have

$$\|\partial_t u(t)\|_{L^\infty(\Omega)} + \|u(t)\|_{W^{2,q}(\Omega)} \le C \tag{8.106}$$

for every trajectory $u(t)$ of equation (8.13) starting from $u_0 \in B$ (and the constant $C = C_q$ depends on $q$ but is independent of the choice of $u_0 \in B$). In particular, we see that $B \subset W^{2,q}(\Omega)$ for every finite $q$.

Thus, we will construct below the exponential attractor $\mathcal{M}$ for the restriction of the semigroup $S(t)$ on the above absorbing set $B$. To this end, we need the following lemma, which gives the key estimate of the form (8.100).

LEMMA 8.1. *Let the above assumptions hold. Then, for every two solutions $u_1(t)$ and $u_2(t)$ such that $u_i(0) \in B$, $i = 1, 2$, the following estimate holds:*

(8.107)

$$\|u_1(t) - u_2(t)\|^2_{W^{1,2}(\Omega)} + \gamma \int_0^t (a(\cdot), |\partial_t u_1(s) - \partial_t u_2(s)|^2)\, ds \\ \le Ce^{-\beta t}\|u_1(0) - u_2(0)\|^2_{W^{1,2}(\Omega)} + Ce^{kt} \int_0^t \|\chi_{\Omega\setminus\Omega_0}(\cdot)(u_1(s) - u_2(s))\|^2_{L^2(\Omega)}\, ds$$

*where $\Omega_0 := \{x \in \Omega,\ a(x) = 0\}$, $\chi_V(\cdot)$ means the characteristic function of the set $V$, and the positive constants $\beta, \gamma$, $C$, and $K$ are independent of $u_1$ and $u_2$.*

PROOF. We first note that (8.89) can be improved as follows:

$$\partial_t(K_{2R}[a(\cdot)]^{1/2}v, v) \le \frac{1}{2}(a(\cdot)\partial_t v, \partial_t v) + 4K^2_{2R}\|\chi_{\Omega\setminus\Omega_0}(\cdot)v\|^2_{L^2(\Omega)} \tag{8.108}$$

and, consequently, (8.90) reads

$$\partial_t \mathcal{G}(v(t)) + \frac{1}{2}(a(x)\partial_t v(t), \partial_t v(t)) + (l_{u_1,u_2}\partial_t v(t), \partial_t v(t)) \\ \le \|\partial_t u_2(t)\|_{L^{3/2}(\Omega)}\|v(t)\|^2_{W^{1,2}(\Omega)} + 4K_{2R}\|\chi_{\Omega\setminus\Omega_0}v(t)\|^2_{L^2(\Omega)} \tag{8.109}$$

where the function $\mathcal{G}$ is defined by (8.91) and

$$l_{u_1,u_2} := \int_0^1 \phi'_{\partial_t u}(x, s\partial_t u_1 + (1-s)\partial_t u_2)\, ds.$$

Now multiplying equation (8.84) by $v(t) := u_1(t) - u_2(t)$,

$$\|\nabla_x v(t)\|^2_{L^2(\Omega)} + (g(\cdot, u_1(t)) - g(\cdot, u_2(t)), v(t)) \\ + (l_{u_1,u_2}(t)\partial_t v(t), v(t)) + \frac{1}{2}\partial_t(a(\cdot)v(t), v(t)) \\ + (g(\cdot, u_1(t)) - g(\cdot, u_2(t)), v(t)) \\ = 0. \tag{8.110}$$

Due to conditions (8.16), we may assume, without loss of generality, that $\phi'_{\partial_t u}(x, 0) = 0$. Then, using estimate (8.106) and the fact that $l_{u_1,u_2}(t) \ge 0$, we have

$$|(l_{u_1,u_2}(t)\partial_t v(t), v(t))| \le \frac{1}{4}(l_{u_1,u_2}\partial_t v(t), \partial_t v(t)) + (l_{u_1,u_2}v(t), v(t)) \\ \le \frac{1}{4}(l_{u_1,u_2}\partial_t v(t), \partial_t v(t)) \\ + C(\|\partial_t u_1(t)\|_{L^{3/2}(\Omega)} + \|\partial_t u_2(t)\|_{L^{3/2}(\Omega)})\|\nabla_x v(t)\|^2_{W^{1,2}(\Omega)}. \tag{8.111}$$

Moreover, due to the uniqueness assumption (8.81),

$$(g(\cdot, u_1(t)) - g(\cdot, u_2(t)), v(t)) \ge -C\|\chi_{\Omega\setminus\Omega_0}v(t)\|^2_{L^2(\Omega)}. \tag{8.112}$$

Inserting estimates (8.111) and (8.112) into (8.110), we arrive at

$$\|\nabla_x v(t)\|^2_{L^2(\Omega)} \le \frac{1}{4}[(l_{u_1,u_2}\partial_t v(t), \partial_t v(t)) + (a(\cdot)\partial_t v(t), \partial_t v(t))] \\ + C(\|\partial_t u_1(t)\|_{L^{3/2}(\Omega)} + \|\partial_t u_2(t)\|_{L^{3/2}(\Omega)})\|\nabla_x v(t)\|^2_{W^{1,2}(\Omega)} \\ + C\|\chi_{\Omega\setminus\Omega_0}v(t)\|^2_{L^2(\Omega)}. \tag{8.113}$$

Taking a sum of (8.109) and (8.113) and using estimates (8.92), we finally infer

$$\begin{aligned}\partial_t\mathcal{G}(v(t)) + [\beta - C(\|\partial_t u_1(t)\|_{L^{3/2}(\Omega)} + \|\partial_t u_2(t)\|_{L^{3/2}(\Omega)})]\mathcal{G}(v(t))\\ + \frac{1}{4}(a(\cdot)\partial_t v(t), \partial_t v(t))\\ \le C\|\chi_{\Omega\backslash\Omega_0} v(t)\|^2_{L^2(\Omega)}\end{aligned} \tag{8.114}$$

for some positive $\beta$, independent of $u_1$ and $u_2$.

In order to deduce estimate (8.107) from (8.115), we note that the existence of a global Lyapunov function (8.18) together with assumptions (8.16) implies that the following dissipation integrals are finite:

$$\int_0^\infty \|\partial_t u_1(s)\|^{p+2}_{L^{p+2}(\Omega)}\,ds + \int_0^\infty \|\partial_t u_2(s)\|^{p+2}_{L^{p+2}(\Omega)}\,ds \le C < \infty \tag{8.115}$$

where the constant $C$ is independent of $u_1$ and $u_2$ (with the initial data belonging to the absorbing set $B$). These integrals imply that

$$\int_0^t (\|\partial_t u_1(s)\|_{L^{3/2}(\Omega)} + \|\partial_t u_2(s)\|_{L^{3/2}(\Omega)})\,ds \le \varepsilon t + C_\varepsilon, \tag{8.116}$$

where $\varepsilon > 0$ can be arbitrary and $C_\varepsilon$ depends only on $\varepsilon$. Now applying the Gronwall inequality to (8.114) and using (8.116) with sufficiently small $\varepsilon$, we deduce the required estimate (8.107) and finish the proof of Lemma 8.1. □

It is now not difficult to finish the proof of the theorem. Indeed, let us fix $T > 0$ in such a way that $Ce^{-\beta T} < 1/2$ where $C$ and $\beta$ are the same as in Lemma 8.1 and let $S = S(T)$.

We also set $H := W^{1,2}(\Omega)$,

$$V := L^2([0,T], L^2(\Omega\backslash\Omega_0)) \tag{8.117}$$

and

$$V_1 := \{u \in L^2([0,T], W^{1,2}(\Omega)),\ \int_0^T (a(\cdot)\partial_t u(s), \partial_t u(s))\,ds < \infty\}. \tag{8.118}$$

Finally, we define the operator $\mathbb{T} : B \to V_1$ as the solving operator of problem (8.95), i.e,

$$\mathbb{T}u_0 := u \in V_1 \quad\text{where } u(t),\ t \in [0,T], \text{ solves (8.13) with } u(0) = u_0. \tag{8.119}$$

We claim that the operator $S : B \to B$, the spaces $H$, $V$, and $V_1$, and the operator $\mathbb{T}$ thus defined satisfy all the assumptions of Proposition 8.6. Indeed, the compactness of the embedding $V_1 \subset V$ is verified in Lemma 1.1 from Chapter 1, the global Lipschitz continuity of $\mathbb{T}$ is an immediate corollary of Theorem 8.4, and estimate (8.100) follows from Lemma 8.1. Thus, due to Proposition 8.6, the semigroup $S(n)$ generated by iterations of the operator $S : B \to B$ possesses an exponential attractor $\mathcal{M}_d$ in $B$ endowed by the topology of $H = W^{1,2}(\Omega)$.

In order to construct the exponential attractor $\mathcal{M}$ for the semigroup $S(t)$ with continuous time, we note that, due to Theorem 8.4, this semigroup is Lipschitz continuous with respect to the initial data in the topology of $H$. Moreover, since the derivative $\partial_t u(t)$ is uniformly bounded for any trajectory $u(t)$ starting from $B$, this semigroup is also uniformly Lipschitz continuous in time in the $L^\infty(\Omega)$-metric. Since $B$ is bounded in $W^{2,q}$, for any finite $q$, the last assertion together with the appropriate interpolation inequality gives the uniform Hölder continuity in time in the metric of $H$. Thus, we have verified that the map $(t, u_0) \to S(t)u_0$ is uniformly Hölder continuous on $[0, T] \times B$ where $B$ is endowed by the $H$-metric. Therefore, the required exponential attractor $\mathcal{M}$ for the case of continuous time can be defined by the standard expression

$$\mathcal{M} := \bigcup_{t \in [0,T]} S(t)\mathcal{M}_d \tag{8.120}$$

where $\mathcal{M}_d$ is the exponential attractor for the discrete semigroup associated with the map $S = S(T)$ constructed above.

So, in order to finish the proof of the theorem, we only need to verify that $\mathcal{M}$ defined by (8.120) will be the exponential attractor for $S(t)$ restricted to $B$ not only in the $H$-metric, but also in a stronger metric of the phase space $\Phi$. But this is an immediate corollary of the fact that $B$ is bounded in $W^{2,q}(\Omega)$ and the interpolation inequality

$$\|w\|_{L^\infty(\Omega)} \le C\|w\|^{\kappa}_{W^{1,2}(\Omega)}\|w\|^{1-\kappa}_{W^{2,q}(\Omega)} \tag{8.121}$$

with the appropriate exponent $0 < \kappa < 1$. Theorem 8.6 is proven. □

REMARK 8.7. Since the global attractor $\mathcal{A}$ is always contained in the exponential attractor $\mathcal{M}$, the proved theorem immediately implies that the fractal dimension of the global attractor $\mathcal{A}$ is also finite.

REMARK 8.8. It is not difficult to show that the global Lyapunov functional is not crucial for the above theory.

CHAPTER 9

# On a class of PDEs with degenerate diffusion and chemotaxis: Autonomous case

We will consider the following model:

$$
\begin{aligned}
&(9.1) \qquad M_t = \nabla\cdot(M^\alpha \nabla M) - \nabla\cdot(M^\gamma \nabla\rho) + f(M,\rho) && \text{in } \Omega\times(0,\infty),\\
&(9.2) \qquad \rho_t = \triangle\rho - g(M,\rho) && \text{in } \Omega\times(0,\infty),\\
&(9.3) \qquad M = 0, \quad \rho = 1 && \text{in } \partial\Omega\times(0,\infty),\\
&(9.4) \qquad M(\cdot,0) = M_0, \quad \rho(\cdot,0) = \rho_0 && \text{in } \Omega,
\end{aligned}
$$

where $\alpha$ and $\gamma$ are given constants satisfying

$$\gamma + 1 \le \alpha \le 2(\gamma - 1). \tag{9.5}$$

REMARK 9.1. (1) We will call a condition of this type for $\alpha$ and $\gamma$ a "balance condition" since it establishes a balance between the diffusion and transport terms, that is, between the porous medium and the chemotaxis effects.

(2) It is clear from (9.5) that $\alpha \ge 4$ and $\gamma \ge 3$ should hold.

(3) We will propose a weaker "balance condition" in Chapter 10. See also Remark 9.2.

Moreover, $\Omega \subset \mathbb{R}^N$ is a bounded domain ($N = 1, 2, 3$), $\rho_0, M_0 \ge 0$ and they are smooth. We assume that the functions $f$ and $g$ satisfy

$$
\begin{aligned}
&(9.6) \qquad -f_1 M^2 \le f(M,\rho) \le f_2 M - f_3 M^2 \quad \text{for } M \ge 0,\ \rho \ge 0,\\
&(9.7) \qquad g(M,\rho) = g_0(\rho)M + g_1\rho \quad \text{for } M \ge 0,\ \rho \ge 0,\\
&(9.8) \qquad 0 \le g_0(\rho) \le g_2 \quad \text{for } \rho \ge 0,\\
&(9.9) \qquad |g_0'(\rho)| \le g_3 \quad \text{for } \rho \ge 0,\ g(0) = 0,\\
&(9.10) \qquad \tilde f(M,\rho) = f(M^{\frac{2}{\alpha+2}},\rho) - f_4 M^{\frac{2}{\alpha+2}} \text{ is a } C^1\text{-function},
\end{aligned}
$$

where $f_1, f_2, f_3, f_4, g_0, g_1, g_2, g_3$ are positive constants.

Functions $f$ and $g$ satisfying conditions (9.6)–(9.10) are, for example,

$$
\begin{aligned}
f(M,\rho) &= M - \frac{M^{\frac{\alpha}{2}+3}}{1+M^{\frac{\alpha}{2}+1}} \quad \text{or} \quad \left(M - \frac{M^{\frac{\alpha}{2}+3}}{1+M^{\frac{\alpha}{2}+1}}\right)(\rho+1),\\
g(M,\rho) &= \frac{\rho}{\rho+1}M + \rho.
\end{aligned}
$$

In this book we treat weak solutions to the system (9.1)–(9.4). The definition is as follows.

DEFINITION 9.1. For $T > 0$, $\alpha > 1$, and $\gamma > 1$, nonnegative functions $(M, \rho)$ defined in $\Omega \times [0, T)$ are said to be a weak solution of (9.1)–(9.4) for $M_0$, $\rho_0 \in C^1(\overline{\Omega})$, if

(i) $M \in L^\infty(0, T; L^2(\Omega))$, $M^\alpha \in L^2(0, T; H^1(\Omega))$, $M_t \in C_w([0, T]; H^{-1}(\Omega))$,
(ii) $\rho \in C([0, T]; H^1(\Omega))$,
(iii) $(M, \rho)$ satisfies the equation in the following sense:

$$\int_\Omega (M_t - \nabla M^\alpha \nabla M + \nabla \cdot (M^\gamma \nabla \rho) - f(M, \rho))\, \varphi dx = 0 \quad \text{for any } \varphi \in \mathcal{D}(\Omega),$$

$$\begin{aligned}(\rho(x,t) - 1) = &\int_\Omega G(x, y, t)(\rho_0(y) - 1)dy \\ &- \int_0^t \int_\Omega G(x, y, t - s) g(M(y, s), \rho(y, s)) dyds \\ &\text{for any } (x, t) \in \Omega \times (0, T),\end{aligned}$$

where $G$ is a heat kernel in $\Omega$ with the homogeneous Dirichlet boundary condition.

This system of partial differential equations, models, for example, a population described in terms of its density $M$, which grows in dependence of a substrate with concentration $\rho$. The substrate is degraded by abiotic decay. The spatial movement of the population is caused by two different effects. Firstly, the model includes a density-dependent diffusion term. This nonlinear diffusion effect becomes stronger as the population grows larger locally, following a power law as in the case of the porous medium equation. Secondly, the population moves towards regions with increased substrate availability, i.e., follows the chemical signal $\rho$. This effect is also controlled by the population density and its intensity increases as the local population density grows. Both effects of population mobility increase/diminish with the population, each following a power law. Thus, the model degenerates for $M = 0$. Finally, our model includes a "source term": a nonlinear reaction-interaction term $f$. As usual, it corresponds to the sink/source density (net number of particles created per unit of time and per unit of volume). This effect also vanishes with the population.

The study of this equation is inspired by a work of the current author (see [43]). Indeed, in [43], we dealt with the biomass spatial spreading mechanisms via nonlinear diffusion (modeling in meso-scale), which leads to a reaction-diffusion equation comprising simultaneously two kinds of degeneracy: porous medium and fast diffusion. More precisely, in [43] we studied existence, uniqueness, the dependence of a global existence in time of the solution on boundary conditions, and the existence of a global attractor for the associated semigroup. It is worth noting that a numerical simulation of this equation leads to the mushroom-shaped patterns observed in an experimental study (see [26]).

The structure of the global attractor which captures all dynamics (long-time behavior of solutions) was given in [33], which shows, in particular, the dominant role of the order of porous medium degeneracy over fast diffusion for the spatial spreading mechanism of biomass. Therefore, we simplify the nonlinear diffusion term and include chemotaxis (see also [48] and the references therein). The main aim of the present study is to prove the well-posedness of the system (9.1)–(9.4) and to study its dynamics in terms of the global attractor. We emphasize the fact that the analysis of equations with a chemotaxis type term even without degeneracy

($\alpha = 0$) is quite difficult (examples, though for somewhat different biological models, can be found in [73], [87], [88], [89], [91]; see also the references therein), and, in our degenerate case, we face significant difficulties. To overcome these difficulties and to obtain the existence of the global attractor, we propose so-called "balance condition"' between the order of porous medium degeneracy and the growth order of the chemotaxis function.

The chapter is organized as follows. In Section 9.1 we obtain several a priori estimates for the solutions of (9.1)–(9.4), which in turn lead to $L^\infty$-bounds for the biomass component. Section 9.2 is devoted to the uniqueness of solutions. In Section 9.3 we obtain dissipative estimates for the solutions of (9.1)–(9.4) and, as a consequence, we prove the existence of the weak global attractor. In the first appendix (Section 9.4) we present some standard ideas that we used proving the well-posedness of solutions (for the convenience of the reader).

## 9.1. Global existence and boundedness

In this section, we obtain some $L^\infty$-estimates for the limiting function of classical solutions to an approximated equation given in A1 of Section 9.4. The limiting function is a weak solution for (9.1)–(9.4). To this end, we multiply (9.1) by $\left(\frac{1}{\alpha+1-\gamma}M^{\alpha+1-\gamma} - \rho\right)$ and integrate over $\Omega$. Then, we obtain

$$\begin{aligned} \mathrm{I} &:= \int_\Omega M_t \left(\frac{M^{\alpha+1-\gamma}}{\alpha+1-\gamma} - \rho\right) dx \\ &= \mathrm{II} + \mathrm{III}, \end{aligned} \tag{9.11}$$

where by II and III we denote the integrals

$$\begin{aligned} \mathrm{II} &:= \int_\Omega \nabla \cdot M^\gamma \nabla \left(\frac{M^{\alpha+1-\gamma}}{\alpha+1-\gamma} - \rho\right) \times \left(\frac{M^{\alpha+1-\gamma}}{\alpha+1-\gamma} - \rho\right) dx, \\ \mathrm{III} &:= \int_\Omega f(M,\rho) \left(\frac{M^{\alpha+1-\gamma}}{\alpha+1-\gamma} - \rho\right) dx, \end{aligned}$$

respectively. Note that

$$\begin{aligned} \mathrm{I} &= \frac{1}{(\alpha+2-\gamma)(\alpha+1-\gamma)} \frac{d}{dt} \int_\Omega M^{\alpha+2-\gamma} dx - \frac{d}{dt} \int_\Omega M\rho dx + \int_\Omega M \rho_t dx \\ &= \mathrm{IV} + \mathrm{V}, \end{aligned} \tag{9.12}$$

where by IV and V we denote the integrals

$$\begin{aligned} \mathrm{IV} &:= \frac{d}{dt} \int_\Omega \left( \frac{1}{(\alpha+2-\gamma)(\alpha+1-\gamma)} M^{\alpha+2-\gamma} - M\rho \right) dx \\ \mathrm{V} &:= \int_\Omega M \rho_t dx. \end{aligned}$$

Multiplying the equation (9.2) by $\rho_t$ using $\rho_t|_{\partial\Omega} = 0$, we get

$$\int_\Omega |\rho_t|^2 dx + \frac{1}{2}\frac{d}{dt} \int_\Omega |\nabla \rho|^2 dx + \frac{1}{2}\frac{d}{dt}\int_\Omega \rho^2 dx \le \int_\Omega (g_2 M + g_1|\rho|)|\rho_t| dx.$$

Hence,

$$\frac{1}{2}\int_\Omega |\rho_t|^2 dx + \frac{1}{2}\frac{d}{dt}\int_\Omega (|\nabla\rho|^2 + \rho^2)dx \le \frac{g_2^2}{2}\int_\Omega M^2 dx + g_1^2|\Omega|\|\rho\|_{L^\infty}^2. \tag{9.13}$$

From (9.6) it follows that $f_2M - f_3M^2 \le F_2 - F_3M^2$ for $M \ge 0$, where $F_2$ and $F_3$ are positive constants. Hence

$$\begin{aligned} \mathrm{III} &\le \int_\Omega (F_2 - F_3M^2)\frac{M^{\alpha+1-\gamma}}{\alpha+1-\gamma}dx + f_1\int_\Omega M^2\rho dx \\ &\le -\frac{F_3}{2(\alpha+1-\gamma)}\int_\Omega M^{\alpha+3-\gamma}dx + C(\|\rho\|_\infty^3 + 1). \end{aligned}$$

Since $\alpha + 3 - \gamma \ge 3$, we have

$$\begin{aligned} \int_\Omega M^2\rho dx &\le \left(\int_\Omega M^3 dx\right)^{\frac{2}{3}}\left(\int_\Omega \rho^3 dx\right)^{\frac{1}{3}} \le \frac{2}{3}\int_\Omega M^3 dx + \frac{1}{3}\|\rho\|_\infty^3 \\ &\le \frac{F_3}{4(\alpha+1-\gamma)}\int_\Omega M^{\alpha+3-\gamma}dx + \frac{1}{3}\|\rho\|_\infty^3 + C, \\ \int_\Omega M^{\alpha+1-\gamma} &\le \left(\int_\Omega M^{\alpha+3-\gamma}dx\right)^{\frac{\alpha+1-\gamma}{\alpha+3-\gamma}}\left(\int_\Omega 1dx\right)^{\frac{2}{\alpha+3-\gamma}} \\ &\le \frac{F_3}{4(\alpha+1-\gamma)}\int_\Omega M^{\alpha+3-\gamma}dx + C. \end{aligned} \tag{9.14}$$

Using the comparison theorem and (9.7) and (9.8), we obtain

$$0 \le \rho \le \max(\|\rho_0\|_\infty, 1). \tag{9.15}$$

By (9.11), (9.12), and (9.14), we have

$$\begin{aligned} &\frac{d}{dt}\int_\Omega \left(\frac{1}{(\alpha+2-\gamma)(\alpha+1-\gamma)}M^{\alpha+2-\gamma} - M\rho\right)dx \\ &\quad = -\int_\Omega M^\gamma\left|\nabla\left(\frac{M^{\alpha+1-\gamma}}{\alpha+1-\gamma} - \rho\right)\right|^2 dx - \int_\Omega M\rho_t dx \\ &\qquad - \frac{F_3}{2(\alpha+1-\gamma)}\int_\Omega M^{\alpha+3-\gamma}dx + C. \end{aligned}$$

Combining this equality with (9.13) and the obvious inequality $M^{\alpha+2-\gamma} \le \varepsilon M^{\alpha+3-\gamma} + C\varepsilon$, we obtain

$$\begin{aligned} &\frac{d}{dt}\int_\Omega \left(\frac{M^{\alpha+2-\gamma}}{(\alpha+2-\gamma)(\alpha+1-\gamma)} - M\rho\right) dx \\ &\quad \le \left(\frac{1}{2} + g_2\right)\int_\Omega M^2 dx - \frac{F_3}{2(\alpha+1-\gamma)}\int_\Omega M^{\alpha+2-\gamma}dx + C \end{aligned} \tag{9.16}$$

where $C$ is a positive constant independent of $t$. ($C$ may not be equal to the other $C$'s.) It follows from (9.15) that

$$\int_\Omega M\rho dx \leq \max(\|\rho_0\|_\infty, 1)\int_\Omega M dx \leq \frac{F_3}{8(\alpha+1-\gamma)}\int_\Omega M^{\alpha+2-\gamma}dx + C. \tag{9.17}$$

We have $(1/2+g_2)M^2 \leq (F_3/[8(\alpha+1-\gamma)])M^{\alpha+2-\gamma} + C$ (by $\alpha \geq \gamma+1$). Then, we observe that

$$\frac{d}{dt}\int_\Omega M^{\alpha+2-\gamma}dx \leq -\frac{F_3(\alpha+2-\gamma)}{2}\int_\Omega M^{\alpha+2-\gamma}dx + C,$$

or

$$\int_\Omega M^{\alpha+2-\gamma}dx \leq C \qquad \text{for } t \geq 0. \tag{9.18}$$

Let $A = -\triangle + I$ with $\cdot\,|_{\partial\Omega} = 0$. Then

$$(\rho-1)_t = \triangle(\rho-1) - (\rho-1) + g_0(\rho)M - 1,$$
$$e^t(\rho-1) = e^{-At}(\rho_0-1) + \int_0^t e^{-A(t-s)}e^A(g_o(\rho)M-1)ds.$$

Hence, for $0 < \varepsilon < 1$,

$$\begin{aligned} e^t\|A^{1-\varepsilon}(\rho-1)\|_{L^{\alpha-\gamma+2}} \leq{}& \|A^{1-\varepsilon}e^{-At}(\rho_0-1)\|_{L^{\alpha-\gamma+2}} \\ &+ \int_0^t e^s\|A^{1-\varepsilon}e^{-A(t-s)}g(M,\rho)\|_{L^{\alpha-\gamma+2}}ds. \end{aligned}$$

Consequently,

$$\begin{aligned} e^t\|A^{1-\varepsilon}(\rho-1)\|_{L^{\alpha-\gamma+2}} &\leq \underbrace{C}_{\text{by } \rho_0\text{: smooth}} + \int_0^t e^s\|A^{1-\varepsilon}e^{-A(t-s)}g\|_{L^{\alpha-\gamma+2}}ds \\ &\leq C + \int_0^t e^s(t-s)^{-1+\varepsilon}C(\|M\|_{L^{\alpha-\gamma+2}}+1)ds \\ &\leq C(1+e^t). \end{aligned}$$

By $\alpha-\gamma+2 \geq 3$ and $D(A^{4/5}) \subset L^6(\Omega)$,

$$\|\rho\|_{L^6} \leq C\|A^{4/5}\rho\|_{L^{\alpha-\gamma+2}} \leq C(\|M\|_{L^{\alpha-\gamma+2}}+1).$$

Multiplying (9.1) by $M^\delta$ and integrating over $\Omega$, we obtain

(9.19)

$$\begin{aligned} &\int_\Omega M_t M^\delta dx \\ &\quad= \int_\Omega \left(\triangle M^{\alpha+1}\right)M^\delta dx - \int_\Omega \nabla(M^\gamma\nabla\rho)M^\delta dx + \int_\Omega f(M,\rho)M^\delta dx \\ &\quad= -\delta(\alpha+1)\int_\Omega M^{\alpha+\delta-1}|\nabla M|^2 + \frac{\gamma+\delta}{\delta}\int_\Omega \nabla\rho\nabla M^{\gamma+\delta}dx + \hat{C}\int_\Omega M^\delta dx \\ &\quad= -\frac{4\delta(\alpha+1)}{(\alpha+\delta+1)^2}\int_\Omega |\nabla M^{\frac{\alpha+\delta+1}{2}}|^2 dx - \frac{\gamma+\delta}{\delta}\int_\Omega \nabla\rho\cdot\nabla M^{\gamma+\delta}dx + \hat{C}\int_\Omega M^\delta dx. \end{aligned}$$

Based on the embedding theorem, we obtain

$$\left(\int_\Omega M^{3(\alpha+\delta+1)}dx\right)^{\frac{1}{3}} \leq \hat{C}\int_\Omega |\nabla M^{\frac{\alpha+\delta+1}{2}}|^2 dx,$$

$$\begin{aligned}\frac{\gamma+\delta}{\delta}\left|\int_\Omega \nabla\rho\cdot\nabla M^{\gamma+\delta}dx\right| &\leq C\|\nabla\rho\|_{L^6}\cdot\|M^{(2\gamma+\delta-1-\alpha)/2}\|_{L^3}\cdot\|\nabla M^{(\alpha+\delta+1)/2}\|_{L^2}\\ &\leq \varepsilon\|\nabla M^{(\alpha+\delta+1)/2}\|_{L^2}^2+\varepsilon\|M^{(2\gamma+\delta-1-\alpha)/2}\|_{L^3}^{2\theta}+C_{\varepsilon,\delta}\\ &\leq \varepsilon\|\nabla M^{(\alpha+\delta+1)/2}\|_{L^2}^2+\varepsilon|\Omega|^{2\theta/(1-\theta)}\|M^{\alpha+\delta+1}\|_{L^3}^2+C_{\varepsilon,\delta}\\ &\leq \varepsilon(1+|\Omega|^{2\theta/(1-\theta)})\|\nabla M^{(\alpha+\delta+1)/2}\|_{L^2}^2+C_{\varepsilon,\delta}\end{aligned}$$

for sufficiently small $\varepsilon>0$ and $\delta>\alpha+1-2\gamma$, where $\theta=2(\alpha+\delta+1)/(2\gamma+\delta-1-\alpha)>1$.

Combining the Hölder inequality with the inequalities above and $\alpha\geq\gamma+1$ implies that

(9.20)
$$\frac{d}{dt}\int_\Omega M^{\delta+1}dx+\int_\Omega M^{\delta+1}dx+\frac{1}{C_\delta}\int_\Omega|\nabla M^{(\alpha+\delta+1)/2}|^2dx\leq C_\delta \quad \text{for some } C_\delta>1.$$

Then, we have

$$\int_\Omega M^{\delta+1}dx+\int_0^t\frac{e^{s-t}}{C_\delta}\int_\Omega|\nabla M^{(\alpha+\delta+1)/2}|^2dxds\leq C_\delta \ \forall\delta\geq 1. \tag{9.21}$$

We see that

$$\|A^{4/5}(\rho-1)\|_{L^7}\leq C\left(\sup_{0\leq t}\|M(t)\|_{L^7}+1\right)\leq C. \tag{9.22}$$

If we regard the operator $A$ in $L^7(\Omega)$, then $D(A^{4/5})\subset C^1(\overline{\Omega})$. Hence, it follows that we have that

$$\|\nabla\rho\|_{L^\infty}\leq C\left(\sup_{t\geq 0}\|M(t)\|_{L^7}+1\right)\leq C.$$

Then, we obtain

$$\begin{aligned}\left|\int_\Omega\nabla\rho\cdot\nabla M^{\gamma+\delta}dx\right| &\leq C\int_\Omega\left|\nabla M^{\gamma+\delta}\right|dx\\ &\leq C\frac{\alpha+\delta+1}{2(\gamma+\delta)}\int_\Omega\left|\nabla M^{(\gamma+\delta+1)/2}|M^{(\frac{2(\gamma+\delta)}{\alpha+\delta+1}-1)\frac{\alpha+\delta+1}{2}}\right|dx\\ &\leq\frac{\delta(\alpha+1)}{(\alpha+\delta+1)^2}\int_\Omega\left|\nabla M^{(\gamma+\delta+1)/2}\right|dx+C(1+\delta^2)\int_\Omega M^{\alpha+\delta-3}dx.\end{aligned} \tag{9.23}$$

Since

$$\|u\|_{L^2}\leq K_2\left(\|\nabla u\|_{L^2}+\|u\|_{L^2}\right)^{1/2}\|u\|_{L^{6/5}}^{1/2},$$

we obtain

(9.24)
$$\int_\Omega M^{\alpha+\delta+1}dx$$
$$\leq K_2\left(\int_\Omega|\nabla M^{(\alpha+\delta+1)/2}|^2dx+\int_\Omega M^{\alpha+\delta+1}dx\right)^{1/2}\cdot\left(\int_\Omega M^{3(\alpha+\delta+1)/5}dx\right)^{5/6}.$$

Then, we have that

$$\begin{aligned}
C(1+\delta^2)\int_\Omega M^{\alpha+\delta-3}dx &\le C(1+\delta^2)\left(\int_\Omega M^{\alpha+\delta+1}dx\right)\\
&\le \frac{\delta(\alpha+1)}{(\alpha+\delta+1)^2}\int_\Omega |\nabla M^{(\alpha+\delta+1)/2}|^2dx\\
&\quad + C(\delta^5+1)\left(\int_\Omega M^{3(\alpha+\delta+1)/5}dx\right)^{5/3} + C(1+\delta^2)\\
&\le \frac{\delta(\alpha+1)}{(\alpha+\delta+1)^2}\int_\Omega |\nabla M^{(\alpha+\delta+1)/2}|^2dx\\
&\quad + C(\delta^5+1)\left(\int_\Omega M^{4(\alpha+1)/5}dx\right)^{5/4}\left(\int_\Omega M^{12\alpha/5}dx\right)^{5/12}\\
&\quad + C(1+\delta^2).
\end{aligned}$$

By using an argument similar to that above, we get that

$$\begin{aligned}
\hat{C}\int_\Omega M^{\delta}dx \le{}& \frac{\delta(\alpha+1)}{(\alpha+\delta+1)^2}\int_\Omega |\nabla M^{(\alpha+\delta+1)/2}|^2dx\\
&+ C(1+\delta)\left(\int_\Omega M^{4(\alpha+1)/5}dx\right)^{5/4}\left(\int_\Omega M^{12\alpha/5}dx\right)^{5/12} + C.
\end{aligned}$$

Combining this estimate with (9.21) and (9.19) with $\delta = (12/5)\alpha - 1$ implies that

$$\begin{aligned}
&\frac{d}{dt}\int_\Omega M^{\delta+1}dx + \frac{\delta(\delta+1)(\alpha+1)}{(\alpha+\delta+1)^2}\int_\Omega |\nabla M^{(\alpha+\delta+1)/2}|^2dx\\
&\quad \le C(1+\delta^6)\left(\int_\Omega M^{4(\alpha+1)/5}dx+1\right)^{5/4}.
\end{aligned}$$

By this and (9.24), we obtain that

$$\frac{d}{dt}\left(\int_\Omega M^{\delta+1}dx+1\right) + \left(\int_\Omega M^{\delta+1}dx+1\right) \le C(1+\delta^6)\left(\int_\Omega M^{4(\alpha+1)/5}dx+1\right)^{5/4}$$

or

$$\begin{aligned}
\sup_{0\le t}\left(\int_\Omega M^{\delta+1}dx+1\right) \le{}& \left(\int_\Omega M_0^{\delta+1}dx+1\right)\\
&+ C(1+\delta^6)\sup_{0\le t}\left(\int_\Omega M^{4(\alpha+1)/5}dx+1\right)^{5/4}.
\end{aligned}$$

Letting

$$A(n) = \sup_{0\le t}\left(\int_\Omega M^{2(5/4)^n}dx+1\right)$$

and $B = \|M_0\|_{L^\infty}|\Omega| + 1$. We get

$$\begin{aligned} A(n) &\le B^{2(5/4)^n} + C\{1 + (5/4)^n\} A(n-1) \\ &\le \sum_{k=1}^n B^{2(5/4)^k} \left\{ \prod_{i+k}^n \left[ 3C\,(5/4)^{n-i} \right] \right\} + \left\{ \prod_{i+k}^n \left[ 3C\,(5/4)^{n-i} \right] \right\} A(0)^{(5/4)^n} \\ &\le \left[ 3C \left( \frac{5}{4} \right)^{n^3} \right] \left( B^{2(5/4)^{n+1}} + A(0)^{(5/4)^n} \right). \end{aligned}$$

Then, we have that

$$\sup_{0\le t} \|M(t)\|_{L^{2(5/4)^n}} \le \left[ 3C \left( \frac{5}{4} \right)^{n^3} \right]^{(4/5)^n/2} \left( B^{5/4} + \sup_{0\le t} \|M(t)\|_{L^2} + 1 \right) \le C.$$

As $n \to \infty$, this leads to

$$\|M\|_{L^\infty} \le \hat{C}(I + \|M_0\|_{L^\infty}).$$

This proves the $L^\infty$-bounds for $M(t,x)$.

To show well-posedness, it remains to prove the uniqueness of the solutions.

## 9.2. Uniqueness

Let $(M_1, \rho_1), (M_2, \rho_2)$ be bounded solutions to (9.1)–(9.4) with $\rho_1(0) = \rho_2(0)$, $M_1(0) = M_2(0)$. Then

$$\begin{aligned} (M_1 - M_2)_t = \triangle(M_1^{\alpha+1} - M_2^{\alpha+1}) - \nabla \cdot (M_1^\gamma \nabla \rho_1 - M_2^\gamma \nabla \rho_2) \\ + f(M_1, \rho_1) - f(M_2, \rho_2). \end{aligned} \tag{9.25}$$

Multiplying equation (9.25) by $(-\triangle)^{-1}(M_1 - M_2)$ and integrating over $\Omega$, we obtain

$$\begin{aligned} &\int_\Omega (M_1 - M_2)_t (-\triangle)^{-1}(M_1 - M_2) dx \\ &\quad = -\int_\Omega (M_1^{\alpha+1} - M_2^{\alpha+1})(M_1 - M_2) dx \\ &\quad\quad + \int_\Omega (M_1^\gamma \nabla \rho_1 - M_2^\gamma \nabla \rho_2)(-\triangle)^{-\frac{1}{2}}(M_1 - M_2) dx \\ &\quad\quad + \int_\Omega (f(M_1, \rho_1) - f(M_2, \rho_2))(-\triangle)^{-1}(M_1 - M_2) dx. \end{aligned}$$

Then,

$$\begin{aligned} \mathrm{VI} &:= \frac{1}{2}\frac{d}{dt} \int_\Omega |(-\triangle)^{-\frac{1}{2}}(M_1 - M_2)|^2 dx \\ &= -\mathrm{VII} + \mathrm{VIII} + \mathrm{IX} + \mathrm{X}, \end{aligned} \tag{9.26}$$

where

$$
\begin{aligned}
\mathrm{VII} &:= \int_\Omega (M_1^{\alpha+1} - M_2^{\alpha+1})(M_1 - M_2)dx, \\
\mathrm{VIII} &:= \int_\Omega (M_1^\gamma - M_2^\gamma)\nabla\rho \cdot (-\triangle)^{-\frac{1}{2}}(M_1 - M_2)dx, \\
\mathrm{IX} &:= \int_\Omega M_2^\gamma \nabla(\rho_1 - \rho_2)(-\triangle)^{-\frac{1}{2}}(M_1 - M_2)dx, \\
\mathrm{X} &:= \int_\Omega (f(M_1, \rho_1) - f(M_2, \rho_2))(-\Delta)^{-1}(M_1 - M_2)dx.
\end{aligned}
$$

We first estimate the integral VIII:

$$
\mathrm{VIII} \le \varepsilon \int_\Omega (M_1^\gamma - M_2^\gamma)^2 dx + C_\varepsilon \|\nabla\rho\|_\infty^2 \int_\Omega |(-\triangle)^{-\frac{1}{2}}(M_1 - M_2)|^2 dx.
$$

Combining this with $\|M_1\|_{L^\infty}, \|M_2\|_{L^\infty} \le C$ and using $\gamma \ge \frac{\alpha}{2} + 1$, we obtain

$$
\begin{aligned}
(M_1^\gamma - M_2^\gamma)^2 &\le C\left(\|M_1\|_{L^\infty}^{\gamma - \frac{\alpha}{2} - 1} + \|M_2\|_{L^\infty}^{\gamma - \frac{\alpha}{2} + 1}\right)^2 (M_1^{\frac{\alpha}{2}+1} - M_2^{\frac{\alpha}{2}+1})^2 \\
&\le C\left(\int_{M_2}^{M_1} \tau^{\frac{\alpha}{2}} d\tau\right)^2 \le C\left|\int_{M_2}^{M_1} \tau^\alpha d\tau\right| \left|\int_{M_2}^{M_1} d\tau\right| \\
&= C(M_1^{\alpha+1} - M_2^{\alpha+1})(M_1 - M_2).
\end{aligned}
$$

Hence, the expression VIII can be estimated by (choosing $\varepsilon$ sufficiently small)

(9.27)

$$
\begin{aligned}
\mathrm{VIII} &\le \frac{1}{2}\int_\Omega (M_1^{\alpha+1} - M_2^{\alpha+1})(M_1 - M_2)dx + C\|\nabla\rho\|_\infty^2 \int_\Omega |(-\triangle)^{-\frac{1}{2}}(M_1 - M_2)|^2 dx \\
&\le \frac{1}{2}\int_\Omega (M_1^{\alpha+1} - M_2^{\alpha+1})(M_1 - M_2)dx + C\int_\Omega |(-\triangle)^{-\frac{1}{2}}(M_1 - M_2)|^2 dx.
\end{aligned}
$$

Here, we used the estimate (9.22). Next, multiplying equation (9.2) by $\rho = \rho_1$ and $\rho = \rho_2$, respectively, and subtracting them from each other, we get

$$
\begin{aligned}
&\frac{1}{2}\frac{d}{dt}\int_\Omega (\rho_1 - \rho_2)^2 dx + \int_\Omega |\nabla(\rho_1 - \rho_2)|^2 dx + \int_\Omega (\rho_1 - \rho_2)^2 dx \\
&\quad = \int_\Omega (g_0(\rho_1)M_1 - g_0(\rho_2)M_2)(\rho_1 - \rho_2)dx \\
&\quad \le \int_\Omega (g_0(\rho_1) - g_0(\rho_2))(\rho_1 - \rho_2)M_1 dx \\
&\qquad + \int_\Omega g_0(\rho_2)(M_1 - M_2)(\rho_1 - \rho_2)dx \\
&\quad \le \underbrace{g_3\|M\|_{L^\infty}\int_\Omega |\rho_1 - \rho_2|^2 dx}_{\text{by } |g_0'(\rho)| \le g_3} + \frac{1}{2}\int_\Omega |\nabla(\rho_1 - \rho_2)|^2 dx \\
&\qquad + \underbrace{\frac{g_2}{2}\int_\Omega |(-\triangle)^{-\frac{1}{2}}(M_1 - M_2)|^2 dx}_{\text{by } |g_0(\rho)| \le g_2}.
\end{aligned}
$$

Then,

$$\begin{aligned}&\frac{1}{2}\frac{d}{dt}\int_\Omega |\rho_1-\rho_2|^2dx+\frac{1}{2}\int_\Omega |\nabla(\rho_1-\rho_2)|^2dx\\&\quad\le C\int_\Omega |\rho_1-\rho_2|^2dx+C\int_\Omega |(-\triangle)^{-\frac{1}{2}}(M_1-M_2)|^2dx\end{aligned}$$

or, equivalently,

$$\begin{aligned}&\frac{d}{dt}\left(e^{-Ct}\int_\Omega |\rho_1-\rho_2|^2dx\right)+\frac{1}{2}e^{-Ct}\int_\Omega |\nabla(\rho_1-\rho_2)|^2dx\\&\quad\le Ce^{-Ct}\int_\Omega |(-\triangle)^{-\frac{1}{2}}(M_1-M_2)|^2dx.\end{aligned}$$

Integrating the last inequality over $[0,t]$, we have

$$\begin{aligned}&e^{-Ct}\int_\Omega |\rho_1(t)-\rho_2(t)|^2dx+\int_0^t e^{-Cs}\int_\Omega |\nabla(\rho_1(s)-\rho_2(s))|^2dxds\\&\quad\le\int_0^t Ce^{-Cs}\int_\Omega |(-\triangle)^{-\frac{1}{2}}(M_1(s)-M_2(s))|^2dxds+\underbrace{\int_\Omega |\rho_1(0)-\rho_2(0)|^2\,dx}_{=0}.\end{aligned}$$

Hence,

$$\begin{aligned}&\int_0^t\int_\Omega |\nabla(\rho_1(s)-\rho_2(s))|^2dxds\\&\quad\le Ce^{-Ct}\int_0^t\int_\Omega |(-\triangle)^{-\frac{1}{2}}(M_1(s)-M_2(s))|^2dxds.\end{aligned}\tag{9.28}$$

The boundedness of $M_2$ guarantees

$$\begin{aligned}|\mathrm{IX}|&\le \|M_2\|_{L^\infty}^\gamma\frac{1}{2}\left(\int_\Omega |\nabla(\rho_1-\rho_2)|^2dx+\int_\Omega |(-\triangle)^{-\frac{1}{2}}(M_1-M_2)|^2dx\right)\\&\le C\left(\int_\Omega |\nabla(\rho_1-\rho_2)|^2dx+\int_\Omega |(-\triangle)^{-\frac{1}{2}}(M_1-M_2)|^2dx\right).\end{aligned}$$

Since $\tilde f(M^{\frac{\alpha}{2}+1},\rho):=f(M^{\frac{\alpha}{2}+1},\rho)-f_4M$ is a $C^1$-function by (9.10),

$$\begin{aligned}&|(f(M_1,\rho_1)-f_4M_1)-(f(M_2,\rho_2)-f_4M_2)|\\&\quad=\left|\tilde f(M_1^{\frac{\alpha}{2}+1},\rho_1)-\tilde f(M_2^{\frac{\alpha}{2}+1},\rho_2)\right|\\&\quad\le C\left(|M_1^{\frac{\alpha}{2}+1}-M_2^{\frac{\alpha}{2}+1}|+|\rho_1-\rho_2|\right).\end{aligned}$$

Therefore,

$$
\begin{aligned}
|\mathrm{X}| &= \left| \int_\Omega \left( f(M_1,\rho_1) - f(M_2,\rho_2) \right) (-\Delta)^{-1}(M_1 - M_2) dx \right| \\
&\le \varepsilon \int_\Omega \left( M_1^{\frac{\alpha}{2}+1} - M_2^{\frac{\alpha}{2}+1} \right)^2 dx + C_\varepsilon \int_\Omega (\rho_1 - \rho_2)^2 dx \\
&\quad + C_\varepsilon \int_\Omega \left| (-\Delta)^{-1}(M_1 - M_2) \right|^2 dx + f_4 \int_\Omega (M_1 - M_2)(-\Delta)^{-1}(M_1 - M_2) dx \\
&\le \varepsilon C \int_\Omega (M_1^{\frac{\alpha}{2}+1} - M_2^{\frac{\alpha}{2}+1})(M_1 - M_2) dx + C_\varepsilon \int_\Omega (\rho_1 - \rho_2)^2 dx \\
&\quad + C_\varepsilon \int_\Omega |(-\Delta)^{\frac{-1}{2}}(M_1 - M_2)|^2 dx.
\end{aligned}
$$

Taking $\varepsilon$ sufficiently small and combining with (9.26) and (9.27), we have

$$
\begin{aligned}
\frac{1}{2}\frac{d}{dt} \int_\Omega |(-\triangle)^{-\frac{1}{2}}(M_1 - M_2)|^2 dx &\le C \int_\Omega |(-\triangle)^{-\frac{1}{2}}(M_1 - M_2)|^2 dx \\
&\quad + C \int_\Omega (\rho_1 - \rho_2)^2 + |\nabla(\rho_1 - \rho_2)|^2 dx. \qquad (9.29)
\end{aligned}
$$

Using (9.28), (9.29), and the Poincaré inequality with $M_1(0) = M_2(0)$, we have

$$
\begin{aligned}
&\int_\Omega |(-\triangle)^{-\frac{1}{2}}(M_1(t) - M_2(t))|^2 dx \\
&\le C(1 + e^{Ct}) \int_0^t \int_\Omega |(-\triangle)^{-\frac{1}{2}}(M_1(s) - M_2(s))|^2 dx ds. \qquad (9.30)
\end{aligned}
$$

Let $R(t) = \int_0^t \int_\Omega |(-\triangle)^{-\frac{1}{2}}(M_1(s) - M_2(s))|^2 dx ds$. Then $R(t)$ satisfies

$$R'(t) \le Ce^{Ct}R(t) \quad \text{or} \quad R(t) \le e^{e^{Ct}} R(0) = 0.$$

Combining this with (9.29) implies that

$$\int_\Omega |(-\triangle)^{-\frac{1}{2}}(M_1(t) - M_2(t))|^2 dx = 0 \quad \text{or} \quad M_1(t) = M_2(t).$$

This proves uniqueness.

Summarizing, we can state the main result of the chapter.

THEOREM 9.1. *Let the functions $f$ and $g$ satisfy assumptions* (9.6)–(9.10) *and let the given constants $\alpha$ and $\gamma$ satisfy $0 \le \alpha \le 2(\gamma - 1)$ and $\alpha \ge \gamma + 1$. Then the initial boundary value problem* (9.1)–(9.4) *is uniquely solvable in terms of Definition* 9.1.

REMARK 9.2. One can show well-posedness under a less restrictive "balance condition" than the one proposed in Section 9.1. The new balance condition reads: $\frac{\alpha}{2} + 1 \le \gamma < \alpha$. We may also assume that the nonlinearities $f$ and $g$ satisfy the more general assumptions than those given in (9.6)–(9.10):

$$|f(M,\rho)| \le F_1(1 + |M|^\xi)^{\frac{1}{2}}, \ 0 \le \xi < \alpha - \gamma + 2, \ F_1 \in \mathbb{R}_0^+, \qquad (9.31)$$

$$f(M,\rho)M \le -F_2M^2 + F_3|M|, \ F_2 \in \mathbb{R}^+, \ F_3 \in \mathbb{R}_0^+, \qquad (9.32)$$

$$g(M,\rho) = G_1\rho + g_2(\rho)M, \ G_1 \in \mathbb{R}_0^+, \qquad (9.33)$$

$$|g_2(\rho)| \le G_3, \ G_3 \in \mathbb{R}_0^+. \qquad (9.34)$$

Well-posedness for the problem (9.1)–(9.4) under such assumptions will be proved in the next chapter for the nonautonomous case (see Theorem 10.1).

## 9.3. Dissipative estimates and the weak attractor in $L^\infty$

In this section we study the dynamics of the semigroup generated by the system (9.1)–(9.4) in terms of its global attractor in the weak topology of $L^\infty(\Omega)$.

To this end we derive dissipative estimates in various phase spaces for the solutions of the problem (9.1)–(9.4), which in turn lead to a dissipative estimate in $L^\infty(\Omega)$ for both $M$ and $\rho$.

Our main result can be summarized as follows:

THEOREM 9.2. *Let the functions $f$ and $g$ satisfy assumptions* (9.31)–(9.33) *and let the given constants $\alpha$ and $\gamma$ satisfy $\gamma \in \left[\frac{\alpha}{2}+1, \alpha\right)$. Then the following dissipative estimate holds for the initial boundary value problem* (10.1)–(10.4)*:*

$$
\begin{aligned}
&||M(t)||_{L^\infty(\Omega)} + ||\rho(t)||_{W^{1,\infty}(\Omega)} \\
&\qquad \le C_\infty \left(||M_0||_{L^\infty(\Omega)} + ||\rho_0||_{W^{1,\infty}(\Omega)}\right)^{r_\infty} \cdot e^{-\omega_\infty t} + D_\infty,
\end{aligned} \tag{9.35}
$$

*where the positive constants $C_\infty, r_\infty, \omega_\infty, D_\infty$ depend only on $\alpha, \gamma, f,$ and $g$ and are independent of $M_0, \rho_0,$ or $t$.*

PROOF. We start with rewriting the equation (9.1) in the following way:

$$
\dot{M} = \nabla \cdot \left(|M|^\gamma \nabla \left(\frac{1}{\alpha-\gamma+1} M|M|^{(\alpha-\gamma+1)-1} - \rho\right)\right) + f(M,\rho). \tag{9.36}
$$

In order to derive our first a priori estimate, we multiply this equation by

$$
\left(\frac{1}{\alpha-\gamma+1} M|M|^{(\alpha-\gamma+1)-1} - \rho\right)
$$

and integrate (formally) over $\Omega$ to get

$$
\begin{aligned}
&\left(\dot{M}, \frac{1}{\alpha-\gamma+1} M|M|^{(\alpha-\gamma+1)-1} - \rho\right) \\
&\quad = -\left(|M|^\gamma, \left|\nabla \left(\frac{1}{\alpha-\gamma+1} M|M|^{(\alpha-\gamma+1)-1} - \rho\right)\right|^2\right) \\
&\qquad + \left(f(M,\rho), \frac{1}{\alpha-\gamma+1} M|M|^{(\alpha-\gamma+1)-1} - \rho\right) \\
&\quad \le \left(f(M,\rho), \frac{1}{\alpha-\gamma+1} M|M|^{(\alpha-\gamma+1)-1} - \rho\right) \\
&\Leftrightarrow \frac{d}{dt}\left(\frac{1}{(\alpha-\gamma+1)(\alpha-\gamma+2)} \left\| |M|^{\frac{\alpha-\gamma+2}{2}} \right\|^2 - (M,\rho)\right) \\
&\quad \le \left(f(M,\rho), \frac{1}{\alpha-\gamma+1} M|M|^{(\alpha-\gamma+1)-1} - \rho\right) - (\dot{\rho}, M)
\end{aligned} \tag{9.37}
$$

and we multiply equation (9.2) by $(\dot\rho+\rho-1)$ in the same sense as above to get

$$\|\dot\rho\|^2+\frac{1}{2}\frac{d}{dt}||\rho-1||^2=-\frac{1}{2}\frac{d}{dt}||\nabla\rho||^2-||\nabla\rho||^2-(g(M,\rho),\dot\rho+\rho-1)$$

$$\Leftrightarrow\frac{1}{2}\frac{d}{dt}\left(||\nabla\rho||^2+||\rho-1||^2\right)=-||\nabla\rho||^2-\|\dot\rho\|^2-(g(M,\rho),\dot\rho+\rho-1). \tag{9.38}$$

Adding the inequalities (9.37) and (9.38) together, we obtain

(9.39)

$$\frac{d}{dt}\left(\frac{1}{(\alpha-\gamma+1)(\alpha-\gamma+2)}\left\||M|^{\frac{\alpha-\gamma+2}{2}}\right\|^2-(M,\rho)+\frac{1}{2}||\nabla\rho||^2+\frac{1}{2}||\rho-1||^2\right)$$
$$\le\left(f(M,\rho),\frac{1}{\alpha-\gamma+1}M|M|^{(\alpha-\gamma+1)-1}-\rho\right)-||\nabla\rho||^2-(\dot\rho,M)-\|\dot\rho\|^2$$
$$-(g(M,\rho),\dot\rho+\rho-1).$$

We first consider the term containing $g(M,\rho)=G_1\rho+g_2(\rho)M$. Then,

$$\begin{aligned}-(G_1\rho,\dot\rho+\rho-1)&=-\frac{1}{2}\frac{d}{dt}\left(G_1||\rho||^2\right)-G_1\left(||\rho||^2-(1,\rho)\right)\\&\le-\frac{1}{2}\frac{d}{dt}\left(G_1||\rho||^2\right)-(1-\varepsilon)G_1||\rho||^2+\frac{1}{4\varepsilon}G_1|\Omega|\end{aligned} \tag{9.40}$$

and

$$\begin{aligned}-(g_2(\rho)M,\dot\rho+\rho-1)&\le\ \varepsilon\|\dot\rho\|^2+\varepsilon||\rho-1||^2+\frac{1}{2\varepsilon}||g_2(\rho)M||^2\\&\underset{(9.34)}{\le}\ \varepsilon\|\dot\rho\|^2+\varepsilon||\rho-1||^2+\frac{1}{2\varepsilon}G_3^2||M||^2,\end{aligned} \tag{9.41}$$

where $|\Omega|$ denotes the volume of $\Omega$. By combining (9.40) and (9.41) with the inequality

$$-(\dot\rho,M)-\|\dot\rho\|^2\le\frac{1}{2}||M||^2-\frac{1}{2}\|\dot\rho\|^2 \tag{9.42}$$

and by choosing $\varepsilon\le\frac{1}{2}$, we have

(9.43)

$$-(\dot\rho,M)-\|\dot\rho\|^2-(g(M,\rho),\dot\rho+\rho-1)$$
$$\le-\frac{1}{2}\frac{d}{dt}\left(G_1||\rho||^2\right)-(1-\varepsilon)G_1||\rho||^2+\varepsilon||\rho-1||^2+\frac{1}{4\varepsilon}G_1|\Omega|-\left(\frac{1}{2}-\varepsilon\right)\|\dot\rho\|^2$$
$$+\left(\frac{1}{2}+\frac{1}{2\varepsilon}G_3^2\right)||M||^2$$
$$\underset{\varepsilon\le\frac{1}{2}}{\le}-\frac{1}{2}\frac{d}{dt}\left(G_1||\rho||^2\right)-(1-\varepsilon)G_1||\rho||^2+\varepsilon||\rho-1||^2+\frac{1}{4\varepsilon}G_1|\Omega|$$
$$+\left(\frac{1}{2}+\frac{1}{2\varepsilon}G_3^2\right)||M||^2.$$

Further, we can estimate the terms with $f$ from (9.39) in the following way:

$$\begin{aligned}\left(f(M,\rho), M|M|^{(\alpha-\gamma+1)-1}\right) &\underset{(9.32)}{\leq} \left(-F_2M^2+F_3|M|, |M|^{(\alpha-\gamma+1)-1}\right)\\ &= -F_2\left\||M|^{\frac{\alpha-\gamma+2}{2}}\right\|^2+F_3\left\||M|^{\frac{\alpha-\gamma+1}{2}}\right\|^2,\end{aligned} \tag{9.44}$$

$$\begin{aligned}-\left(f(M,\rho),\rho\right) &\underset{(9.31)}{\leq} \varepsilon||\rho||^2+\frac{1}{4\varepsilon}F_1^2\left(|\Omega|+\left\||M|^{\frac{\xi}{2}}\right\|^2\right)\\ &\leq 2\varepsilon||\rho-1||^2+\left(2\varepsilon+\frac{1}{4\varepsilon}F_1^2\right)|\Omega|+\frac{1}{4\varepsilon}F_1^2\left\||M|^{\frac{\xi}{2}}\right\|^2.\end{aligned} \tag{9.45}$$

Using the inequalities (9.43)–(9.45), we conclude from (9.39) that

$$\begin{aligned}&\frac{d}{dt}\left(\frac{1}{(\alpha-\gamma+1)(\alpha-\gamma+2)}\left\||M|^{\frac{\alpha-\gamma+2}{2}}\right\|^2-(M,\rho)\right.\\ &\qquad\left.+\frac{1}{2}||\nabla\rho||^2+\frac{1}{2}||\rho-1||^2+\frac{1}{2}G_1||\rho||^2\right)\\ &\leq -F_2\left\||M|^{\frac{\alpha-\gamma+2}{2}}\right\|^2+F_3\left\||M|^{\frac{\alpha-\gamma+1}{2}}\right\|^2+\frac{1}{4\varepsilon}F_1^2\left\||M|^{\frac{\xi}{2}}\right\|^2+\left(\frac{1}{2}+\frac{1}{2\varepsilon}G_3^2\right)||M||^2\\ &\quad -||\nabla\rho||^2-(1-\varepsilon)G_1||\rho||^2+3\varepsilon||\rho-1||^2+\left(2\varepsilon+\frac{1}{4\varepsilon}G_1+\frac{1}{4\varepsilon}F_1^2\right)|\Omega|.\end{aligned} \tag{9.46}$$

In order to shorten the formulas, we introduce a new variable:

$$\begin{aligned}\varphi := &\frac{1}{(\alpha-\gamma+1)(\alpha-\gamma+2)}\left\||M|^{\frac{\alpha-\gamma+2}{2}}\right\|^2-(M,\rho)\\ &+\frac{1}{2}||\nabla\rho||^2+\frac{1}{2}||\rho-1||^2+\frac{1}{2}G_1||\rho||^2.\end{aligned} \tag{9.47}$$

$|M|^{\frac{\alpha-\gamma+2}{2}}$ is the leading $M$-power in the expression (9.47) due to the assumptions made on $\alpha, \gamma$, and $\xi$, and we also have the estimate

$$(M,\rho) \leq \varepsilon||\rho||^2+\frac{1}{4\varepsilon}||M||^2 \tag{9.48}$$

for all $\varepsilon > 0$. Moreover, applying the Poincaré and Hölder inequalities and adjusting the constant $\varepsilon$, we can deduce from (9.46) the inequality

$$\dot{\varphi} \leq -A_1\varphi + A_2 \tag{9.49}$$

for some $A_1 \in \mathbb{R}^+$, $A_2 \in \mathbb{R}_0^+$ and finally obtain our first dissipative estimate: setting for short

$$\begin{aligned}y_{\delta_0} &:= ||M||_{\delta_0}^{\delta_0}+1+||\nabla\rho||^2,\\ \delta_0 &:= \alpha-\gamma+2 > 2,\end{aligned}$$

then with (9.49)

$$y_{\delta_0}(t) \leq C_{y_{\delta_0}} y_{\delta_0}(0)e^{-\omega_{y_{\delta_0}}t}+D_{y_{\delta_0}} \tag{9.50}$$

for some $C_{y_{\delta_0}}, \omega_{y_{\delta_0}}, D_{y_{\delta_0}}$.

In what follows, we use (9.50) to obtain several intermediate dissipative estimates for $M$ and $\rho$, which in turn lead to an $L^\infty$-dissipative estimate.

NOTATION 9.1. *For the sake of convenience, we assume that the constants $B_i$ (below) are only dependent upon the parameters of the problem (9.31)–(9.33), that is, upon the constants $\alpha, \gamma, F_2, F_3, G_1, G_3$ and the domain $\Omega$, and* **not** *upon the initial data $M_0, \rho_0$, or $t$, or (unless stated otherwise) any other parameters.*

The following observation, which is an implication from the theory of abstract parabolic evolution equations (see [95]), will be helpful in further discussion.

Having a $\delta \in (2, \infty)$ fixed, consider the unbounded operator

$$\Delta : \; L^{\delta}(\Omega) \to L^{\delta}(\Omega)$$

equipped with the domain

$$D(\Delta) := \left\{ u \in W_0^{1,\delta}(\Omega) : \; \Delta u \in L^{\delta}(\Omega) \right\}.$$

It is known (see [95]) that this operator generates an analytic semigroup $e^{t\Delta}$ and its spectrum lies entirely in $\{\lambda \in \mathbb{R} : \lambda \leq -\beta\}$ for some $\beta > 0$. As such it has the following properties:

$$(-\Delta)^{\mu} e^{t\Delta} = e^{t\Delta}(-\Delta)^{\mu}, \tag{9.51}$$

$$||e^{t\Delta}(-\Delta)^{\mu}||_{\delta} \leq A^{\mu,\delta} e^{-\beta t} t^{-\mu} \tag{9.52}$$

for all $t > 0$ and $\mu > 0$ for some constants $A^{\mu,\delta}$ that depend only on $\mu, \delta$, and the domain $\Omega$. Now, equation (9.2) can be rewritten as

$$\frac{d}{dt}(\rho - 1) = \Delta(\rho - 1) - g(M, \rho)$$

and can thus be regarded as an abstract parabolic evolution equation with respect to $\rho - 1$. Therefore, for all $t > 0$,

$$\rho(t) - 1 = e^{t\Delta}(\rho_0 - 1) - \int_0^t e^{(t-s)\Delta} g(M(s), \rho(s))\, ds \tag{9.53}$$

and applying the operator $\nabla$ to both sides of (9.53) and making use of property (9.51), we obtain

$$\nabla \rho(t) = e^{t\Delta} \nabla \rho_0 - \int_0^t \nabla \left( e^{(t-s)\Delta} g(M(s), \rho(s)) \right) ds. \tag{9.54}$$

The initial value $\rho_0$ is assumed to be sufficiently smooth, so

$$||\nabla \rho_0||_{\delta} < \infty. \tag{9.55}$$

What remains is to estimate the $\delta$-norm of the integral from (9.54) with the help of (9.52) and the assumptions on $g$. Choosing $\mu \in \left(\frac{1}{2}, 1\right)$ and $\hat{\delta} \geq 1$ such that $W^{2\mu,\hat{\delta}}(\Omega) \hookrightarrow W^{1,\delta}(\Omega)$, we obtain

$$\begin{aligned} &\left\| \int_0^t \nabla \left( e^{(t-s)\Delta} g(M(s), \rho(s)) \right) ds \right\|_{\delta} \\ &\leq \int_0^t \left\| (-\Delta)^{\mu} \left( e^{(t-s)\Delta} g(M(s), \rho(s)) \right) \right\|_{\hat{\delta}} ds \\ &\leq A^{\mu,\hat{\delta}} \int_0^t e^{-\beta(t-s)} (t-s)^{-\mu} \left( G_1 ||\rho(s)||_{\hat{\delta}} + G_3 ||M(s)||_{\hat{\delta}} \right) ds. \end{aligned} \tag{9.56}$$

Leaving this result for a moment and returning to equation (9.1), we multiply this equation by $M|M|^{\delta-1}$ for an arbitrary $\delta \geq \alpha - \gamma + 1$, so that all occurring powers remain nonnegative, and we (formally) integrate over $\Omega$:

$$\left(\dot{M}, M|M|^{\delta-1}\right) = \left(\nabla \cdot (|M|^{\alpha} \nabla M) - \nabla \cdot (|M|^{\gamma} \nabla \rho) + f(M, \rho), M|M|^{\delta-1}\right).$$

It follows that

$$\begin{aligned} \frac{1}{\delta+1} \frac{d}{dt} \left\| |M|^{\frac{\delta+1}{2}} \right\|^2 = & - \frac{4\delta}{(\alpha+\delta+1)^2} \left\| \nabla |M|^{\frac{\alpha+\delta+1}{2}} \right\|^2 \\ & + \frac{2\delta}{\alpha+\delta+1} \left( \nabla |M|^{\frac{\alpha+\delta+1}{2}}, |M|^{\gamma - \frac{\alpha}{2} + \frac{\delta-1}{2}} \nabla \rho \right) \\ & + (f(M, \rho), M|M|^{\delta-1}). \end{aligned} \tag{9.57}$$

Denote $\vartheta(\delta) := \frac{\gamma - \frac{\alpha}{2} + \frac{\delta-1}{2}}{\frac{\alpha+\delta+1}{2}}$. Then $\vartheta(\delta) < 1$ holds due to the assumption $\alpha > \gamma$ that we made. Applying the Hölder inequality we obtain

$$\begin{aligned} \left( \nabla |M|^{\frac{\alpha+\delta+1}{2}}, |M|^{\gamma - \frac{\alpha}{2} + \frac{\delta-1}{2}} \nabla \rho \right) & = \left( \nabla |M|^{\frac{\alpha+\delta+1}{2}}, |M|^{\vartheta(\delta) \frac{\alpha+\delta+1}{2}} \nabla \rho \right) \\ & \leq \|1\|_{\frac{6}{1-\theta(\delta)}} \left\| \nabla |M|^{\frac{\alpha+\delta+1}{2}} \right\| \left\| |M|^{\frac{\alpha+\delta+1}{2}} \right\|_6^{\vartheta(\delta)} \|\nabla \rho\|_3 \\ & \leq B_1 \left\| \nabla |M|^{\frac{\alpha+\delta+1}{2}} \right\|^{1+\vartheta(\delta)} \|\nabla \rho\|_3 . \end{aligned} \tag{9.58}$$

For the last inequality the embedding $H^1(\Omega) \hookrightarrow L^6(\Omega)$ has been used. Further, we use once more the Hölder inequality and the assumptions on the function $f$ and write

$$(f(M, \rho), M|M|^{\delta-1}) \leq - F_2 \left\| |M|^{\frac{\delta+1}{2}} \right\|^2 + F_3 \left\| |M|^{\frac{\delta}{2}} \right\|^2 \tag{9.59}$$

$$\leq - F_2 \left\| |M|^{\frac{\delta+1}{2}} \right\|^2 + F_3 \|1\|_{\delta+1} \left( \left\| |M|^{\frac{\delta+1}{2}} \right\|^2 \right)^{\frac{\delta}{\delta+1}} . \tag{9.60}$$

We can conclude from (9.57) using (9.58) and (9.60) that

$$\begin{aligned} \frac{1}{\delta+1} \frac{d}{dt} \left\| |M|^{\frac{\delta+1}{2}} \right\|^2 \leq & - \frac{4\delta}{(\alpha+\delta+1)^2} \left\| \nabla |M|^{\frac{\alpha+\delta+1}{2}} \right\|^2 \\ & + \frac{2\delta}{\alpha+\delta+1} B_1 \left\| \nabla |M|^{\frac{\alpha+\delta+1}{2}} \right\|^{1+\vartheta(\delta)} \|\nabla \rho\|_3 \\ & - F_2 \left\| |M|^{\frac{\delta+1}{2}} \right\|^2 + F_3 \|1\|_{\delta+1} \left( \left\| |M|^{\frac{\delta+1}{2}} \right\|^2 \right)^{\frac{\delta}{\delta+1}} . \end{aligned}$$

Since $1 + \vartheta(\delta) < 2$, it follows from the Young inequality that

$$\begin{aligned} \frac{1}{\delta+1} \frac{d}{dt} \left\| |M|^{\frac{\delta+1}{2}} \right\|^2 \leq & -F_2 \left\| |M|^{\frac{\delta+1}{2}} \right\|^2 + F_3 \|1\|_{\delta+1} \left( \left\| |M|^{\frac{\delta+1}{2}} \right\|^2 \right)^{\frac{\delta}{\delta+1}} \\ & + B_2(\delta) \|\nabla \rho\|_3^{\frac{2}{1-\vartheta(\delta)}} , \end{aligned} \tag{9.61}$$

where $B_2(\delta) = \frac{1-\vartheta(\delta)}{2} \left( \frac{2\delta}{\alpha+\delta+1} B_1 \right)^{\frac{2}{1-\vartheta(\delta)}} \left( \frac{4\delta}{(\alpha+\delta+1)^2} \frac{2}{1+\vartheta(\delta)} \right)^{-\frac{1+\vartheta(\delta)}{1-\vartheta(\delta)}}$; therefore, this constant depends only on $\delta$ and the parameters of the problem.

Next, we return to equality (9.57) to repeat the whole procedure once more but this time being more precise about the estimates being made and using the regularity achieved up to this point. First, due to (9.59) and two obvious inequalities, we have

$$
\begin{aligned}
\frac{d}{dt}\left\||M|^{\frac{\delta+1}{2}}\right\|^2 =& -\frac{4\delta(\delta+1)}{(\alpha+\delta+1)^2}\left\|\nabla|M|^{\frac{\alpha+\delta+1}{2}}\right\|^2 \\
&+\frac{2\delta(\delta+1)}{\alpha+\delta+1}\left(\nabla|M|^{\frac{\alpha+\delta+1}{2}}, |M|^{\gamma-\frac{\alpha}{2}+\frac{\delta-1}{2}}\nabla\rho\right) \\
&+(\delta+1)(f(M,\rho), M|M|^{\delta-1}) \\
\le& -B_3\left\|\nabla|M|^{\frac{\alpha+\delta+1}{2}}\right\|^2 \\
&+(\delta+1)B_4||\nabla\rho||_\infty\left\|\nabla|M|^{\frac{\alpha+\delta+1}{2}}\right\|\left\||M|^{\frac{\alpha+\delta+1}{2}}\right\|^{\vartheta(\delta)} \\
&-(\delta+1)F_2\left\||M|^{\frac{\delta+1}{2}}\right\|^2+(\delta+1)B_5F_3\left\||M|^{\frac{\alpha+\delta+1}{2}}\right\|^{2\zeta}
\end{aligned}
\tag{9.62}
$$

for $\delta \ge \alpha-\gamma+1$ with $\zeta = \frac{\delta}{\alpha+\delta+1}$. Taking into account a special case of the interpolation inequality for Sobolev spaces (see [2]):

$$||v|| \le C_\kappa||\nabla v||^{\frac{3}{5}}||v||_1^{\frac{2}{5}},$$

where the constant $C_\kappa$ depends only on the domain $\Omega$, we obtain with the help of the Young inequality

$$
\begin{aligned}
&(\delta+1)||\nabla v||||v||^{\vartheta(\delta)} \\
&\quad\le (\delta+1)C_\kappa^{\vartheta(\delta)}||\nabla v||^{1+\vartheta(\delta)\frac{3}{5}}||v||_1^{\vartheta(\delta)\frac{2}{5}} \\
&\quad\le C_\kappa^{\vartheta(\delta)}\left(\varepsilon||\nabla v||^2+B_6(\varepsilon)(\delta+1)^{\frac{2}{1-\vartheta(\delta)\frac{3}{5}}}||v||_1^{\frac{2\vartheta(\delta)\frac{2}{5}}{1-\vartheta(\delta)\frac{3}{5}}}\right)
\end{aligned}
\tag{9.63}
$$

and

$$
\begin{aligned}
(\delta+1)F_3||v||^{2\zeta} &\le (\delta+1)F_3C_\kappa^{2\zeta}||\nabla v||^{2\zeta\frac{3}{5}}||v||_1^{2\zeta\frac{2}{5}} \\
&\le C_\kappa^{2\zeta}\left(\varepsilon||\nabla v||^2+B_7(\varepsilon)\left(F_3(\delta+1)\right)^{\frac{1}{1-\zeta\frac{3}{5}}}||v||_1^{\frac{2\zeta\frac{2}{5}}{1-\zeta\frac{3}{5}}}\right),
\end{aligned}
\tag{9.64}
$$

where $B_6(\varepsilon)$ and $B_7(\varepsilon)$ depend only on $\varepsilon$ and the parameters of the problem. With the Hölder inequality we also have

$$\left\||M|^{\frac{\alpha+\delta+1}{2}}\right\|_1 \le \left\||M|^{\frac{\alpha}{2}}\right\|_{\frac{q}{q-1}}\left\||M|^{\frac{\delta+1}{2}}\right\|_q \tag{9.65}$$

for some $q \in (1,2)$ independent of $\delta$. Combining (9.63)–(9.64) for $v := |M|^{\frac{\alpha+\delta+1}{2}}$ with (9.65) and choosing $\varepsilon$ small enough, depending only on $C_\kappa$ and $B_3$ (thus it depends only on the parameters of the problem), we can conclude from (9.62) that

$$\begin{aligned}\frac{d}{dt}\left\||M|^{\frac{\delta+1}{2}}\right\|^2 &\leq B_8\left(\|\nabla\rho\|_\infty(\delta+1)\right)^{\frac{2}{1-\vartheta(\delta)\frac{3}{5}}}\left(\left\||M|^{\frac{\alpha}{2}}\right\|_{\frac{q}{q-1}}\left\||M|^{\frac{\delta+1}{2}}\right\|_q\right)^{\frac{2\vartheta(\delta)\frac{2}{5}}{1-\vartheta(\delta)\frac{3}{5}}}\\ &\quad+B_8\left(F_3(\delta+1)\right)^{\frac{1}{1-\zeta\frac{3}{5}}}\left(\left\||M|^{\frac{\alpha}{2}}\right\|_{\frac{q}{q-1}}\left\||M|^{\frac{\delta+1}{2}}\right\|_q\right)^{\frac{2\zeta\frac{2}{5}}{1-\zeta\frac{3}{5}}}\\ &\quad-F_2(\delta+1)\left\||M|^{\frac{\delta+1}{2}}\right\|^2\end{aligned}$$

for $\delta\geq\alpha-\gamma+1$. Since $\vartheta(\delta),\zeta\in(0,1)$, it follows for all $\delta\geq\alpha-\gamma+2$ that

$$\begin{aligned}\frac{d}{dt}\left(\|M\|_\delta^\delta+1\right)&\leq B_8\delta^5\left(\|\nabla\rho(s)\|_\infty+1\right)^5\|M\|^\alpha_{\frac{\alpha}{2}\frac{q}{q-1}}\left(\|M\|_{q\delta/2}^{q\delta/2}+1\right)^{\frac{2}{q}}\\ &\quad-F_2\delta\left(\|M\|_\delta^\delta+1\right)\end{aligned}$$

and once more we get an integral inequality for $\|M(t)\|_\delta^\delta+1$:

$$\begin{aligned}\|M(t)\|_\delta^\delta+1\leq B_8&\int_0^t e^{-\delta F_2(t-s)}\delta^5\left(\|\nabla\rho(s)\|_\infty+1\right)^5\|M(s)\|^\alpha_{\frac{\alpha}{2}\frac{q}{q-1}}\\ &\cdot\left(\|M(s)\|_{q\delta/2}^{q\delta/2}+1\right)^{\frac{2}{q}}ds+e^{-\delta F_2 t}\left(\|M_0\|_\delta^\delta+1\right).\end{aligned}\tag{9.66}$$

Now we are ready to derive more dissipative estimates for the problem (9.1)–(9.4). We will extensively use the following.

LEMMA 9.1. *Let $z_1,z_2,z_3:[0,+\infty)\to[0,+\infty)$ be functions such that*

$$\begin{aligned}&z_1(t)\leq\psi_1(z_1(0))e^{-\omega_1 t}+D_1,\\ &z_2(t)\leq\psi_2(z_2(0))e^{-\omega_2 t}+D_2,\\ &z_3(t)\leq z_3(0)e^{-\omega_3 t}+\int_0^t e^{-\omega_3(t-s)}d_3(t,s)z_1(s)\,ds,\\ &z_1(0),z_2(0),z_3(0)\geq 1,\end{aligned}\tag{9.67}$$

*for some constants $\omega_1,\omega_2,\omega_3>0$ and $D_1,D_2\geq 1$, some nondecreasing functions $\psi_1,\psi_2:[1,+\infty)\to[1,+\infty)$, and some $d_3\in L^\infty(\mathbb{R}_0^+,L_b^1(\mathbb{R}_0^+))$, where*

$$L_b^1(\mathbb{R}_0^+)=\left\{u\in L^1_{loc}(\mathbb{R}_0^+):||u||_{L_b^1(\mathbb{R}_0^+)}:=\sup_{x_0\in\mathbb{R}_0^+}||u||_{L^1([x_0,x_0+1])}<\infty\right\}.$$

*Then:*

(1) $(z_1+z_2)(t)\leq(\psi_1+\psi_2)((z_1+z_2)(0))e^{-\min\{\omega_1,\omega_2\}t}+D_1+D_2$.
(2) $z_1z_2(t)\leq 3D_1D_2\psi_1\psi_2(z_1z_2(0))e^{-\min\{\omega_1,\omega_2\}t}+D_1D_2$.
(3) $z_1^\sigma(t)\leq\max\left\{1,2^{\sigma-1}\right\}\left(\psi_1^\sigma(z_1(0))e^{-\sigma\omega_1 t}+D_1^\sigma\right)\quad\forall\sigma>0$.
(4) *For $\omega_1\neq\omega_3$*

$$\begin{aligned}z_3(t)\leq&\left(\psi_1(z_1(0))\frac{1}{1-e^{-|\omega_1-\omega_3|}}e^{-\min\{\omega_1,\omega_3\}t}+D_1\frac{1}{1-e^{-\omega_3}}\right)\\ &\cdot||d_3||_{L^\infty(\mathbb{R}_0^+,L_b^1(\mathbb{R}_0^+))}+z_3(0)e^{-\omega_3 t}\end{aligned}\tag{9.68}$$

*and for* $\omega_3 = \omega_1$

$$z_3(t) \leq \left(\psi_1(z_1(0)) \lceil t \rceil e^{-\omega_1 t} + D_1 \frac{1}{1-e^{-\omega_1}}\right) ||d_3||_{L^\infty(\mathbb{R}_0^+, L_b^1(\mathbb{R}_0^+))} + z_3(0)e^{-\omega_1 t}.$$

*For* $\omega_1 < \omega_3$ *we also have*

$$z_3(t) \leq z_3(0)e^{-\omega_3 t} + z_1(t) \int_0^t e^{-(\omega_3-\omega_1)(t-s)} d_3(t,s)\, ds. \tag{9.69}$$

(See Lemma 9.2 from the second appendix (Section 9.5) for some details regarding the proof.)

Lemma 9.2 is very useful in our situation. It actually shows that the "dissipative property" is preserved under standard operations (addition, multiplication, raising to a power, and integration).

To shorten the formulas, let us set

$$h_1 := ||\nabla\rho||_3 + 1,$$
$$h_2 := ||\nabla\rho||_\infty + 1,$$
$$u_\delta := ||M||_\delta^\delta + 1, \quad \delta \in [1,\infty).$$

Observe that particular powers of $y_{\delta_0}$ and $h_1$, $h_2$ and $u_\delta$ for sufficiently large $\delta$, $u_{\widetilde{\delta_1}}$ for some $\widetilde{\delta_1} > 6$ and $h_2$ can be connected with one another by the inequalities of the type (9.67) in the same manner that $z_1$ and $z_3$ from Lemma 9.1 are (see (9.50), (9.54)–(9.56), (9.61)). From Lemma 9.1 we can conclude that all of them dissipate exponentially with $t$:

$$h_1(t) \leq C_{h_1}(h_1 + y_{\delta_0})^{r_{h_1}}(0)e^{-\omega_{h_1} t} + D_{h_1}, \tag{9.70}$$

$$h_2(t) \leq C_{h_2}(h_2 + u_{\widetilde{\delta_1}})^{r_{h_2}}(0)e^{-\omega_{h_2} t} + D_{h_2}, \tag{9.71}$$

$$u_\delta(t) \leq U\left(u_\delta(0) + C_{u_\delta}(h_1 + y_{\delta_0})^{r\delta}(0)\right) e^{-\frac{F_2}{2}\delta t} + D_{u_\delta} =: \widetilde{u}_\delta(t), \tag{9.72}$$

where the coefficients that appear depend on the parameters of the problem and only the coefficients $C_{u_\delta}$ and $D_{u_\delta}$ depend on $\delta$ as well. We especially emphasize that $r$ is independent of $\delta$ (it will be crucial for the existence of the uniform dissipative estimate). Indeed, for $h_1$ and $h_2$ the statement follows without great difficulty due to the fact that for the function $d_\mu(t,s) := (t-s)_+^{-\mu}$ for $\mu \in \left(\frac{1}{2}, 1\right)$ the condition $\sup_{t>0} ||d_\mu(t,\cdot)||_{L_b^1(\mathbb{R}_0^+)} < \infty$ is satisfied. Further, with (9.61) we have

$$\frac{1}{\delta}\dot{u}_\delta \leq -F_2 u_\delta + F_3|\Omega| u_\delta^{\frac{\delta-1}{\delta}} + B_2(\delta) h_1^{\frac{2}{1-\vartheta(\delta)}}. \tag{9.73}$$

Recall that $\vartheta(\delta) = \frac{\gamma - \frac{\alpha}{2} + \frac{\delta-2}{2}}{\frac{\alpha+\delta}{2}}$ and consequently $\frac{2}{1-\vartheta(\delta)} = \frac{\alpha+\delta}{\alpha-\gamma+1} \leq B_9\delta$ for some $B_9$ and $\delta \geq \delta_*$ sufficiently large. Now, the Young inequality yields

$$u_\delta^{\frac{\delta-1}{\delta}} = (\varepsilon u_\delta)^{\frac{\delta-1}{\delta}} \varepsilon^{-\frac{\delta-1}{\delta}} \leq \frac{\delta-1}{\delta}\varepsilon u_\delta + \frac{1}{\delta}\varepsilon^{-(\delta-1)};$$

therefore, it follows from (9.73) that

$$\begin{aligned}\dot{u}_\delta \leq & -\delta\left(F_2-\varepsilon F_3|\Omega|\frac{\delta-1}{\delta}\right)u_\delta+\varepsilon^{-(\delta-1)}F_3|\Omega|+\delta B_2(\delta)h_1^{B_9\delta} \\ \leq & -\delta\frac{F_2}{2}u_\delta+\varepsilon^{-(\delta-1)}F_3|\Omega|+\delta B_2(\delta)h_1^{B_9\delta}\end{aligned}$$

for $\varepsilon$ small (depends only on the parameters of the problem). Gronwall's lemma then yields

$$\begin{aligned}u_\delta(t) \leq & \int_0^t e^{-\delta\frac{F_2}{2}(t-s)}\left(\varepsilon^{-(\delta-1)}F_3|\Omega|+\delta B_2(\delta)h_1^{B_9\delta}(s)\right)ds \\ & + e^{-\delta\frac{F_2}{2}t}u_\delta(0).\end{aligned} \tag{9.74}$$

The dissipative estimate (9.72) now follows from (9.68) and the dissipative estimate (9.70). Now, from inequality (9.66), we can conclude that

$$u_\delta(t) \leq e^{-\delta F_2 t}u_\delta(0)+B_8\int_0^t e^{-\delta F_2(t-s)}\delta^5 h_2^5(s)\widetilde{u}^{\frac{2q}{q-1}}_{\frac{\alpha}{2}\frac{q}{q-1}}(s)\widetilde{u}^{\frac{2}{q}}_{\frac{q}{2}\delta}(s)\,ds. \tag{9.75}$$

Taking into account that $u^{\frac{2}{q}}_{\frac{q}{2}\delta}$ dissipates with $e^{-\delta\frac{F_2}{2}t}$ and that $h_2^5\widetilde{u}^{\frac{2q}{q-1}}_{\frac{\alpha}{2}\frac{q}{q-1}}$ dissipates with an exponent independent of $\delta$, we consecutively apply (9.69) to (9.75) and get

$$\begin{aligned}u_\delta(t) \leq & e^{-\delta\frac{F_2}{2}t}u_\delta(0)+B_8\widetilde{u}^{\frac{2}{q}}_{\frac{q}{2}\delta}(t)\int_0^t e^{-\delta\frac{F_2}{2}(t-s)}\delta^5 h_2^5(s)\widetilde{u}^{\frac{2q}{q-1}}_{\frac{\alpha}{2}\frac{q}{q-1}}(s)\,ds \\ \leq & e^{-\delta\frac{F_2}{2}t}u_\delta(0)+\frac{2}{F_2}B_8\delta^4 h_2^5(t)\widetilde{u}^{\frac{2q}{q-1}}_{\frac{\alpha}{2}\frac{q}{q-1}}(t)\widetilde{u}^{\frac{2}{q}}_{\frac{q}{2}\delta}(t)\end{aligned}$$

for $\delta \geq \delta_{**}$ sufficiently large. The bound $\delta_{**}$ depends only on the parameters of the problem. Therefore, we may assume that

$$\widetilde{u}_\delta(t) = e^{-\delta\frac{F_2}{2}t}u_\delta(0)+B_{10}\delta^4 h_2^5(t)\widetilde{u}^{\frac{2q}{q-1}}_{\frac{\alpha}{2}\frac{q}{q-1}}(t)\widetilde{u}^{\frac{2}{q}}_{\frac{q}{2}\delta}(t). \tag{9.76}$$

Since

$$u_\delta(0) = ||M_0||_\delta^\delta+1 \leq ||M_0||_\infty^\delta|\Omega|+1,$$

we conclude from (9.76) that for

$$A_\delta(t) := \widetilde{u}_\delta(t)\left(\frac{e^{\frac{F_2}{2}t}}{||M_0||_\infty+1}\right)^\delta+1, \tag{9.77}$$

it is true that

$$A_\delta(t) \leq B_{11}\delta^4 h_2^5(t)\widetilde{u}^{\frac{2q}{q-1}}_{\frac{\alpha}{2}\frac{q}{q-1}}(t)A^{\frac{2}{q}}_{\frac{q}{2}\delta}(t).$$

One can then show by induction that

$$\begin{aligned}A^{\left(\frac{q}{2}\right)^n}_{\left(\frac{2}{q}\right)^n\delta_{**}}(t) \leq & \left(B_{11}\delta_{**}^4 h_2^5(t)\widetilde{u}^{\frac{2q}{q-1}}_{\frac{\alpha}{2}\frac{q}{q-1}}(t)\right)^{\sum_{k=1}^n\left(\frac{q}{2}\right)^k}\left(\frac{q}{2}\right)^{4\sum_{k=1}^n k\left(\frac{q}{2}\right)^k}A_{\delta_{**}}(t) \\ \underset{n\to\infty}{\to} & \left(B_{11}\delta_{**}^4 h_2^5(t)\widetilde{u}^{\frac{2q}{q-1}}_{\frac{\alpha}{2}\frac{q}{q-1}}(t)\right)^{\frac{\frac{q}{2}}{1-\frac{q}{2}}}\left(\frac{q}{2}\right)^{2q\left(\frac{1}{1-\frac{q}{2}}\right)^2}A_{\delta_{**}}(t) \\ =: & \ H^{\delta_{**}}(t)A_{\delta_{**}}(t).\end{aligned}$$

Therefore, we get

$$\limsup_{\delta\to\infty} A_\delta^{\frac{1}{\delta}}(t) \le H(t) A_{\delta_{**}}^{\frac{1}{\delta_{**}}}(t). \tag{9.78}$$

Combining (9.78) with (9.77), we finally arrive at an estimate for $||M(t)||_\infty$:

$$\begin{aligned} ||M(t)||_\infty + 1 &= \lim_{\delta\to\infty} u_\delta^{\frac{1}{\delta}}(t) \\ &\le \limsup_{\delta\to\infty} \widetilde{u}_\delta^{\frac{1}{\delta}}(t) \\ &\le H(t)\left(\widetilde{u}_{\delta_{**}}^{\frac{1}{\delta_{**}}}(t) + (||M_0||_\infty + 1)\, e^{-\frac{F_2}{2}t}\right). \end{aligned} \tag{9.79}$$

Now, since the functions $h_2, \widetilde{u}_{\frac{\alpha}{2}\frac{q}{q-1}}, \widetilde{u}_{\delta_{**}}$ dissipate exponentially (recall (9.71) and (9.72) and the definition of $H$), we apply Lemma 9.1 to (9.79) and conclude that $||M||_\infty$ dissipates exponentially as well. Moreover, there exists a dissipative estimate for $||M||_\infty$ of the form given in (9.35). This is a consequence of Lemma 9.1, the estimates (9.71) and (9.72), and the fact that we only used the initial data norms $||M_0||_\infty$ and $||\nabla\rho_0||_\infty$ throughout the proof. All other parameters depended only upon the parameters of the problem (9.1)–(9.4). The dissipative estimate for $||\nabla\rho||_\infty + 1 = h_2$ is given in (9.71) and Theorem 9.2 is thus proved. □

**Weak attractor in $L^\infty$**

We showed in Sections 9.1 and 9.2 that the problem (9.1)–(9.4) if considered as an equation with respect to $(M,\rho)$ in the Banach space $L^\infty(\Omega)\times W^{1,\infty}(\Omega)$ is well-posed: for each pair $(M_0,\rho_0)\in L^\infty(\Omega)\times W^{1,\infty}(\Omega)$ there exist a unique solution $(M(t),\rho(t))_{t\in\mathbb{R}_0}$ in terms of Definition 9.1. Therefore, we can define a semigroup in $L^\infty(\Omega)\times W^{1,\infty}(\Omega)$ through

$$S(t)(M_0,\rho_0) := (M(t),\rho(t)) \quad \text{for all } t\ge 0.$$

The dissipative estimate (9.35) provides the existence of a uniformly absorbing set $\mathcal{B}$ for the semigroup $S(t)$ which is bounded in $L^\infty(\Omega)\times W^{1,\infty}(\Omega)$. Moreover, since

$$\Delta\rho(s) = \Delta e^{(s-0)\Delta}\rho_0 - \int_0^s \Delta e^{(s-\omega)\Delta} g(M(\omega),\rho(\omega))\, d\omega,$$

we get

$$\begin{aligned} ||\Delta\rho(s)||_6 &\le \left\| \Delta e^{(s-0)\Delta}\rho_0 - \int_0^s \Delta e^{(s-\omega)\Delta} g(M(\omega),\rho(\omega))\, d\omega \right\|_6 \\ &\le C_1\frac{1}{\sqrt{s}}\,||\nabla\rho_0||_6 + C_1\int_0^s \frac{1}{\sqrt{s-\omega}}\,||g(M(\omega),\rho(\omega))\, d\omega||_6 . \end{aligned}$$

Thus, due to the assumptions on $g$, there exists a nonnegative function $\Phi_\rho(s,x,y)$ which is nondecreasing with respect to $s, x$, and $y$, independent of $M_0$ and $\rho_0$, and such that the following smoothing estimate holds:

$$\sqrt{s}\,||\Delta\rho(s)||_6 \le \Phi_\rho\left(s, ||M_0||_\infty, ||\nabla\rho_0||_\infty\right). \tag{9.80}$$

With (9.80) and the compact embeddings

$$\begin{aligned} L^\infty(\Omega) &\subset\subset L^\infty_{w-*}(\Omega), \\ W^{2,6}(\Omega) &\subset\subset W^{1,\infty}(\Omega) \end{aligned}$$

($L^\infty_{w-*}(\Omega)$ denotes the space $L^\infty(\Omega)$ equipped with the weak-$*$ topology of $L^\infty(\Omega)$), it follows that the set $S(t)\mathcal{B}$ is a relatively compact absorbing set for the semigroup $S(t)$ in the (Fréchet) space $L^\infty_{w-*}(\Omega) \times W^{1,\infty}(\Omega)$.

In Section 10.2 we derive (see (10.80)) local Lipschitz continuity for the solutions of (9.1)–(9.4) in the following sense: for all $T, R > 0$,

$$\begin{aligned}
&\left\| S(t)\left(M_0^{(1)}, \rho_0^{(1)}\right) - S(t)\left(M_0^{(2)}, \rho_0^{(2)}\right)\right\|_{H^{-1}(\Omega)\times L^2(\Omega)} \\
&\qquad \le L(t,R)\left\| \left(M_0^{(1)}, \rho_0^{(1)}\right) - \left(M_0^{(2)}, \rho_0^{(2)}\right)\right\|_{H^{-1}(\Omega)\times L^2(\Omega)}
\end{aligned} \tag{9.81}$$

for all $\left\|\left(M_0^{(i)}, \rho_0^{(i)}\right)\right\|_{L^\infty(\Omega)\times W^{1,\infty}(\Omega)} \le R$, $i = 1, 2$, and some nonnegative nondecreasing in both $t$ and $R$ function $L(t, R)$ independent of $M, \rho$, and $0$.

Recall that due to embedding theorems for Sobolev spaces we have

$$W^{1,\infty}(\Omega) \subset L^2(\Omega) \tag{9.82}$$

and

$$L^\infty(\Omega) \subset\subset H^{-1}(\Omega). \tag{9.83}$$

Since compact operators between two Banach spaces are weak-$*$-to-norm continuous, property (9.83) allows the interpretation

$$L^\infty_{w-*}(\Omega) \subset H^{-1}(\Omega). \tag{9.84}$$

Combining (9.82) with (9.84), we obtain

$$L^\infty_{w-*}(\Omega) \times W^{1,\infty}(\Omega) \subset H^{-1}(\Omega) \times L^2(\Omega). \tag{9.85}$$

Let $\left(M_0^{(n)}, \rho_0^{(n)}\right)$ be a sequence of initial data convergent to some $(M_0, \rho_0)$ in $L^\infty_{w-*}(\Omega) \times W^{1,\infty}(\Omega)$. This sequence converges in $H^{-1}(\Omega) \times L^2(\Omega)$ to the same limit $(M_0, \rho_0)$ due to the continuous embedding (9.85). From property (9.81) we deduce that the sequences $\left(S(t)\left(M_0^{(n)}, \rho_0^{(n)}\right)\right)$ converge to $S(t)(M_0, \rho_0)$ in $H^{-1}(\Omega) \times L^2(\Omega)$ for all $t \ge 0$. Let us further assume that for some $t \ge 0$ the sequence $\left(S(t)\left(M_0^{(n)}, \rho_0^{(n)}\right)\right)$ is convergent in $L^\infty_{w-*}(\Omega) \times W^{1,\infty}(\Omega)$. Due to the continuity of the embedding (9.85), the limit is $S(t)(M_0, \rho_0)$. Therefore, we can conclude that the operators $S(t)$ if considered as mapping in $L^\infty_{w-*}(\Omega) \times W^{1,\infty}(\Omega)$ are closed operators. Since any (nonlinear) closed compact operator is completely continuous (that is, continuous and compact), we get the continuity of the operators $S(t)$ in $L^\infty_{w-*}(\Omega) \times W^{1,\infty}(\Omega)$.

Applying the general theorem from [9] to our case, we deduce the existence of the global attractor $\mathcal{A}$ for the semigroup $\{S(t)\}_{t\ge 0}$:

THEOREM 9.3. *Let the functions $f$ and $g$ satisfy the assumptions* (9.31)–(9.33) *and let the given constants $\alpha$ and $\gamma$ satisfy $\gamma \in \left[\frac{\alpha}{2} + 1, \alpha\right)$. Further let the solution of the problem* (9.1)–(9.4) *be described by the semigroup $\{S(t)\}_{t\ge 0}$. Then there exists the global attractor for $\{S(t)\}_{t\ge 0}$ in the (Fréchet) space $L^\infty_{w-*}(\Omega) \times W^{1,\infty}(\Omega)$.*

## 9.4. Appendix (Existence)

**A1.**

$$M_{\varepsilon t} = \nabla(M_\varepsilon+\varepsilon)^\alpha \nabla M_\varepsilon - \nabla(M_\varepsilon+\varepsilon)^\gamma \nabla \rho_\varepsilon + f(M_\varepsilon+\varepsilon,\rho_\varepsilon) \quad \text{in } \Omega, \tag{9.86}$$

$$\rho_{\varepsilon t} = \Delta\rho_\varepsilon - g(M_\varepsilon+\varepsilon,\rho_\varepsilon) \quad \text{in } \Omega. \tag{9.87}$$

By a standard theorem, we get a unique classical solution to the above system. Moreover, we obtain a uniform bound for the solution (which is independent of $\varepsilon>0$, by the argument in Section 9.1).

The following estimates are independent of $\varepsilon \in (0,1]$. Then, the limiting function $(M,\rho) = \lim\limits_{\varepsilon\to 0}(M_\varepsilon,\rho_\varepsilon)$ satisfies the same estimates.

**A2.**
In this section we show that any solution $(M_\varepsilon,\rho_\varepsilon)$ of (9.86), (9.87), (9.3), and (9.4) satisfies

$$\|\rho_{\varepsilon t}\|_{L^\delta(\Omega)},\quad \|\Delta\rho_\varepsilon\|_{L^\delta(\Omega)} \le C \quad \text{for any } \delta>1. \tag{9.88}$$

PROOF. Indeed, let $A=-\Delta+I$ with $\cdot\,|_{\partial\Omega}=0$. Then

$$(\rho_\varepsilon-1)_t = \Delta(\rho_\varepsilon-1)-(\rho_\varepsilon-1)+g_0(\rho_\varepsilon)(M_\varepsilon+\varepsilon)-1,$$

$$e^t(\rho_\varepsilon-1) = e^{-At}(\rho_0-1)+\int_0^t e^{-A(t-s)}e^A(g_o(\rho_\varepsilon)(M_\varepsilon+\varepsilon)-1)ds.$$

Hence, for any $\delta\ge 1$

$$e^t\|A(\rho_\varepsilon-1)\|_{L^\delta} \le \|Ae^{-At}(\rho_0-1)\|_{L^\delta}+\int_0^t e^s\|Ae^{-A(t-s)}g((M_\varepsilon+\varepsilon),\rho_\varepsilon)\|_{L^\delta}ds. \tag{9.89}$$

Since the domain of $A^\eta$ is $D(A^\eta)\subset L^{2\delta}$ $\quad(\eta=\eta(\delta)>0)$, it follows that

$$\|A^\eta g\|_{L^\delta} \le C(\|(M_\varepsilon+\varepsilon)\|_{L^{2\delta}}+1). \tag{9.90}$$

Consequently,

$$e^t\|A(\rho_\varepsilon-1)\|_{L^\delta} \le \underbrace{C}_{\text{by } \rho_0\text{: smooth}} + \int_0^t e^s\|A^{1-\varepsilon}e^{-A(t-s)}A^\varepsilon g\|_{L^\delta}ds$$

$$\le C+\int_0^t e^s(t-s)^{-1+\varepsilon}C(\|(M_\varepsilon+\varepsilon)\|_{L^{2\delta}}+1)ds. \tag{9.91}$$

Note that from (9.22) we obtain for any $\delta\ge 1$

$$\int_\Omega (M_\varepsilon+\varepsilon)^{\delta+1}dx + C_3^\star\int_0^t\int_\Omega (M_\varepsilon+\varepsilon)^{\delta+1}dx \le C_4^\star,$$

where $C_3^\star$ and $C_4^\star$ depend on $\delta$. We set

$$Q(t) := \int_\Omega (M_\varepsilon+\varepsilon)^{\delta+1}dx - \frac{C_4^\star}{C_3^\star}. \tag{9.92}$$

Then $Q(t)$ satisfies $Q(t)+\int_0^t Q(s)ds\le 0$. Consequently, for any $\delta\ge 1$

$$Q(t)\le C_5^\star \quad\text{and}\quad \int_\Omega (M_\varepsilon+\varepsilon)^{\delta+1}dx \le C_6^\star,$$

where $C_5^\star, C_6^\star$ depend on $\delta$. Thus

$$e^t \|A(\rho_\varepsilon - 1)\|_{L^\delta(\Omega)} \le C_7^\star, \tag{9.93}$$

where $C_7^\star$ also depends on $\delta$. Hence, we have

$$\|\Delta\rho_\varepsilon\|_{L^\delta(\Omega)} \le C_7^\star \quad \text{for any } \delta \ge 1. \tag{9.94}$$

Combining this with (9.92) implies

$$\|(\rho_\varepsilon)_t\|_{L^\delta(\Omega)} \le C_8^\star.$$

Then, we obtain (9.88). □

**A3.**
In this section we show that any solution $(M_\varepsilon, \rho_\varepsilon)$ of (9.86), (9.87), (9.3), and (9.4) satisfies

$$\int_0^T \int_\Omega |(M_\varepsilon + \varepsilon)_t^{(\alpha+2)/2}|^2 dxdt \le C, \quad M_\varepsilon \in H^1((0,T); H^{-1}(\Omega)),$$
$$\rho_\varepsilon \in C([0,T]; W^{1,\delta}(\Omega)).$$

PROOF. We multiply (9.86) by $[(M_\varepsilon + \varepsilon)^{\alpha+1}]_t$ and integrate over $\Omega$. Noting that $M_{\varepsilon t} = 0$ on $\partial\Omega$, we have that

$$\begin{aligned}
&(\alpha+1)\left(\frac{\alpha+2}{2}\right)^2 \int_\Omega \left|\left\{(M_\varepsilon+\varepsilon)^{(\alpha+2)/2}\right\}_t\right|^2 dx \\
&+ \frac{1}{2(\alpha+1)} \frac{d}{dt} \int_\Omega |\nabla(M_\varepsilon+\varepsilon)^{\alpha+1}|^2 dx \\
&\quad = -\int_\Omega \left(\nabla\cdot(M_\varepsilon+\varepsilon)^\gamma \nabla\rho_\varepsilon + f(M_\varepsilon+\varepsilon,\rho_\varepsilon)\right)(M_\varepsilon^{\alpha+1})_t dx \\
&\quad = -\frac{2\gamma}{2\gamma+\alpha}\frac{2(\alpha+1)}{\alpha+2} \int_\Omega \left\{\nabla(M_\varepsilon+\varepsilon)^{(2\gamma+\alpha)/2}\nabla\rho_\varepsilon\right\}\left\{(M_\varepsilon+\varepsilon)^{(\alpha+2)/2}\right\}_t dx \\
&\qquad - \frac{2(\alpha+1)}{\alpha+2} \int_\Omega \left\{(M_\varepsilon+\varepsilon)^{(2\gamma+\alpha)/2}\Delta\rho_\varepsilon\right\}\left\{(M_\varepsilon+\varepsilon)^{(\alpha+2)/2}\right\}_t dx \\
&\qquad + \int_\Omega f(M_\varepsilon+\varepsilon,\rho_\varepsilon)\left\{(M_\varepsilon+\varepsilon)^{(\alpha+2)/2}\right\}_t dx.
\end{aligned}$$

Hence,

$$\begin{aligned}
&\int_0^t \int_\Omega \left|\left\{(M_\varepsilon+\varepsilon)^{(\alpha+2)/2}\right\}_s\right|^2 dx + \int_\Omega |\nabla(M_\varepsilon+\varepsilon)^{\alpha+1}|^2 dx \\
&\qquad \le C\|\nabla\rho_\varepsilon\|_{L^\infty}^2 \int_\Omega \left|\nabla(M_\varepsilon+\varepsilon)^{(2\gamma+\alpha)/2}\right|^2 dxds \\
&\qquad\quad + C\int_0^t \left(\int_\Omega (M_\varepsilon+\varepsilon)^{2(2\gamma+\alpha)} dx\right)^{1/2} \left(\int_\Omega |\Delta\rho_\varepsilon|^4 dx\right)^{1/2} ds \\
&\qquad\quad + \int_0^t \int_\Omega |f(M_\varepsilon+\varepsilon,\rho_\varepsilon)|^2 dxds.
\end{aligned}$$

Combining this with **A2** implies that

$$\int_0^t \int_\Omega \left|\left\{(M_\varepsilon+\varepsilon)^{(\alpha+2)/2}\right\}_s\right|^2 dxds \le C.$$

We can get the same estimate as (9.20) for $M_\varepsilon + \varepsilon$. Then,

$$(M_\varepsilon + \varepsilon)^{\delta+1} \in C([0,T]; L^2(\Omega))$$

and

$$(M_\varepsilon + \varepsilon)^{(\alpha+\delta+1)/2} \in L^2((0,T); H^1(\Omega))$$

for $\delta \geq 1$. Since, for any $\delta \geq 1$, $\|M_\varepsilon + \varepsilon\|_{L^\delta}$ is bounded, then $\rho_\varepsilon \in C([0,T]; W^{1,\delta}(\Omega))$ for any $\delta \geq 1$, by the standard argument. By all of this, we have

$$M_\varepsilon \in H^1((0,T); H^{-1}(\Omega)). \qquad \square$$

## 9.5. Appendix (Proof of the auxiliary lemma)

Consider first the differential inequality

$$\dot{y} \leq -\omega_y y + d_y y^{\zeta_y} \tag{9.95}$$

assuming that $y \geq 1$, $\zeta_y \in (0,1)$, $d_y \in L^1_b(\mathbb{R}^+_0)$ so that with some computation the estimate

$$(y(t))^{1-\zeta_y} \leq (y(0))^{1-\zeta_y} e^{-\omega_y(1-\zeta_y)t} + (1-\zeta_y)\int_0^t e^{-\omega_y(1-\zeta_y)(t-s)} d_y(s)\, ds$$

follows.

Lemma 9.2. *Let $z_1, z_2, z_3 : [0,+\infty) \to [0,+\infty)$ be functions such that*

$$\begin{aligned} &z_1(t) \leq \psi_1(z_1(0))e^{-\omega_1 t} + D_1,\\ &z_2(t) \leq \psi_2(z_2(0))e^{-\omega_2 t} + D_2,\\ &z_3(t) \leq z_3(0)e^{-\omega_3 t} + \int_0^t e^{-\omega_3(t-s)} d_3(t,s) z_1(s)\, ds,\\ &z_1(0), z_2(0), z_3(0) \geq 1, \end{aligned} \tag{9.96}$$

*for some constants $\omega_1, \omega_2, \omega_3 > 0$ and $D_1, D_2 \geq 1$, some nondecreasing functions $\psi_1, \psi_2 : [1,+\infty) \to [1,+\infty)$, and some $d_3 \in L^\infty(\mathbb{R}^+_0, L^1_b(\mathbb{R}^+_0))$, where*

$$L^1_b(\mathbb{R}^+_0) = \left\{ u \in L^1_{loc}(\mathbb{R}^+_0) : ||u||_{L^1_b(\mathbb{R}^+_0)} := \sup_{x_0 \in \mathbb{R}^+_0} ||u||_{L^1([x_0, x_0+1])} < \infty \right\}.$$

*Then:*

(1) $(z_1 + z_2)(t) \leq (\psi_1 + \psi_2)((z_1+z_2)(0))e^{-\min\{\omega_1,\omega_2\}t} + D_1 + D_2$.
(2) $z_1 z_2(t) \leq 3D_1 D_2 \psi_1\psi_2(z_1 z_2(0))e^{-\min\{\omega_1,\omega_2\}t} + D_1 D_2$.
(3) $z_1^\sigma(t) \leq \max\{1, 2^{\sigma-1}\}\left(\psi_1^\sigma(z_1(0))e^{-\sigma\omega_1 t} + D_1^\sigma\right) \quad \forall \sigma > 0$.
(4) *For $\omega_1 \neq \omega_3$*

$$\begin{aligned} z_3(t) \leq &\left(\psi_1(z_1(0))\frac{1}{1-e^{-|\omega_1-\omega_3|}} e^{-\min\{\omega_1,\omega_3\}t} + D_1 \frac{1}{1-e^{-\omega_3}}\right)\\ &\cdot ||d_3||_{L^\infty(\mathbb{R}^+_0, L^1_b(\mathbb{R}^+_0))} + z_3(0)e^{-\omega_3 t} \end{aligned} \tag{9.97}$$

*and for $\omega_3 = \omega_1$*

$$\begin{aligned} z_3(t) \leq &\left(\psi_1(z_1(0))\lceil t\rceil e^{-\omega_1 t} + D_1\frac{1}{1-e^{-\omega_1}}\right) ||d_3||_{L^\infty(\mathbb{R}^+_0, L^1_b(\mathbb{R}^+_0))}\\ &+ z_3(0)e^{-\omega_1 t}. \end{aligned}$$

*For $\omega_1 < \omega_3$ we also have*

$$z_3(t) \leq z_3(0)e^{-\omega_3 t} + z_1(t)\int_0^t e^{-(\omega_3-\omega_1)(t-s)} d_3(t,s)\,ds. \tag{9.98}$$

PROOF. We only prove (4) for $\omega_1 \neq \omega_3$. Since

$$\begin{aligned}
&\int_0^t e^{-\omega_3(t-s)}e^{-\omega_1 s} d_3(t,s)\,ds \\
&\quad = e^{-\min\{\omega_1,\omega_3\}t}\begin{cases} \int_0^t e^{-|\omega_1-\omega_3|(t-s)} d_3(t,s)\,ds & \text{if } \omega_1 < \omega_3, \\ \int_0^t e^{-|\omega_1-\omega_3|s} d_3(t,s)\,ds & \text{if } \omega_1 > \omega_3 \end{cases} \\
&\quad \leq \frac{1}{1-e^{-|\omega_1-\omega_3|}} e^{-\min\{\omega_1,\omega_3\}t} ||d_3||_{L^\infty(\mathbb{R}_0^+, L_b^1(\mathbb{R}_0^+))} \qquad (9.99)
\end{aligned}$$

we conclude from (9.96) that

$$\begin{aligned}
&\int_0^t e^{-\omega_3(t-s)} d_3(t,s) z_1(s)\,ds \\
&\quad \leq \int_0^t e^{-\omega_3(t-s)} d_3(t,s)\left(\psi_1(z_1(s))e^{-\omega_1(t-s)} + D_1\right) ds \\
&\quad \leq \left(\psi_1(z_1(0))\frac{1}{1-e^{-|\omega_1-\omega_3|}} e^{-\min\{\omega_1,\omega_3\}t} + D_1\frac{1}{1-e^{-\omega_3}}\right) \\
&\qquad \cdot ||d_3||_{L^\infty(\mathbb{R}_0^+, L_b^1(\mathbb{R}_0^+))}
\end{aligned}$$

and the statement follows. □

CHAPTER 10

# On a class of PDEs with degenerate diffusion and chemotaxis: Nonautonomous case

This study aims to consider the following model:

$$\dot{M} = \nabla\cdot(|M|^{\alpha}\nabla M) - \nabla\cdot(|M|^{\gamma}\nabla\rho) + f(t,M,\rho) \text{ in } (0,\infty)\times\Omega, \tag{10.1}$$

$$\dot{\rho} = \Delta\rho - g(t,M,\rho) \quad \text{in } (0,\infty)\times\Omega, \tag{10.2}$$

$$M = 0, \quad \rho = 1 \quad \text{in } (0,\infty)\times\partial\Omega, \tag{10.3}$$

$$M(\cdot,0) = M_0, \quad \rho(\cdot,0) = \rho_0 \quad \text{in } \Omega, \tag{10.4}$$

where $\alpha$ and $\gamma$ are given constants satisfying a balance condition (see Remark 9.2)

$$\frac{\alpha}{2} + 1 \le \gamma < \alpha. \tag{10.5}$$

REMARK 10.1. (1) This new balance condition (10.5) is clearly weaker than condition (9.5) proposed in the previous chapter. Note: The condition might not be optimal.

(2) It follows from (10.5) that $\alpha, \gamma > 2$ should hold.

Moreover, $\Omega \subset \mathbb{R}^N$ is a smooth bounded domain ($N = 1,2,3$) and $M_0 \in L^{\infty}(\Omega)$, $\rho_0 \in W^{1,\infty}(\Omega)$ with $\rho_0$, $M_0 \ge 0$. We assume the functions $f$ and $g$ satisfy the following assumptions: for all $M,\rho \in \mathbb{R}$, $t \in \mathbb{R}_0^+$ let

$$|f(t,M,\rho)| \le f_1(t)(1+|M|^{\xi})^{\frac{1}{2}},\ f_1 \in L_b^2(\mathbb{R}_0^+),\ 0 \le \xi < \alpha - \gamma + 2, \tag{10.6}$$

$$f(t,M,\rho)M \le -F_2 M^2 + f_3(t)|M|,\ f_3 \in L_b^{\kappa}(\mathbb{R}_0^+),\ \kappa > 1, \tag{10.7}$$

$$g(t,M,\rho) = g_1(t)\rho + g_2(t,\rho)M,\ g_1 \in C^1[0,+\infty),\ \dot{g_1}(t) \le 0, \tag{10.8}$$

$$|g_2(t,\rho)| \le g_3(t),\ g_3 \in L_b^{\eta}(\mathbb{R}_0^+),\ \eta > 4, \tag{10.9}$$

and, in order to ensure the uniqueness and the nonnegativity of solution,

$$\widetilde{f}(t,M,\rho) := f\left(t, M^{\frac{2}{2+\alpha}}, \rho\right) - F_4 M^{\frac{2}{2+\alpha}},\ \frac{\partial \widetilde{f}}{\partial M} \in L_{loc}^{\infty}(\mathbb{R}_0^+ \times \mathbb{R}_0^+ \times \mathbb{R}_0^+), \tag{10.10}$$

$$\frac{\partial f}{\partial M}, \frac{\partial f}{\partial \rho} \in L_{loc}^{\infty}(\mathbb{R}_0^+ \times \mathbb{R} \times \mathbb{R}),\ \frac{\partial g_2}{\partial \rho} \in L_{loc}^{\infty}(\mathbb{R}_0^+ \times \mathbb{R}),\ f(t,0,\rho) = 0,\ g_2(t,0) = 0, \tag{10.11}$$

where $f_1, f_3, g_1, g_3$ are nonnegative functions and $F_2, F_4$ are some constants, $F_2$ is strictly positive, and for $p \in [1, \infty]$, $Q \subset \mathbb{R}^m$

$$L^p_{loc}(Q) = \{u : Q \to \mathbb{R} : u \in L^p(K) \text{ for all compact sets } K \subset Q\},$$

$$L^p_b(Q) = \left\{u \in L^p_{loc}(Q) : ||u||_{L^p_b(Q)} := \sup_{x_0 \in \mathbb{R}^m} ||u||_{L^p(Q \cap B^1_{x_0})} < \infty\right\}$$

where $B^1_{x_0}$ is a ball of unit radius centered at $x_0$. The following example of functions $f$ and $g$ satisfies the conditions (10.6)–(10.11):

EXAMPLE 10.1.

$$f(t, M, \rho) = -M + \frac{M_+^{\frac{2+\alpha}{2}}}{M_+^{\frac{2+\alpha}{2}} + 1} \sin(t),$$

$$g(t, M, \rho) = \frac{1}{1+t}\rho + M \frac{\rho_+}{\rho_+ + 1} \cos(t),$$

where $M_+ = \max\{M, 0\}$.

In this book we treat weak solutions of the system (10.1)–(10.4). The definition is as follows:

DEFINITION 10.1. For $T > 0$, $\alpha > 1$, and $\gamma > 1$, a pair of functions $(M, \rho)$ defined in $[0, T) \times \Omega$ is said to be a weak solution of (10.1)–(10.4) for $M_0 \in L^\infty(\Omega)$, $\rho_0 \in W^{1,\infty}(\Omega)$, if

(i) $M \in L^\infty((0, T) \times \Omega)$, $|M|^{\frac{\alpha}{2}} M \in L^2\left(0, T; H^1_0(\Omega)\right)$, $\dot{M} \in L^2\left(0, T; H^{-1}(\Omega)\right)$,

(ii) $\rho - 1 \in C([0, T]; H^1_0(\Omega))$,

(iii) $(M, \rho)$ satisfies the equation in the following sense:

$$\int_0^T (M, v)\, \dot{\varphi} - (|M|^\alpha \nabla M - |M|^\gamma \nabla \rho, \nabla v)\, \varphi + (f(\tau, M, \rho), v)\, \varphi \, d\tau = 0$$

for any $v \in H^1_0(\Omega)$, $\varphi \in C_0^\infty(0, T)$,

$$\rho(x, t) - 1 = \int_\Omega G(x, y, t)(\rho_0(y) - 1)\, dy - \int_0^t \int_\Omega G(x, y, t - s) g(s, M(y, s), \rho(y, s))\, dyds$$

for a.e. $(x, t) \in \Omega \times (0, T)$, where $G$ is a heat kernel in $\Omega$ with the homogeneous Dirichlet boundary condition and the initial conditions hold: $\rho(0) = \rho_0$ and, in the $C_w([0, T]; L^2(\Omega))$ sense, $M(0) = M_0$.

REMARK 10.2. From $M \in L^\infty(0, T; L^2(\Omega))$ and $\dot{M} \in L^2\left(0, T; H^{-1}(\Omega)\right)$ it follows that $M \in C_w([0, T]; L^2(\Omega))$ (see [9]); therefore, the initial condition for $M$ does make sense.

The model (10.1)–(10.4) expands the model (9.1)–(9.4) from the previous chapter, where the autonomous case was studied. In this chapter, we generalize the model

(9.1)–(9.4) to include the nonautonomous case, in which the source term may now be time dependent; additionally, its dependence on population density is more general. We also relax the balance condition.

The main focus of the present study is on proving the well-posedness of (10.1)–(10.4). We emphasize the fact that the analysis of equations with a chemotaxis type term even in the autonomous case and without degeneracy ($\alpha = 0$) is quite difficult (see [73], [87], [88], [89], [91] and the references therein), and in proving the well-posedness in our degenerate case, we face significant difficulties. In order to overcome these difficulties, we impose the so-called "balance condition" between the order of porous medium degeneracy and the growth order of the chemotaxis function: $\frac{\alpha}{2} + 1 \leq \gamma < \alpha$. We show that our model is a well-posed model and that it exhibits no singular behavior. For each pair of starting values the solution is uniformly bounded in time and space. Recall that this is not the case for the models that contain the chemotaxis effect alone (the solution may blow up; see [56]).

The condition $\alpha > \gamma$ (an improvement over the condition $\alpha \geq \gamma + 1$ imposed in Chapter 9) reads: the density-dependent diffusion coefficient "dominates" the intensity of response to the chemical signal as the population density grows. This, as we will see later, results in the uniform boundedness of $M$ and $\rho$.

On the other hand, even in the areas with low population density the porous medium effect is due to $\frac{\alpha}{2} + 1 \leq \gamma$ being strong enough to keep the population from spreading without vanishing locally, which means that the support of $M(t, \cdot)$, the set $\overline{\{x \in \Omega | M(t,x) > 0\}}$, is expanding in $t$. In this study we use the condition $\frac{\alpha}{2} + 1 \leq \gamma$ to establish the well-posedness for our model.

Our main result can be summarized as follows:

THEOREM 10.1. *Let the functions $f$ and $g$ satisfy the assumptions* (10.6)–(10.11) *and let the given constants $\alpha$ and $\gamma$ satisfy $\gamma \in \left[\frac{\alpha}{2} + 1, \alpha\right)$. Then the initial boundary value problem* (10.1)–(10.4) *has at most one nonnegative solution.*

The chapter is organized as follows: in Section 10.1 we obtain several a priori estimates for the solutions of (10.1)–(10.4), which in turn lead to $L^\infty$-bounds for both the biomass component and the substrate concentration. Section 10.2 is devoted to the uniqueness of solutions. For the convenience of the reader, we present some standard ideas which we used while proving the well-posedness of solutions in the first appendix (Section 10.3) and we show the positivity of total biomass in the second appendix (Section 10.4).

**Notation**: $||\cdot||$ stands for the $||\cdot||_{L^2(\Omega)}$-norm and $(u,v)$ for $\int_\Omega u(x)v(x)\,dx$ or, more generally (in the case of distributional derivatives, for instance), for $\langle u, v\rangle$.

## 10.1. A priori estimates

In this section we derive several *a priori* estimates for the solutions of the problem (10.1)–(10.4), which in turn lead to $L^\infty$-bounds for both $M$ and $\rho$. The existence of a solution for this problem can then be proved by means of a standard approximation procedure (see the first appendix of this chapter).

We start with rewriting equation (10.1) in the following way:

$$\dot{M} = \nabla \cdot \left( |M|^{\gamma} \nabla \left( \frac{1}{\alpha - \gamma + 1} M|M|^{(\alpha-\gamma+1)-1} - \rho \right) \right) + f(t, M, \rho). \tag{10.12}$$

In order to derive our first *a priori* estimate, we multiply this equation by the expression $\left(\frac{1}{\alpha-\gamma+1} M|M|^{(\alpha-\gamma+1)-1} - \rho\right)$ and integrate (formally) over $\Omega$ to get

$$\begin{aligned}
&\left(\dot{M}, \frac{1}{\alpha-\gamma+1} M|M|^{(\alpha-\gamma+1)-1} - \rho\right) \\
&\quad = -\left(|M|^{\gamma}, \left|\nabla\left(\frac{1}{\alpha-\gamma+1} M|M|^{(\alpha-\gamma+1)-1} - \rho\right)\right|^2\right) \\
&\qquad + \left(f(t, M, \rho), \frac{1}{\alpha-\gamma+1} M|M|^{(\alpha-\gamma+1)-1} - \rho\right) \\
&\quad \leq \left(f(t, M, \rho), \frac{1}{\alpha-\gamma+1} M|M|^{(\alpha-\gamma+1)-1} - \rho\right)
\end{aligned}$$

which is equivalent to

$$\begin{aligned}
&\frac{d}{dt}\left(\frac{1}{(\alpha-\gamma+1)(\alpha-\gamma+2)} \left\| |M|^{\frac{\alpha-\gamma+2}{2}} \right\|^2 - (M, \rho)\right) \\
&\qquad \leq \left(f(t, M, \rho), \frac{1}{\alpha-\gamma+1} M|M|^{(\alpha-\gamma+1)-1} - \rho\right) - (\dot{\rho}, M)
\end{aligned} \tag{10.13}$$

and we multiply equation (10.2) by $(\dot{\rho} + \rho - 1)$ (which we find more appropriate, in contrast to the previous study in Chapter 9) in the same sense as above to get

$$\begin{aligned}
&\|\dot{\rho}\|^2 + \frac{1}{2}\frac{d}{dt}||\rho - 1||^2 = -\frac{1}{2}\frac{d}{dt}||\nabla\rho||^2 - ||\nabla\rho||^2 - (g(t, M, \rho), \dot{\rho} + \rho - 1) \\
\Leftrightarrow\ &\frac{1}{2}\frac{d}{dt}\left(||\nabla\rho||^2 + ||\rho - 1||^2\right) = -||\nabla\rho||^2 - \|\dot{\rho}\|^2 - (g(t, M, \rho), \dot{\rho} + \rho - 1).
\end{aligned} \tag{10.14}$$

Adding the inequalities (10.13) and (10.14) together, we obtain

(10.15)
$$\begin{aligned}
&\frac{d}{dt}\left(\frac{1}{(\alpha-\gamma+1)(\alpha-\gamma+2)} \left\| |M|^{\frac{\alpha-\gamma+2}{2}} \right\|^2 - (M, \rho) + \frac{1}{2}||\nabla\rho||^2 + \frac{1}{2}||\rho - 1||^2\right) \\
&\quad \leq \left(f(t, M, \rho), \frac{1}{\alpha-\gamma+1} M|M|^{(\alpha-\gamma+1)-1} - \rho\right) - ||\nabla\rho||^2 - (\dot{\rho}, M) - \|\dot{\rho}\|^2 \\
&\quad - (g(t, M, \rho), \dot{\rho} + \rho - 1).
\end{aligned}$$

We first consider the term containing $g(t, M, \rho) = g_1(t)\rho + g_2(t,\rho)M$. It holds true that

$$
\begin{aligned}
-(g_1\rho, \dot\rho + \rho - 1) &= -\frac{1}{2}\frac{d}{dt}\left(g_1||\rho||^2\right) + \frac{1}{2}\dot g_1||\rho||^2 - g_1\left(||\rho||^2 - (1,\rho)\right)\\
&\underset{(10.8)}{\le} -\frac{1}{2}\frac{d}{dt}\left(g_1||\rho||^2\right) - g_1\left(||\rho||^2 - (1,\rho)\right)\\
&\le -\frac{1}{2}\frac{d}{dt}\left(g_1||\rho||^2\right) - (1-\varepsilon)g_1||\rho||^2 + \frac{1}{4\varepsilon}g_1\\
&\underset{(10.8)}{\le} -\frac{1}{2}\frac{d}{dt}\left(g_1||\rho||^2\right) - (1-\varepsilon)g_1||\rho||^2 + \frac{1}{4\varepsilon}g_1(0)
\end{aligned}
\tag{10.16}
$$

and

$$
\begin{aligned}
-(g_2(t,\rho)M, \dot\rho + \rho - 1) \le{} & \varepsilon\,\|\dot\rho\|^2 + \varepsilon||\rho - 1||^2 + \frac{1}{2\varepsilon}||g_2(t,\rho)M||^2\\
\underset{(10.9)}{\le}{} & \varepsilon\,\|\dot\rho\|^2 + \varepsilon||\rho - 1||^2 + \frac{1}{2\varepsilon}|g_3|^2||M||^2.
\end{aligned}
\tag{10.17}
$$

By combining (10.16) and (10.17) with the inequality

$$
-(\dot\rho, M) - \|\dot\rho\|^2 \le \frac{1}{2}||M||^2 - \frac{1}{2}\,\|\dot\rho\|^2 \tag{10.18}
$$

and by choosing $\varepsilon \le \frac{1}{2}$, we have

$$
\begin{aligned}
&- (\dot\rho, M) - \|\dot\rho\|^2 - (g(t,M,\rho), \dot\rho + \rho - 1)\\
&\quad\le -\frac{1}{2}\frac{d}{dt}\left(g_1||\rho||^2\right) - (1-\varepsilon)g_1||\rho||^2 + \varepsilon||\rho - 1||^2 + \frac{1}{4\varepsilon}g_1(0) - \left(\frac{1}{2} - \varepsilon\right)\|\dot\rho\|^2\\
&\qquad + \left(\frac{1}{2} + \frac{1}{2\varepsilon}|g_3|^2\right)||M||^2\\
&\quad\underset{\varepsilon\le\frac{1}{2}}{\le} -\frac{1}{2}\frac{d}{dt}\left(g_1||\rho||^2\right) - (1-\varepsilon)g_1||\rho||^2 + \varepsilon||\rho - 1||^2 + \frac{1}{4\varepsilon}g_1(0)\\
&\qquad + \left(\frac{1}{2} + \frac{1}{2\varepsilon}|g_3|^2\right)||M||^2.
\end{aligned}
\tag{10.19}
$$

Further, we can estimate the terms with $f$ from (10.15) in the following way:

$$
\begin{aligned}
\left(f(t,M,\rho), M|M|^{(\alpha-\gamma+1)-1}\right) &\underset{(10.7)}{\le} \left(-F_2M^2 + f_3|M|, |M|^{(\alpha-\gamma+1)-1}\right)\\
&= -F_2\left\||M|^{\frac{\alpha-\gamma+2}{2}}\right\|^2 + f_3\left\||M|^{\frac{\alpha-\gamma+1}{2}}\right\|^2,
\end{aligned}
\tag{10.20}
$$

$$
\begin{aligned}
-\left(f(t,M,\rho),\rho\right) &\underset{(10.6)}{\le} \varepsilon||\rho||^2 + \frac{1}{4\varepsilon}f_1^2\left(1 + \left\||M|^{\frac{\xi}{2}}\right\|^2\right)\\
&\le 2\varepsilon||\rho - 1||^2 + 2\varepsilon + \frac{1}{4\varepsilon}f_1^2 + \frac{1}{4\varepsilon}f_1^2\left\||M|^{\frac{\xi}{2}}\right\|^2.
\end{aligned}
\tag{10.21}
$$

Using the inequalities (10.19)–(10.21), we conclude from (10.15) that

(10.22)

$$\frac{d}{dt}\left(\frac{1}{(\alpha-\gamma+1)(\alpha-\gamma+2)}\left\||M|^{\frac{\alpha-\gamma+2}{2}}\right\|^2-(M,\rho)+\frac{1}{2}||\nabla\rho||^2+\frac{1}{2}||\rho-1||^2+\frac{1}{2}g_1||\rho||^2\right)$$
$$\leq -F_2\left\||M|^{\frac{\alpha-\gamma+2}{2}}\right\|^2+f_3\left\||M|^{\frac{\alpha-\gamma+1}{2}}\right\|^2+\frac{1}{4\varepsilon}f_1^2\left\||M|^{\frac{\xi}{2}}\right\|^2+\left(\frac{1}{2}+\frac{1}{2\varepsilon}|g_3|^2\right)||M||^2$$
$$-||\nabla\rho||^2-(1-\varepsilon)g_1||\rho||^2+3\varepsilon||\rho-1||^2+2\varepsilon+\frac{1}{4\varepsilon}g_1(0)+\frac{1}{4\varepsilon}f_1^2.$$

In order to shorten the formulas, we introduce a new variable:

$$\varphi:=\frac{1}{(\alpha-\gamma+1)(\alpha-\gamma+2)}\left\||M|^{\frac{\alpha-\gamma+2}{2}}\right\|^2-(M,\rho)$$
$$+\frac{1}{2}||\nabla\rho||^2+\frac{1}{2}||\rho-1||^2+\frac{1}{2}g_1||\rho||^2+C \tag{10.23}$$

where the constant $C$ can be chosen in such a way that $\varphi\geq 1$ holds. Indeed, $|M|^{\frac{\alpha-\gamma+2}{2}}$ is the leading $M$-power present in the expression (10.23) due to the assumptions made on $\alpha,\gamma$, and $\xi$, and we also have the estimate

$$(M,\rho)\leq\varepsilon||\rho||^2+\frac{1}{4\varepsilon}||M||^2 \tag{10.24}$$

valued for all $\varepsilon>0$. Moreover, applying the Poincaré and the Hölder inequalities and adjusting the constants $C$ and $\varepsilon$, we can deduce from (10.22) the inequality

$$\dot{\varphi}\leq -A_1\varphi+a_2\varphi^\theta \tag{10.25}$$

for some $A_1\in\mathbb{R}_+$ and $a_2\in L_b^1(\mathbb{R}_0^+)$, $a_2\geq 0$, and

$$\theta:=\frac{\max\left\{\frac{\alpha-\gamma+1}{2},\frac{\xi}{2}\right\}}{\frac{\alpha-\gamma+2}{2}}\in(0,1).$$

A simple calculation shows that any solution $\varphi$ of the inequality (10.25) satisfies the inequality

$$\varphi(t)\leq\left(\varphi_0^{1-\theta}e^{-A_1(1-\theta)t}+(1-\theta)\int_0^t e^{-A_1(1-\theta)(t-\tau)}a_2\,d\tau\right)^{\frac{1}{1-\theta}}. \tag{10.26}$$

In particular, it is bounded because $a_2\in L_b^1(\mathbb{R}_0^+)$ holds and as a consequence of that and the inequality (10.24), we finally achieve our first uniform estimate:

$$\sup_{t>0}\|M(t)\|_{\alpha-\gamma+2},\quad \sup_{t>0}||\rho(t)||_{H_0^1(\Omega)}<\infty. \tag{10.27}$$

REMARK 10.3. We emphasize the fact that, unlike as in Chapter 9, the nonlinearity $g$ is not necessarily nonnegative. Thus we cannot rely on the comparison principle and we have to work with both equations (10.1) and (10.2) simultaneously to get the desired estimates.

In what follows, we use (10.27) to obtain some intermediate estimates for $M$ and $\rho$, which in turn lead to $L^\infty$-bounds in both time and space. The following observation, which is an implication from the theory of abstract parabolic evolution equations (see [95]), will be helpful in the future.

Having a $\delta \in (2, \infty)$ fixed, consider the unbounded operator

$$\Delta : \; L^{\delta}(\Omega) \to L^{\delta}(\Omega)$$

equipped with the domain

$$D(\Delta) := \left\{ u \in W_0^{1,\delta} : \; \Delta u \in L^{\delta}(\Omega) \right\}.$$

It is known (see [95]) that this operator generates an analytic semigroup $e^{t\Delta}$ and its spectrum lies entirely in the left half-plane $\{\lambda \in \mathbb{R} : \lambda \leq -\beta\}$ for some $\beta > 0$. As such it has the following properties:

$$(-\Delta)^{\mu} e^{t\Delta} = e^{t\Delta}(-\Delta)^{\mu}, \tag{10.28}$$

$$||e^{t\Delta}(-\Delta)^{\mu}||_{\delta} \leq A^{\mu,\delta} e^{-\beta t} t^{-\mu} \tag{10.29}$$

for all $t > 0$ and $\mu > 0$ for some constants $A^{\mu,\delta}$. Now, equation (10.2) can be rewritten in the following way:

$$\frac{d}{dt}(\rho - 1) = \Delta(\rho - 1) - g(t, M, \rho)$$

and can thus be regarded as an abstract parabolic evolution equation with respect to $\rho - 1$. Therefore, for all $t > 0$,

$$\rho(t) - 1 = e^{t\Delta}(\rho(0) - 1) - \int_0^t e^{(t-\tau)\Delta} g(\tau, M(\tau), \rho(\tau))\, d\tau, \tag{10.30}$$

and applying the operator $\nabla$ to both sides of (10.30) and making use of property (10.28), we obtain

$$\nabla\rho(t) = e^{t\Delta}\nabla\rho_0 - \int_0^t \nabla\left(e^{(t-\tau)\Delta} g(\tau, M(\tau), \rho(\tau))\right) d\tau. \tag{10.31}$$

The initial value $\rho(0)$ is assumed to be sufficiently smooth, so that

$$||\nabla\rho_0||_{\delta} < \infty. \tag{10.32}$$

What remains is to estimate the $\delta$-norm of the integral from (10.31) with the help of (10.29) and the assumptions on $g$. Choosing $\mu \in \left(0, \frac{1}{2}\right)$ and $\hat{\delta} \geq 1$ such that $W^{2\mu,\hat{\delta}}(\Omega) \hookrightarrow W^{1,\delta}(\Omega)$, we obtain

$$\begin{aligned} &\left\| \int_0^t \nabla\left(e^{(t-\tau)\Delta} g(\tau, M(\tau), \rho(\tau))\right) d\tau \right\|_{\delta} \\ &\quad \leq \int_0^t \left\|(-\Delta)^{\mu}\left(e^{(t-\tau)\Delta} g(\tau, M(\tau), \rho(\tau))\right)\right\|_{\hat{\delta}} d\tau \\ &\quad \leq A^{\mu,\hat{\delta}} \int_0^t e^{-\beta t}(t-\tau)^{-\mu}\left(g_1(0)||\rho(\tau)||_{\hat{\delta}} + g_3(\tau)||M(\tau)||_{\hat{\delta}}\right) d\tau. \end{aligned} \tag{10.33}$$

Suppose for a moment that

$$\sup_{t>0}||\rho(t)||_{\hat{\delta}}, \; \sup_{t>0}||M(t)||_{\hat{\delta}} < \infty. \tag{10.34}$$

Due to (10.9), we have $g_3 \in L^\eta_b(\mathbb{R}^+_0)$ and can therefore conclude from (10.32) and (10.33) with the Hölder inequality

$$\sup_{t>0} \|\nabla\rho(t)\|_\delta \leq \widetilde{A}^{\mu,\hat{\delta}} \left(\sup_{t>0}||\rho(t)||_{\hat{\delta}} + \sup_{t>0}||M(t)||_{\hat{\delta}}\right), \quad \mu \in \left(0, 1-\frac{1}{\eta}\right). \tag{10.35}$$

For instance, $\tilde{\delta_0} := \min\{\alpha - \gamma + 2, 6\}$ and $\frac{3}{4}$ are an appropriate choice of $\hat{\delta}$ and $\mu$, respectively. This is a consequence of the estimate (10.27), the fact that $H^1(\Omega) \hookrightarrow L^6(\Omega)$, $\eta > 4$, and that due to $\alpha > \gamma$ we have $\tilde{\delta_0} \geq 2$, so that $2\mu - \frac{3}{\tilde{\delta_0}} \geq 0 = 1 - \frac{3}{3}$. Therefore, it follows from (10.35) that

$$B_1 := \sup_{t>0} \|\rho(t)\|_{W^{1,3}(\Omega)} < \infty. \tag{10.36}$$

It is worth mentioning that in contrast to the previous study (Chapter 9) we have to be more precise in our further derivations due to a more general assumption (10.9) on the nonlinearity $g_2$ which, at this point, has led to a $W^{1,3}$-estimate for $\rho$ instead of a $W^{1,6}$-estimate.

Getting back to equation (10.1), we multiply this equation by $M|M|^{\delta-1}$ for an arbitrary $\delta \geq \alpha - \gamma + 1$, so that all occurring powers remain nonnegative (note: the assumption $\gamma \geq \frac{\alpha}{2} + 1$ will not be necessary here), and (formally) integrate over $\Omega$:

$$\left(\dot{M}, M|M|^{\delta-1}\right) = (\nabla \cdot (|M|^\alpha \nabla M) - \nabla \cdot (|M|^\gamma \nabla\rho) + f(t, M, \rho), M|M|^{\delta-1}).$$

It follows that

$$\begin{aligned} \frac{1}{\delta+1}\frac{d}{dt}\left\||M|^{\frac{\delta+1}{2}}\right\|^2 = &-\frac{4\delta}{(\alpha+\delta+1)^2}\left\|\nabla|M|^{\frac{\alpha+\delta+1}{2}}\right\|^2 \\ &+\frac{2\delta}{\alpha+\delta+1}\left(\nabla|M|^{\frac{\alpha+\delta+1}{2}}, |M|^{\gamma-\frac{\alpha}{2}+\frac{\delta-1}{2}}\nabla\rho\right) \\ &+ (f(t, M, \rho), M|M|^{\delta-1}). \end{aligned} \tag{10.37}$$

Let $\vartheta(\delta) := \frac{\gamma - \frac{\alpha}{2} + \frac{\delta-1}{2}}{\frac{\alpha+\delta+1}{2}}$. Then $\vartheta(\delta) < 1$ holds due to the assumption $\alpha > \gamma$ that we made. Applying the Hölder inequality, we obtain

$$\begin{aligned} \left(\nabla|M|^{\frac{\alpha+\delta+1}{2}}, |M|^{\gamma-\frac{\alpha}{2}+\frac{\delta-1}{2}}\nabla\rho\right) &= \left(\nabla|M|^{\frac{\alpha+\delta+1}{2}}, |M|^{\vartheta(\delta)\frac{\alpha+\delta+1}{2}}\nabla\rho\right) \\ &\leq \|1\|_{\frac{6}{1-\vartheta(\delta)}}\left\|\nabla|M|^{\frac{\alpha+\delta+1}{2}}\right\|\left\||M|^{\frac{\alpha+\delta+1}{2}}\right\|_6^{\vartheta(\delta)}\|\nabla\rho\|_3 \\ &\leq B_2\left\|\nabla|M|^{\frac{\alpha+\delta+1}{2}}\right\|^{1+\vartheta(\delta)}\|\nabla\rho\|_3. \end{aligned} \tag{10.38}$$

For the last inequality the embedding $H^1(\Omega) \hookrightarrow L^6(\Omega)$ has been used. Further, we use the Hölder inequality once more and the assumptions on the function $f$ and write

$$(f(t, M, \rho), M|M|^{\delta-1}) \leq -F_2\left\||M|^{\frac{\delta+1}{2}}\right\|^2 + f_3\left\||M|^{\frac{\delta}{2}}\right\|^2 \tag{10.39}$$

$$\leq -F_2\left\||M|^{\frac{\delta+1}{2}}\right\|^2 + f_3\|1\|_{\delta+1}\left(\left\||M|^{\frac{\delta+1}{2}}\right\|^2\right)^{\frac{\delta}{\delta+1}}. \tag{10.40}$$

We can conclude from (10.37) using (10.38) and (10.40) that

$$\begin{aligned}\frac{1}{\delta+1}\frac{d}{dt}\left\||M|^{\frac{\delta+1}{2}}\right\|^2 \leq &-\frac{4\delta}{(\alpha+\delta+1)^2}\left\|\nabla|M|^{\frac{\alpha+\delta+1}{2}}\right\|^2\\ &+\frac{2\delta}{\alpha+\delta+1}B_2\left\|\nabla|M|^{\frac{\alpha+\delta+1}{2}}\right\|^{1+\vartheta(\delta)}\|\nabla\rho\|_3\\ &-F_2\left\||M|^{\frac{\delta+1}{2}}\right\|^2+f_3\|1\|_{\delta+1}\left(\left\||M|^{\frac{\delta+1}{2}}\right\|^2\right)^{\frac{\delta}{\delta+1}}.\end{aligned}$$

Since $1+\vartheta(\delta)<2$, it follows from the Young inequality that

$$\begin{aligned}\frac{1}{\delta+1}\frac{d}{dt}\left\||M|^{\frac{\delta+1}{2}}\right\|^2 \leq\; &-F_2\left\||M|^{\frac{\delta+1}{2}}\right\|^2+f_3\|1\|_{\delta+1}\left(\left\||M|^{\frac{\delta+1}{2}}\right\|^2\right)^{\frac{\delta}{\delta+1}}\\ &+B(\delta)\|\nabla\rho\|_3^{\frac{2}{1-\vartheta(\delta)}} \qquad (10.41)\\ \underset{(10.36)}{\leq}\; &-F_2\left\||M|^{\frac{\delta+1}{2}}\right\|^2+f_3\|1\|_{\delta+1}\left(\left\||M|^{\frac{\delta+1}{2}}\right\|^2\right)^{\frac{\delta}{\delta+1}}\\ &+\tilde{B}(\delta). \qquad (10.42)\end{aligned}$$

Here, then, we fall upon an inequality of the same kind as (10.25) with $\varphi = \left\||M|^{\frac{\delta+1}{2}}\right\|^2$. Approaching it in the same way as there, we derive finally

$$\sup_{t>0}||M(t)||_\delta<\infty \quad \forall\delta\geq 0. \tag{10.43}$$

As an immediate consequence of this, we gain by means of the regularity rule (10.35) and the Sobolev inequality, the uniform boundedness of $\rho$, i.e.,

$$\|\rho\|_\infty<\infty, \tag{10.44}$$

and by taking $\hat{\delta}:=\widetilde{\delta_1}>6$ and $\mu:=\mu_1$ with $\frac{1}{2}+\frac{3}{2\widetilde{\delta_1}}<\mu_1\leq\frac{3}{4}$, we further get

$$\|\nabla\rho\|_\infty<\infty. \tag{10.45}$$

Next, we return to the equality (10.37) to repeat the whole procedure once more but this time being more precise about the estimates being made and using the regularity achieved up to this point. First, due to (10.39) and two obvious inequalities, we have

$$\begin{aligned}\frac{d}{dt}\left\||M|^{\frac{\delta+1}{2}}\right\|^2 &= -\frac{4\delta(\delta+1)}{(\alpha+\delta+1)^2}\left\|\nabla|M|^{\frac{\alpha+\delta+1}{2}}\right\|^2\\ &\quad+\frac{2\delta(\delta+1)}{\alpha+\delta+1}\left(\nabla|M|^{\frac{\alpha+\delta+1}{2}},|M|^{\gamma-\frac{\alpha}{2}+\frac{\delta-1}{2}}\nabla\rho\right)\\ &\quad+(\delta+1)(f(t,M,\rho),M|M|^{\delta-1})\\ &\leq -B_3\left\|\nabla|M|^{\frac{\alpha+\delta+1}{2}}\right\|^2\\ &\quad+(\delta+1)B_4||\nabla\rho||_\infty\left\|\nabla|M|^{\frac{\alpha+\delta+1}{2}}\right\|\left\||M|^{\frac{\alpha+\delta+1}{2}}\right\|^{\vartheta(\delta)}\\ &\quad-(\delta+1)F_2\left\||M|^{\frac{\delta+1}{2}}\right\|^2+(\delta+1)B_5f_3\left\||M|^{\frac{\alpha+\delta+1}{2}}\right\|^{2\zeta} \qquad (10.46)\end{aligned}$$

for $\delta \geq \alpha-\gamma+1$ with $\zeta=\frac{\alpha+\delta+1}{\delta}$. Recall that $f_3 \in L^\kappa_b(\mathbb{R}^+_0)$ and $\kappa>1$. Taking into account a special case of the interpolation inequality for Sobolev spaces (see [2]):

$$||v|| \leq C_\kappa ||\nabla v||^{1-\frac{1}{\kappa}} ||v||_p^{\frac{1}{\kappa}}, \quad p=\frac{6}{1+2\kappa},$$

we obtain with the help of the Young inequality

$$\begin{aligned}(\delta+1)&||\nabla v|| ||v||^{\vartheta(\delta)}\\ &\leq (\delta+1) C_\kappa^{\vartheta(\delta)} ||\nabla v||^{1+\vartheta(\delta)\left(1-\frac{1}{\kappa}\right)} ||v||_p^{\vartheta(\delta)\frac{1}{\kappa}}\\ &\leq C_\kappa^{\vartheta(\delta)} B_4 \left(\varepsilon ||\nabla v||^2 + B_1(\varepsilon)(\delta+1)^{\frac{2}{1-\vartheta(\delta)\left(1-\frac{1}{\kappa}\right)}} ||v||_p^{\frac{2\vartheta(\delta)\frac{1}{\kappa}}{1-\vartheta(\delta)\left(1-\frac{1}{\kappa}\right)}}\right)\end{aligned} \tag{10.47}$$

and

$$\begin{aligned}(\delta+1) f_3 ||v||^{2\zeta} &\leq (\delta+1) f_3 C_\kappa^{2\zeta} ||\nabla v||^{2\zeta\left(1-\frac{1}{\kappa}\right)} ||v||_p^{2\zeta\frac{1}{\kappa}}\\ &\leq C_\kappa^{2\zeta} B_5 \left(\varepsilon ||\nabla v||^2 + B_2(\varepsilon)\left(f_3(\delta+1)\right)^{\frac{1}{1-\zeta\left(1-\frac{1}{\kappa}\right)}} ||v||_p^{\frac{2\zeta\frac{1}{\kappa}}{1-\zeta\left(1-\frac{1}{\kappa}\right)}}\right).\end{aligned} \tag{10.48}$$

With the Hölder inequality we also have

$$\left\| |M|^{\frac{\alpha+\delta+1}{2}} \right\|_p \leq \left\| |M|^{\frac{\delta+1}{2}} \right\|_s \left\| |M|^{\frac{\alpha}{2}} \right\|_{\frac{sp}{s-p}} \leq B_6 \left\| |M|^{\frac{\delta+1}{2}} \right\|_s, \quad s>\frac{6}{1+2\kappa}. \tag{10.49}$$

Since $\kappa>1$, $\frac{6}{1+2\kappa}<2$ holds; therefore, we can assume that $s<2$. Combining (10.47)–(10.48) for $v:=|M|^{\frac{\alpha+\delta+1}{2}}$ with (10.49) and choosing $\varepsilon$ small enough, we can conclude from (10.46) that

$$\begin{aligned}\frac{d}{dt}\left\| |M|^{\frac{\delta+1}{2}} \right\|^2 \leq\, & B_7 \left(\|\nabla\rho\|_\infty (\delta+1)\right)^{\frac{2}{1-\vartheta(\delta)\left(1-\frac{1}{\kappa}\right)}} \left\| |M|^{\frac{\delta+1}{2}} \right\|_s^{\frac{2\vartheta(\delta)\frac{1}{\kappa}}{1-\vartheta(\delta)\left(1-\frac{1}{\kappa}\right)}}\\ &+ B_7 \left(f_3(\delta+1)\right)^{\frac{1}{1-\zeta\left(1-\frac{1}{\kappa}\right)}} \left\| |M|^{\frac{\delta+1}{2}} \right\|_s^{\frac{2\zeta\frac{1}{\kappa}}{1-\zeta\left(1-\frac{1}{\kappa}\right)}}\\ &- F_2(\delta+1) \left\| |M|^{\frac{\delta+1}{2}} \right\|^2\end{aligned}$$

for $\delta \geq \alpha-\gamma+1$. Since $\vartheta(\delta), \zeta \in (0,1)$, it follows for all $\delta \geq \alpha-\gamma+2$ that

$$\begin{aligned}\frac{d}{dt}\left(\|M\|_\delta^\delta+1\right) \leq\, & B_8\left(\|\nabla\rho\|_\infty^{2\kappa}\delta^{2\kappa}+\delta^\kappa(f_3+1)^\kappa\right)\left(\|M\|_{s\delta/2}^{s\delta/2}+1\right)^{\frac{2}{s}}\\ &- F_2\delta\left(\|M\|_\delta^\delta+1\right)\end{aligned}$$

and once more we get an integral inequality for $\|M(t)\|_\delta^\delta+1$:

$$\begin{aligned}\|M(t)\|_\delta^\delta+1 \leq & B_8\int_0^t e^{-\delta F_2(t-\tau)}\left(\|\nabla\rho(\tau)\|_\infty^{2\kappa}\delta^{2\kappa}+\delta^\kappa(f_3(\tau)+1)^\kappa\right)\\ &\cdot\left(\|M(\tau)\|_{s\delta/2}^{s\delta/2}+1\right)^{\frac{2}{s}} d\tau + e^{-\delta F_2 t}\left(\|M_0\|_\delta^\delta+1\right).\end{aligned} \tag{10.50}$$

To shorten the formulas, we introduce

$$A(\delta) := \sup_{t>0} \|M(t)\|_\delta^\delta + 1.$$

Hence it follows from (10.50) and (10.45) that

$$
\begin{aligned}
A(\delta) \leq\ & B_9 \left(A\left(\delta \frac{s}{2}\right)\right)^{\frac{2}{s}} \int_0^t e^{-\delta F_2(t-\tau)} \delta^{2\kappa} (f_3(\tau)+1)^{\kappa}\, d\tau \\
& + e^{-\delta F_2 t} \left(\|M_0\|_\delta^\delta + 1\right) \\
\leq\ & B_{10} \delta^{2\kappa-1} \left(A\left(\delta \frac{s}{2}\right)\right)^{\frac{2}{s}} \left(1 + ||f_3||^{\kappa}_{L^{\kappa}_b(\mathbb{R}_0^+)}\right) + \left(\|M_0\|_\delta^\delta + 1\right) \\
\leq\ & B_{11} \delta^{2\kappa-1} \left(A\left(\delta \frac{s}{2}\right)\right)^{\frac{2}{s}} + \left(\|M_0\|_\delta^\delta + 1\right). \qquad (10.51)
\end{aligned}
$$

Suppose $\lim\limits_{\delta\to\infty} (A(\delta))^{\frac{1}{\delta}} = +\infty$. Since $\|M_0\|_\delta \leq \|M_0\|_\infty \mu_\lambda(\Omega)$ for all $\delta \geq 1$, there exists a $\delta_1 > 1$ with the property

$$
A(\delta) = \sup_{t>0} \|M(t)\|_\delta^\delta + 1 \geq 2^\delta \left(\|M_0\|_\delta^\delta + 1\right) \quad \forall \delta \geq \delta_1. \qquad (10.52)
$$

Therefore, we may consider the following inequality instead of (10.51):

$$
A(\delta) \leq B_{12} \delta^{2\kappa-1} \left(A\left(\delta \frac{s}{2}\right)\right)^{\frac{2}{s}}. \qquad (10.53)
$$

One can then show by induction that

$$
A\left(\delta_1 \left(\frac{2}{s}\right)^n\right) \leq \left(B_{12}\delta_1^{2\kappa-1}\right)^{\sum_{k=0}^{n-1}\left(\frac{2}{s}\right)^k} \left(\frac{2}{s}\right)^{(2\kappa-1)\sum_{k=0}^{n-1}(n-k)\left(\frac{2}{s}\right)^k} (A(\delta_1))^{\left(\frac{2}{s}\right)^n}
$$

so that we obtain

$$
\begin{aligned}
\left(A\left(\delta_1\left(\frac{2}{s}\right)^n\right)\right)^{\left(\frac{s}{2}\right)^n} &\leq \left(B_{12}\delta_1^{2\kappa-1}\right)^{\sum_{k=1}^{n}\left(\frac{s}{2}\right)^k} \left(\frac{s}{2}\right)^{(2\kappa-1)\sum_{k=1}^{n} k\left(\frac{s}{2}\right)^k} A(\delta_1) \\
&\underset{n\to\infty}{\to} \left(B_{12}\delta_1^{2\kappa-1}\right)^{\frac{\frac{s}{2}}{1-\frac{s}{2}}} \left(\frac{s}{2}\right)^{(2\kappa-1)\frac{s}{2}\left(\frac{1}{1-\frac{s}{2}}\right)^2} A(\delta_1). \qquad (10.54)
\end{aligned}
$$

Consequently, the sequence $\left(\left(A\left(\delta_1\left(\frac{2}{s}\right)^n\right)\right)^{\left(\frac{s}{2}\right)^n}\right)_{n\in\mathbb{N}}$ is bounded and it also follows (see the definition of $A(\delta)$) that $\limsup\limits_{\delta\to\infty} (A(\delta))^{\frac{1}{\delta}} < \infty$ so that our assumption made above proved to be false and $\sup\limits_{t>0} \|M(t)\|_\delta$, $\delta \geq \alpha - \gamma + 1$, appears to be bounded. Since

$$
\|M(t)\|_\delta \underset{\delta\to\infty}{\to} \|M(t)\|_\infty \leq \sup_{t>0,\delta\geq 1} \|M(t)\|_\delta < \infty, \qquad (10.55)
$$

we finally obtain the uniform boundedness of $M$.

In what follows, we will obtain more regularity of solutions, which we will use to prove the existence of solutions (see the first appendix of this chapter). First, due to the assumptions on $g$ and the uniform boundedness of both $M$ and $\rho$,

$$
|g(t,M,\rho)| = |g_1\rho + g_2(t,\rho)M| \leq g_1(0)||\rho||_\infty + g_3||M||_\infty \in L^2_b(\mathbb{R}_0^+).
$$

With the help of Theorem 3.3 from [93] we obtain the following regularity result from equation (10.2):

$$\dot{\rho}, \Delta\rho \in L^2_{loc}(\mathbb{R}^+_0, L^2(\Omega)),$$
$$\rho \in C\left(\mathbb{R}^+_0; H^1(\Omega)\right).$$

Moreover, based on the proof of this theorem we have an estimate

$$\begin{aligned}\sup_{n\in\mathbb{N}} \int_n^{n+1} \|\Delta\rho\|^2 &\le \sup_{n\in\mathbb{N}} \|\nabla\rho(n)\|^2 + \int_n^{n+1} \|g(\tau, M(\tau), \rho(\tau))\|^2 \, d\tau \\ &\le \sup_{t>0} \|\nabla\rho(t)\|^2 + ||g||_{L^2_b(\mathbb{R}^+_0, L^2(\Omega))}.\end{aligned}$$

Together with (10.45), we arrive at a somewhat better regularity for $\rho$:

$$\dot{\rho}, \Delta\rho \in L^2_b(\mathbb{R}^+_0, L^2(\Omega)), \quad \rho \in C_b\left(\mathbb{R}^+_0; H^1(\Omega)\right). \tag{10.56}$$

Secondly, we can make use of $\gamma \ge \frac{\alpha}{2} + 1$ so that $\gamma - \frac{\alpha}{2} + \frac{\delta-1}{2} = \gamma - 1 - \frac{\alpha}{2} + \frac{\delta+1}{2}$ is positive for all $\delta > 0$ and from the inequality (10.46) it follows that

(10.57)

$$\begin{aligned}&\frac{d}{dt} \left\| |M|^{\frac{\delta+1}{2}} \right\|^2 \le -B_{13} \left\| \nabla |M|^{\frac{\alpha+\delta+1}{2}} \right\|^2 + B_{14} f_3 \\ &\quad \Rightarrow \sup_{n\in\mathbb{N}} \int_n^{n+1} \left\| \nabla |M|^{\frac{\alpha+\delta+1}{2}} \right\|^2 d\tau \le B_{15} \sup_{n\in\mathbb{N}} \left\| |M|^{\frac{\delta+1}{2}}(n) \right\|^2 + B_{16} ||f_3||_{L^1_b(\mathbb{R}^+_0)} < \infty.\end{aligned}$$

Hence

$$|M|^{\frac{\alpha+\delta+1}{2}} \in L^2_b(\mathbb{R}^+_0, H^1(\Omega))$$

holds for all $\delta > 0$. With $\frac{\alpha+1}{2} < \gamma$ we conclude that

$$|M|^\gamma \in L^2_b(\mathbb{R}^+_0, H^1(\Omega)). \tag{10.58}$$

Combining (10.56) with (10.58) and using the uniform boundedness of $M$ and $\nabla\rho$ one more time, we then get

$$\nabla \cdot (|M|^\gamma \nabla\rho) = (\nabla|M|^\gamma, \nabla\rho) + (|M|^\gamma, \Delta\rho) \in L^2_b(\mathbb{R}^+_0, L^2(\Omega)).$$

Equation (10.1) can thus be regarded as a classical porous medium equation (i.e., without chemotaxis effect):

$$\dot{M} = \nabla \cdot (|M|^\alpha \nabla M) + \tilde{f} \tag{10.59}$$

where

$$\tilde{f}(t) := -\nabla \cdot (|M|^\gamma \nabla\rho) + f(t, M, \rho) \in L^2_b(\mathbb{R}^+_0, L^2(\Omega)) \tag{10.60}$$

and we therefore conclude that

$$\dot{M} \in L^2_b(\mathbb{R}^+_0, H^{-1}(\Omega)).$$

REMARK 10.4. It is not difficult to see that for $\rho_0 \ge 0$, $M_0 \ge 0$ any solution remains nonnegative. (See [34] for a general result.)

## 10.2. Uniqueness

To show uniqueness of solution, we need some preparation first. Consider the operator

$$\nabla : \ H_0^1(\Omega) \to \left(L^2(\Omega)\right)^N .$$

The corresponding adjoint operator

$$\nabla^* : \ \left(L^2(\Omega)\right)^N \to H^{-1}(\Omega)$$

coincides with the operator $-(\nabla \cdot)$. That is why

$$-\Delta = -\nabla \cdot \nabla = \nabla^* \nabla.$$

Denote by

$$\nabla^+ : \ \left(L^2(\Omega)\right)^N \to H_0^1(\Omega)$$

the Moore-Penrose pseudoinverse (see [52]) of the operator $\nabla$. It exists because the range of $\nabla$ is closed and it can be expressed by

$$\nabla^+ = (\nabla)^{-1} P_{Ran(\nabla)},$$

where $P_{Ran(\nabla)}$ means the orthogonal projection on the range of $\nabla$. It follows directly from the properties of a pseudoinverse that

$$(-\Delta)^{-1} = \nabla^+ \nabla^{+*},$$

$$\nabla(-\Delta)^{-1} = \nabla^{+*}, \tag{10.61}$$

$$\nabla^+ \nabla = \text{Id}. \tag{10.62}$$

Now we are ready to prove the uniqueness of solutions for the problem (10.1)–(10.4). Suppose it has two solutions $(M_1, \rho_1)$, $(M_2, \rho_2)$ with the same initial data:

$$M_1(0) = M_2(0) \geq 0, \quad \rho_1(0) = \rho_2(0) \geq 0.$$

Since both $(M_1, \rho_1)$ and $(M_2, \rho_2)$ are nonnegative (see Remark 10.1) solutions of equation (10.1), we get

$$\begin{aligned}
\frac{d}{dt}(M_1 - M_2) &= \frac{1}{\alpha+1} \Delta \left(M_1^{\alpha+1} - M_2^{\alpha+1}\right) - \nabla \cdot \left(M_1^\gamma \nabla \rho_1 - M_2^\gamma \nabla \rho_2\right) \\
&\quad + \left(f(t, M_1, \rho_1) - f(t, M_2, \rho_2)\right) \\
&= \frac{1}{\alpha+1} \Delta \left(M_1^{\alpha+1} - M_2^{\alpha+1}\right) \\
&\quad - \nabla \cdot \left(M_1^\gamma \nabla (\rho_1 - \rho_2)\right) - \nabla \cdot \left(\left(M_1^\gamma - M_2^\gamma\right) \nabla \rho_2\right) \\
&\quad + \left(f(t, M_1, \rho_1) - f(t, M_1, \rho_2)\right) \\
&\quad + \left(f(t, M_1, \rho_2) - f(t, M_2, \rho_2)\right).
\end{aligned} \tag{10.63}$$

We want to estimate the difference $M_1 - M_2$ and we choose to do so in the $||\nabla^{+*}(\cdot)||$-norm. For this purpose we multiply (10.63) by $(-\Delta)^{-1}(M_1 - M_2)$ and integrate over $\Omega$:

$$\begin{aligned}\left(\frac{d}{dt}(M_1 - M_2), (-\Delta)^{-1}(M_1 - M_2)\right)\\ &= \left(\Delta\left(M_1^{\alpha+1} - M_2^{\alpha+1}\right), (-\Delta)^{-1}(M_1 - M_2)\right)\\ &\quad + \left(-\nabla\cdot\left(M_1^{\gamma}\nabla(\rho_1 - \rho_2)\right), (-\Delta)^{-1}(M_1 - M_2)\right)\\ &\quad + \left(-\nabla\cdot\left(\left(M_1^{\gamma} - M_2^{\gamma}\right)\nabla\rho_2\right), (-\Delta)^{-1}(M_1 - M_2)\right)\\ &\quad + \left(f(t, M_1, \rho_1) - f(t, M_1, \rho_2), (-\Delta)^{-1}(M_1 - M_2)\right)\\ &\quad + \left(f(t, M_1, \rho_2) - f(t, M_2, \rho_2), (-\Delta)^{-1}(M_1 - M_2)\right). \end{aligned} \tag{10.64}$$

On the left side of the resulting equation we find

$$\left(\frac{d}{dt}(M_1 - M_2), (-\Delta)^{-1}(M_1 - M_2)\right) = \frac{1}{2}\frac{d}{dt}\left\|\nabla^{+*}(M_1 - M_2)\right\|^2. \tag{10.65}$$

Suitable estimates for the terms on the right side of (10.64) are required now. The operator $\Delta$ is selfadjoint; therefore,

$$\left(\Delta\left(M_1^{\alpha+1} - M_2^{\alpha+1}\right), (-\Delta)^{-1}(M_1 - M_2)\right) = -\left(M_1^{\alpha+1} - M_2^{\alpha+1}, M_1 - M_2\right). \tag{10.66}$$

We already know (see Section 10.1) that

$$\|M_i\|_\infty, \|\nabla\rho_i\|_\infty, \|\rho_i\|_\infty < \infty \tag{10.67}$$

holds. This and (10.61) lead to

$$\begin{aligned}\left(-\nabla\cdot\left(M_1^{\gamma}\nabla(\rho_1 - \rho_2)\right), (-\Delta)^{-1}(M_1 - M_2)\right) &= \left(\left(M_1^{\gamma}\nabla(\rho_1 - \rho_2)\right), \nabla^{+*}(M_1 - M_2)\right)\\ &\leq U_1\left\|\nabla(\rho_1 - \rho_2)\right\|\left\|\nabla^{+*}(M_1 - M_2)\right\|,\end{aligned} \tag{10.68}$$

$$\begin{aligned}\left(-\nabla\cdot\left(\left(M_1^{\gamma} - M_2^{\gamma}\right)\nabla\rho_2\right), (-\Delta)^{-1}(M_1 - M_2)\right) &= \left(\left(M_1^{\gamma} - M_2^{\gamma}\right)\nabla\rho_2, \nabla^{+*}(M_1 - M_2)\right)\\ &\leq U_2\left\|M_1^{\gamma} - M_2^{\gamma}\right\|\left\|\nabla^{+*}(M_1 - M_2)\right\|\end{aligned} \tag{10.69}$$

for some $U_1$, $U_2 > 0$. Finally, combining (10.67) with the assumptions made on $f$ and fixing an (arbitrary) $T > 0$, we obtain

$$\begin{aligned}|f(s, M_1, \rho_1) - f(s, M_1, \rho_2)| &= \left|\int_0^1 \frac{\partial f}{\partial \rho}(s, M_1, \rho_1 + \tau(\rho_2 - \rho_1))\, d\tau\right| |\rho_1 - \rho_2|\\ &\underset{(10.11)}{\leq} U_3(T)\,|\rho_1 - \rho_2|\end{aligned}$$

and

$$\begin{aligned}&|f(s, M_1, \rho_2) - f(s, M_2, \rho_2)|\\ &\quad\underset{(10.10)}{=} \left|\widetilde{f}\left(s, M_1^{\frac{\alpha}{2}+1}, \rho_2\right) - \widetilde{f}\left(s, M_2^{\frac{\alpha}{2}+1}, \rho_2\right) + F_4\left(M_1^{\frac{\alpha}{2}+1} - M_2^{\frac{\alpha}{2}+1}\right)\right|\\ &\quad\leq \left(\left|\int_0^1 \frac{\partial \widetilde{f}}{\partial M}\left(s, M_1^{\frac{\alpha}{2}+1} + \tau\left(M_2^{\frac{\alpha}{2}+1} - M_1^{\frac{\alpha}{2}+1}\right), \rho_2\right) d\tau\right| + |F_4|\right)\\ &\qquad\cdot\left|M_1^{\frac{\alpha}{2}+1} - M_2^{\frac{\alpha}{2}+1}\right|\\ &\quad\underset{(10.10)}{\leq} U_4(T)\left|M_1^{\frac{\alpha}{2}+1} - M_2^{\frac{\alpha}{2}+1}\right|\end{aligned} \tag{10.70}$$

for some constants $U_3 = U_3(T)$, $U_4 = U_4(T)$ for all $s \in [0,T]$. Having the estimates (10.65)–(10.70), it is possible to conclude from (10.64) that

$$\begin{aligned}
&\frac{1}{2}\frac{d}{dt}\left\|\nabla^{+*}(M_1 - M_2)\right\|^2 \\
&\qquad \le -\left(M_1^{\alpha+1} - M_2^{\alpha+1}, M_1 - M_2\right) + U_5(T)\left\|\nabla^{+*}(M_1 - M_2)\right\| \\
&\qquad\quad \cdot \left(\|M_1^\gamma - M_2^\gamma\| + \|\nabla(\rho_1 - \rho_2)\| + \left\|M_1^{\frac{\alpha}{2}+1} - M_2^{\frac{\alpha}{2}+1}\right\|\right). \qquad (10.71)
\end{aligned}$$

With $\lim_{x\to 1}\frac{\left(x^{\frac{\alpha}{2}+1}-1\right)^2}{(x^{\alpha+1}-1)(x-1)} = \frac{\left(\frac{\alpha}{2}+1\right)^2}{\alpha+1} < \infty$, it follows that

$$\left(M_1^{\frac{\alpha}{2}+1} - M_2^{\frac{\alpha}{2}+1}\right)^2 \le L_1\left(M_1^{\alpha+1} - M_2^{\alpha+1}\right)(M_1 - M_2) \qquad (10.72)$$

and with $\lim_{x\to 1}\frac{x^{\frac{\alpha}{2}+1}-1}{x^\gamma-1} = \frac{\frac{\alpha}{2}+1}{\gamma} < \infty$ and $\gamma \ge \frac{\alpha}{2}+1$, it follows that

$$|M_1^\gamma - M_2^\gamma| \le L_2\left(||M_1||_\infty^{\gamma-\frac{\alpha}{2}+1} + ||M_2||_\infty^{\gamma-\frac{\alpha}{2}+1}\right)\left|M_1^{\frac{\alpha}{2}+1} - M_2^{\frac{\alpha}{2}+1}\right|. \qquad (10.73)$$

Applying (10.72), (10.73) together with the Young inequality to (10.71) yields finally

$$\frac{1}{2}\frac{d}{dt}\left\|\nabla^{+*}(M_1 - M_2)\right\|^2 \le U_6(T)\left\|\nabla^{+*}(M_1 - M_2)\right\|^2 + U_6(T)\left\|\nabla(\rho_1 - \rho_2)\right\|^2. \qquad (10.74)$$

Now we turn to equation (10.2). Both $(M_1, \rho_1)$ and $(M_2, \rho_2)$ solve it. Hence

$$\begin{aligned}
\frac{d}{dt}(\rho_1 - \rho_2) &= \Delta(\rho_1 - \rho_2) - (g(t, M_1, \rho_1) - g(t, M_2, \rho_2)) \\
&= \Delta(\rho_1 - \rho_2) - g_1(\rho_1 - \rho_2) - (g_2(t,\rho_1)M_1 - g_2(t,\rho_2)M_2) \\
&= \Delta(\rho_1 - \rho_2) - g_1(\rho_1 - \rho_2) - (g_2(t,\rho_1) - g_2(t,\rho_2))M_1 \\
&\quad - g_2(t,\rho_2)(M_1 - M_2). \qquad (10.75)
\end{aligned}$$

As usual, we multiply (10.75) by $\rho_1 - \rho_2$ and integrate over $\Omega$ so that we get

$$\begin{aligned}
\frac{1}{2}\frac{d}{dt}\|\rho_1 - \rho_2\|^2 = &- \|\nabla(\rho_1 - \rho_2)\|^2 - g_1\|\rho_1 - \rho_2\|^2 \\
&- ((g_2(t,\rho_1) - g_2(t,\rho_2))M_1, \rho_1 - \rho_2) \\
&- (g_2(t,\rho_2)(M_1 - M_2), \rho_1 - \rho_2). \qquad (10.76)
\end{aligned}$$

Due to property (10.62) of $\nabla^+$ and the assumptions made on $g_2$ together with the Poincaré inequality and the uniform boundedness of $\nabla\rho_2$, $M_1$, we obtain the estimates

$$
\begin{aligned}
(-g_2(s,\rho_2)(M_1-M_2),\rho_1-\rho_2) = & -(g_2(s,\rho_2)(\rho_1-\rho_2),M_1-M_2) \\
= & \left(\nabla(g_2(s,\rho_2)(\rho_1-\rho_2)),\nabla^{+*}(M_1-M_2)\right) \\
= & \left(g_2(s,\rho_2)\nabla(\rho_1-\rho_2),\nabla^{+*}(M_1-M_2)\right) \\
& + \left(\frac{\partial g_2}{\partial \rho}(s,\rho_2)(\rho_1-\rho_2)\nabla\rho_2,\nabla^{+*}(M_1-M_2)\right) \\
\underset{(10.11)}{\leq} & \; g_3\|\nabla(\rho_1-\rho_2)\|\left\|\nabla^{+*}(M_1-M_2)\right\| + \|\nabla\rho_2\|_\infty \\
& \cdot\left\|\frac{\partial g_2}{\partial\rho}(s,\rho_2)|\rho_1-\rho_2|\right\|\left\|\nabla^{+*}(M_1-M_2)\right\| \\
\underset{(10.11)}{\leq} & \; (g_3+U_7(T))\|\nabla(\rho_1-\rho_2)\|\left\|\nabla^{+*}(M_1-M_2)\right\| \qquad (10.77)
\end{aligned}
$$

and

$$
\begin{aligned}
|(g_2(s,\rho_1)-g_2(s,\rho_2))M_1| \underset{(10.11)}{\leq} & \left|\int_0^1 \frac{\partial g_2}{\partial\rho}(s,\rho_1+\tau(\rho_2-\rho_1))\,d\tau\right|\|M_1\|_\infty|\rho_1-\rho_2| \\
\leq & \; U_8(T)|\rho_1-\rho_2|. \qquad (10.78)
\end{aligned}
$$

Applying estimates (10.77) and (10.78) to (10.76) yields

$$
\begin{aligned}
\frac{1}{2}\frac{d}{dt}\|\rho_1-\rho_2\|^2 \leq & -\|\nabla(\rho_1-\rho_2)\|^2 + U_8(T)\|\rho_1-\rho_2\|^2 \\
& + (g_3+U_7(T))\|\nabla(\rho_1-\rho_2)\|\left\|\nabla^{+*}(M_1-M_2)\right\|. \qquad (10.79)
\end{aligned}
$$

Adding (10.74) and (10.79) together and using the Young inequality one more time, we finally come to the inequality

$$
\begin{aligned}
& \frac{d}{dt}\left(\left\|\nabla^{+*}(M_1-M_2)\right\|^2 + U_6(T)\|\rho_1-\rho_2\|^2\right) \\
& \quad \leq U_9(T)(g_3+1)\left(\left\|\nabla^{+*}(M_1-M_2)\right\|^2 + U_6(T)\|\rho_1-\rho_2\|^2\right)
\end{aligned}
$$

and consequently

(10.80)

$$
\begin{aligned}
& \left\|\nabla^{+*}(M_1(t)-M_2(t))\right\|^2 + U_6(T)\|\rho_1(t)-\rho_2(t)\|^2 \\
& \quad \leq \left(\left\|\nabla^{+*}(M_1(0)-M_2(0))\right\|^2 + U_6(T)\|\rho_1(0)-\rho_2(0)\|^2\right) e^{U_9(T)\int_0^t g_3(s)+1\,ds}.
\end{aligned}
$$

This proves uniqueness for the problem (10.1)–(10.4) since the solutions $(M_1,\rho_1)$, $(M_2,\rho_2)$ coincide at $t=0$.

## 10.3. Appendix (Existence of solution)

We start with nondegenerate approximations of the problem (10.1)–(10.4) in $Q_{inf} := (0,\infty)\times\Omega$, namely:

$$
\dot{M}_\varepsilon = \varepsilon\Delta M_\varepsilon + \nabla\cdot(|M_\varepsilon|^\alpha\nabla M_\varepsilon) - \nabla\cdot(|M_\varepsilon|^\gamma\nabla\rho_\varepsilon) + f(t,M_\varepsilon,\rho_\varepsilon) \text{ in } Q_{inf}, \qquad (10.81)
$$

$$
\dot{\rho}_\varepsilon = \Delta\rho_\varepsilon - g(t,M_\varepsilon,\rho_\varepsilon) \qquad \text{in } Q_{inf} \qquad (10.82)
$$

with the same initial and boundary conditions as before:

$$M_\varepsilon = 0, \qquad \rho_\varepsilon = 1 \quad \text{in } (0,\infty)\times\partial\Omega, \tag{10.83}$$

$$M_\varepsilon(\cdot,0) = M_0, \quad \rho_\varepsilon(\cdot,0) = \rho_0 \text{ in } \Omega \tag{10.84}$$

for $\varepsilon \in (0,1)$. In the same manner as in Section 10.1 one can show that

$$\|M_\varepsilon\|_\infty + \|\nabla\rho_\varepsilon\|_\infty + \left\||M_\varepsilon|^{\frac{\alpha}{2}+1}\right\|_{L^2_b(\mathbb{R}_0^+,H^1(\Omega))} + \left\|\dot{M}_\varepsilon - \varepsilon\Delta M_\varepsilon\right\|_{L^2_b(\mathbb{R}_0^+,H^{-1}(\Omega))} \le C_1,$$

where the constant $C_1$ is independent of $\varepsilon$. From $\left\|\dot{M}_\varepsilon - \varepsilon\Delta M_\varepsilon\right\|_{L^2_b(\mathbb{R}_0^+,H^{-1}(\Omega))} \le C_1$ it follows that

$$\left\|\dot{M}_\varepsilon\right\|_{L^2_b(\mathbb{R}_0^+,H^{-1}(\Omega))}, \ \varepsilon\left\|\Delta M_\varepsilon\right\|_{L^2_b(\mathbb{R}_0^+,H^{-1}(\Omega))} \le C_2 \tag{10.85}$$

and again the constant does not depend on $\varepsilon$. These nondegenerate problems are, therefore, uniquely solvable in the class of functions which is defined in Definition 10.1 (see [68] or [92]). We only need the compactness argument then to show the existence of solutions for the problem (10.1)–(10.4): we use a lemma of Dubinskii (see [69] for an application of this lemma to the porous medium equation).

## 10.4. Appendix (Total biomass)

Let us now assume the starting value $M_0$ to be separated from 0 in some ball $B \subseteq \Omega$:

$$\operatorname*{ess\,inf}_B M_0 \ge \varepsilon > 0.$$

This time we multiply equation (10.1) with $-\varphi M^{-1}$ choosing a smooth cut-off function

$$\varphi \in C_0^\infty(B), \quad 0 \le \varphi \le 1, \quad \varphi = 1 \text{ in a ball } B_1 \subset B$$

and we integrate over $\Omega$:

$$\left(\dot{M}, -\varphi M^{-1}\right) = \left(\nabla\cdot(M^\alpha\nabla M) - \nabla\cdot(M^\gamma\nabla\rho) + f(t,M,\rho), -\varphi M^{-1}\right) \tag{10.86}$$

Integrating (10.86) by parts, we get

$$\begin{aligned}\frac{d}{dt}\int_B -\varphi\ln M(\cdot,x)\,dx = &-\frac{4}{\alpha^2}\left\|\varphi^{\frac12}\nabla M^{\frac{\alpha}{2}}\right\|^2 + \frac{4}{\alpha}\left(\varphi^{\frac12}\nabla M^{\frac{\alpha}{2}}, M^{\frac{\alpha}{2}}\nabla\varphi^{\frac12}\right)\\ &+\frac{2}{\alpha}\left(\varphi^{\frac12}\nabla M^{\frac{\alpha}{2}}, \varphi^{\frac12}M^{\gamma-\frac{\alpha}{2}-1}\nabla\rho\right) - \left(M^{\gamma-1}, \nabla\varphi\nabla\rho\right)\\ &-\int_B \varphi\frac{f(t,M,\rho)}{M}\,dx.\end{aligned} \tag{10.87}$$

The expression (10.87) is quite similar to what we had in (10.37). In the same manner it can be shown that the right side of (10.87) is bounded from above by some positive constant $C_1(T)$. This is due to the assumptions made on $f$, the uniform boundedness of $\nabla\rho$, and the cut-off function, but also to the fact that $\frac{\alpha}{2}+1 \le \gamma$ (and $\gamma > 1$ of course). Thus we get

$$\frac{d}{dt}\int_B -\varphi\ln M(t,x)\,dx \le C_1(T)$$

and, consequently,

$$\int_B -\varphi \ln M(t,x)\, dx \le C_2(T), \quad t \in [0,T],$$

since $M_0$ is nonzero in $B$. It follows from $M$ being uniformly bounded that

$$\int_{B_1 \cap \{M<1\}} - \ln M(t,x)\, dx \le C_2(T) + \int_{B \cap \{M \ge 1\}} \ln M(t,x) \le C_3(T). \tag{10.88}$$

Finally, applying the Jensen inequality to the left side of (10.88) (note: the function $-\ln$ is convex), we get

$$- \ln \int_{B_1 \cap \{M<1\}} M(t,x)\, dx \le C_4(T).$$

Thus

$$\int_\Omega M(t,x)\, dx \ge \int_{B_1 \cap \{M<1\}} M(t,x)\, dx \ge C_5(T) > 0$$

and the total mass is therefore nonzero for all $t > 0$.

CHAPTER 11

# ODE-PDE coupling arising in the modelling of a forest ecosystem

We study the coupled ODE-PDE system

$$(11.1)\qquad \begin{cases} \partial_t^2 v + \varphi(v)\partial_t v + f(v) = \alpha w, \\ \partial_t w - \Delta_x w + w = v, \quad \partial_n w\big|_{\partial\Omega} = 0 \end{cases}$$

in a bounded smooth domain $\Omega \subset \mathbb{R}^n$, $n \leq 3$. Here $(v, w) = (v(t,x), w(t,x))$ are unknown functions, $\Delta_x$ is a Laplacian with respect to $x$, $\alpha > 0$ is a given parameter, and $f$ and $\varphi$ are given nonlinearities which are assumed to satisfy some natural assumptions formulated in Section 11.1.

Our interest in that problem is motivated by the following system arising in mathematical biology:

$$(11.2)\qquad \begin{cases} \partial_t u = \beta\delta w - \gamma(v)u - fu, \\ \partial_t v = fu - hv, \\ \partial_t w - d\Delta_x w + \beta w = \alpha v, \quad \partial_n w\big|_{\partial\Omega} = 0, \end{cases}$$

where $\alpha, \beta, \delta, d, f, h$ are given positive parameters and $\gamma(v)$ is a given nonlinearity. This system was introduced in [65] in order to describe the development of a forest ecosystem (the unknown functions $u$, $v$, and $w$ are the densities of yang trees, old trees, and the seeds, respectively, and the given nonlinearity $\gamma(v)$ describes the mortality of yang trees in dependence upon the density of the old ones) and has been recently studied analytically and numerically in [22], [23], [77]. Expressing $u = (\partial_t v + hv)/f$ from the second equation of (11.2) and inserting it into the first one, we end up with the system of the type (11.1) with respect to the variables $(v, w)$.

The main aim of the present chapter is to study the long-time behavior of solutions of (11.1) using the ideas and methods of the theory of attractors. From the mathematical point of view, the problem considered is a coupled system of a second-order ODE with a linear PDE (heat-like equation). Heuristically, it is clear that the dynamics of such coupled dissipative systems should depend drastically on the monotonicity properties of the ODE component. In the case where this ODE is "monotone", i.e., it cannot produce internal instability (and all of the instability is driven by the coupling with the PDE component), one expects asymptotic compactness and the existence of a smooth finite-dimensional global attractor with "good" properties. In contrast to that, in the nonmonotone case, the ODE instability may produce the asymptotic discontinuities and even may completely destroy the initially smooth spatial profile. Thus, in that case, the smoothing effect from the PDE component is not strong enough to suppress the development of discontinuities provided by the internal instabilities of the ODE component and, as a result, spatial discontinuities and extremely complicated (in a sense, pathological) spatial structures may appear (see, for instance, [8] and [84] for the analysis of similar effects in one-dimensional finite visco-elasticity). We also mention that, although the existence and uniqueness of a solution of (11.2) has been rigorously proved in the above-mentioned papers [22], [23], very little has been done concerning the asymptotic behavior of solutions as $t \to \infty$. To be more precise, different types of $\omega$-limit sets of a single trajectory were considered there (associated with the different choice of the topology in the phase space) and their simplest properties were formulated, but even the question of whether or not they are empty remained open. As we will see below, some of them are indeed empty for the greater part of the trajectories if the monotonicity assumption is essentially violated; see Remark 11.4

In the present chapter, we justify the above heuristics in a mathematically rigorous way with the example of the ODE-PDE coupled system (11.1). In particular, we show (in Section 11.2) that the monotonicity arguments work perfectly if

$$f'(v) \geq \kappa_0 > 0, \quad v \in \mathbb{R}. \tag{11.3}$$

In this case, problem (11.1) indeed possesses a smooth global attractor $\mathcal{A}$ of finite fractal dimension in the proper phase space $\Phi_\infty$. Moreover, as we have already mentioned in Chapter 2, due to the presence of a global Lyapunov function, this attractor generically has very nice properties (it is the so-called regular attractor in the terminology of Babin and Vishik [9]). Namely, it is a finite collection of the finite-dimensional unstable manifolds associated with the equilibria:

$$\mathcal{A} = \bigcup_{u_0 \in \mathcal{R}} \mathcal{M}^+_{u_0}$$

(where $\mathcal{R}$ denotes a (generically finite) set of equilibria of problem (11.1) and $\mathcal{M}^+_{u_0}$ is an unstable manifold associated with the equilibrium $u_0 \in \mathcal{R}$; see Section 11.2). Moreover, every trajectory of (11.1) converges *exponentially* to one of these equilibria. We also mention that the first equation of (11.1) is a *second*-order ODE and, therefore, the monotonicity of $f$ does not automatically imply the absence of the internal instability. For instance, the ODE

$$y'' + \varphi(y)y' + f(y) = h(t)$$

may produce the nontrivial dynamics even if $f$ is monotone and $\varphi$ is strictly positive, say for the case of a given periodic-in-time external force $h$. For this reason, our proof of the monotonicity of the ODE component is based on rather delicate arguments related to the existence of the global Lyapunov function and associated dissipative integrals; see Section 11.2.

The case where the monotonicity assumption is violated (which is considered in Section 11.3) occurs to be (as predicted by the heuristics) much more complicated. In contrast to the monotone case, there is very little hope of developing a reasonable global attractor theory here (no matter whether in a strong or weak topology of the phase space), since, as a rule, even the equilibria set $\mathcal{R}$ is already not compact in the strong topology of the phase space and is not closed in the weak topology. In addition, we indeed see a huge (uncountable) number of well-separated essentially discontinuous equilibria here; see Section 11.3.

Nevertheless, in the particular case of the small coupling constant $\alpha$, we succeed in giving a complete description of the equilibria set $\mathcal{R} = \mathcal{R}_\alpha$ and verify that every trajectory of (11.1) converges as $t \to \infty$ to one of these equilibria. We mention that the standard Lojasiewicz technique for proving stabilization seems to be nonapplicable here even in the case of analytic nonlinearities, since the equilibria set is not compact in any reasonable topology and the alternative technique of [8] and [84] (see also [63] where the pointwise stabilization for the nonsmooth temperature-driven phase separation model is proved) also does not work here since it is essentially based on the fact that the corresponding nonmonotone ODE is a *first*-order scalar ODE and cannot be generalized to the case of higher-order equations.

For this reason, we develop a new method of proving stabilization, based on the theory of nonautonomous perturbations of regular attractors; see Proposition 2.1

We mention also that, as pointed out in [22], the solutions with discontinuous densities are rather expected in view of the forest ecosystem, and a curve in $\Omega$ where the density has discontinuities is called the ecotone boundary. However, as our result shows, this "curve" is typically not smooth (and even not continuous) and may have an extremely complicated structure.

The chapter is organized as follows. Section 11.1 is devoted to the study of the analytical properties of problem (11.1) such as existence and uniqueness, dissipative estimates in different norm, etc. The case of monotone nonlinearity $f$ is considered in Section 11.2; in particular, the existence of a smooth regular attractor is proved here. In Section 11.3 we deal with the nonmonotone case and, in particular, prove here the above-mentioned stabilization result for the weakly coupled case.

## 11.1. A priori estimates, existence, and uniqueness

We consider the coupled system of a second-order ODE with a heat equation

$$\begin{cases} \partial_t^2 v + \varphi(v)\partial_t v + f(v) = \alpha w, \ v(0) = v_0, \ \partial_t v(0) = v_0', \\ \partial_t w - \Delta_x w + w = v, \ \partial_n w\big|_{\partial\Omega} = 0, \ \ w\big|_{t=0} = w_0 \end{cases} \tag{11.4}$$

in a bounded three-dimensional domain $\Omega \subset \mathbb{R}^3$ with a smooth boundary. Here, $(v, w) = (v(t, x), w(t, x))$ are unknown functions, $\Delta_x$ is a Laplacian with respect to the variable $x$, $\alpha > 0$ is a given constant, and $\varphi$ and $f$ are given nonlinearities which satisfy the following assumptions:

$$\begin{cases} (1) \ \ \varphi, f \in C^2(\mathbb{R}), \\ (2) \ \ \varphi(v) \geq \beta_0 > 0, \\ (3) \ \ f(v)v \geq -C + \gamma_0 |v|^{2+\delta}, \\ (4) \ \ f'(v) \geq -K \end{cases} \tag{11.5}$$

for some positive constants $C$, $K$, $\beta_0$, $\delta$, and $\gamma_0$.

Finally, we assume that the initial data $(v_0, v_0', w_0)$ is taken from $L^\infty(\Omega)$:

$$(v_0, v_0', w_0) \in \Phi_\infty := [L^\infty(\Omega)]^2 \times [L^\infty(\Omega) \cap H^1(\Omega)]. \tag{11.6}$$

The aim of that section is to establish a number of basic a priori estimates for that system which will allow us to verify the existence and uniqueness of a solution and to study its behavior as $t \to \infty$. We start with the following lemma, which gives the global Lyapunov function for that problem.

PROPOSITION 11.1. *Let the above assumptions hold and let $(v(t), w(t))$ be a sufficiently regular solution of problem* (11.4)*. Introduce a functional*

$$\mathcal{L}(v, w) := \|\partial_t v\|_{L^2}^2 + 2(F(v), 1) - 2\alpha(v, w) + \alpha\|\nabla_x w\|^2 + \alpha\|w\|_{L^2}^2, \tag{11.7}$$

*where* $F(v) := \int_0^v f(s)\,ds$ *and* $(\cdot,\cdot)$ *is used for the inner product in* $L^2$. *Then, the following equality holds:*

$$\frac{d}{dt}\mathcal{L}(v(t),w(t)) = -2(\varphi(v(t))\partial_t v(t), \partial_t v(t)) - \alpha\|\partial_t w(t)\|^2_{L^2}. \tag{11.8}$$

Indeed, multiplying the first and the second equations of (11.4) by $\partial_t v$ and $\alpha\partial_t w$, respectively, taking a sum, and integrating over $\Omega$, we arrive at (11.8).

COROLLARY 11.1. *Let the above assumptions hold and let* $(v(t),w(t))$ *be a solution of* (11.4). *Then, the following estimate holds:*

$$\|\partial_t v(t)\|_{L^2} + \|v(t)\|_{L^2} + \|w(t)\|_{H^1} \le Q(\|(v,\partial_t v, w)\|_{\Phi_\infty}) \tag{11.9}$$

*for some monotone function* $Q$ *independent of* $t$ *and the solution.*

Indeed, according to our dissipativity assumption (11.5)(3),

$$F(u) \ge -C_1 + \gamma_1|u|^{2+\delta} \tag{11.10}$$

for some new constants $C_1$ and $\gamma_1$. Using this inequality, we easily check that

$$\gamma_2(\|\partial_t v\|^2_{L^2} + \|v\|^2_{L^2} + \|w\|^2_{H^1}) - C_2 \le \mathcal{L}(v,w) \le Q(\|(v,\partial_t v,w)\|_{\Phi_\infty}) \tag{11.11}$$

for some constants $\gamma_2, C_2 > 0$ and some monotone function $Q$. Now integrating equation (11.8) by $t$ and using that $\varphi(v) \ge 0$ and $\alpha > 0$, we arrive at (11.9).

The next corollary gives the $L^2$-dissipation integral for that problem.

COROLLARY 11.2. *Let the above assumptions hold. Then,*

$$\int_0^\infty \|\partial_t v(t)\|^2_{L^2} + \|\partial_t w(t)\|^2_{L^2}\,dt \le Q(\|(v_0, v_0', w_0)\|_{\Phi_\infty}) \tag{11.12}$$

*for some monotone function* $Q$.

Indeed, this estimate is an immediate corollary of (11.8), (11.11), and the assumption that $\varphi(v) \ge \beta_0 > 0$.

We are now going to verify that the solution is globally bounded in $\Phi_\infty$.

PROPOSITION 11.2. *Let the above assumptions hold. Then, the following estimate is valid:*

$$\|v(t)\|_{L^\infty} + \|\partial_t v(t)\|_{L^\infty} + \|w(t)\|_{L^\infty\cap H^1} \le Q(\|(v_0, v_0', w_0)\|_{\Phi_\infty}) \tag{11.13}$$

*for some monotone function* $Q$ *independent of* $t$ *and the solution.*

PROOF. We first establish the $L^\infty$-bound for the $w$-component. Indeed, according to Corollary 11.1, the right-hand side $v$ of the second equation of (11.4) is bounded in $L^\infty(\mathbb{R}_+, L^2(\Omega))$. Consequently, the standard regularity result for the heat equation gives

$$\|w(t)\|_{L^\infty} \le C\|w(0)\|_{L^\infty}e^{-t} + C\|v\|_{L^\infty(\mathbb{R}_+,L^2)} \le Q(\|(v_0,v_0',w_0)\|_{\Phi_\infty}) \tag{11.14}$$

(here we have implicitly used the restriction on the space dimension).

Thus, we only need to establish the $L^\infty$-bounds for the $v$-component. To this end, we will use the $L^\infty$-bounds for the $w$-component obtained earlier and will consider the equation for the $v$-component as an ODE for every (almost every, as a mathematician) *fixed* $x \in \Omega$. Indeed, let $y(t) := v(t,x_0)$. Then, this function solves

$$y''(t) + \varphi(y)y' + f(y) = h(t) = h_{w,x_0}(t) := \alpha w(t,x_0). \tag{11.15}$$

Multiplying this equation by $y' + \varepsilon y$, we have

(11.16)
$$[(y')^2 + 2F(y) + 2\varepsilon yy' + 2\varepsilon R(y)]' + 2(\varphi(y) - \varepsilon)(y')^2 + 2\varepsilon f(y)y = 2h(y' + \varepsilon y),$$

where $R(y) := \int_0^y \varphi(s)s\,ds$. Using the fact that $\varphi(v)$ is strictly positive and $f$ is dissipative, we deduce from this equation that, for sufficiently small $\varepsilon > 0$,

$$\frac{d}{dt}S(y, y') + \gamma((y')^2 + y^2) \le C(|h(t)|^2 + 1), \tag{11.17}$$

where

$$S(y, y') := (y')^2 + 2F(y) + 2\varepsilon yy' + 2\varepsilon R(y)$$

and $\gamma$ is positive. Moreover,

$$\varepsilon_0(y^2 + (y')^2) - C \le S(y, y') \le Q(y^2 + (y')^2)$$

for some positive $\varepsilon_0$ and $C$ and monotone $Q$. Applying the Gronwall lemma to inequality (11.17), we conclude that

$$y(t)^2 + (y'(t))^2 \le Q(y(0)^2 + y'(0)^2 + \|h\|^2_{L^\infty(\mathbb{R}_+)}) \tag{11.18}$$

for some monotone function $Q$ which is independent of $t$ and $y$; see, e.g., [10]. Now taking the supremum with respect to all $x_0 \in \Omega$ and using (11.14) for estimating $h$, we deduce estimate (11.13) and finish the proof of the proposition. □

We are now ready to verify the existence and uniqueness of a solution for the problem (11.4).

DEFINITION 11.1. A pair of functions $(v(t), w(t))$ is a solution of problem (11.4) if

$$(v(t), \partial_t v(t), w(t)) \in \Phi_\infty$$

for every $t \ge 0$ and (11.4) is satisfied in the sense of distributions.

Note that, from the first equation of (11.4), we see that $\partial_t^2 v(t) \in L^\infty(\Omega)$. Therefore, $v(t) \in W^{2,\infty}([0, T], L^\infty(\Omega))$ and the initial data for $v$ is well-defined. Analogously, the $w$-component is continuous as a function with values, say, in $L^2(\Omega)$ and the initial data is again well-defined.

THEOREM 11.1. *Let the above assumptions hold. Then, for every* $(v_0, v_0', w_0) \in \Phi_\infty$, *problem* (11.4) *possesses a unique solution in the sense of Definition* (11.1) *and this solution satisfies estimate* (11.13). *Moreover, any two solutions* $(v_1(t), w_1(t))$ *and* $(v_2(t), w_2(t))$ *satisfy the following estimate:*

$$\begin{aligned} &\|(v_1(t), \partial_t v_1(t), w_1(t)) - (v_2(t), \partial_t v_2(t), w_2(t))\|_{\Phi_\infty} \\ &\quad \le Ce^{Kt}\|(v_1(0), \partial_t v_1(0), w_1(0)) - (v_2(0), \partial_t v_2(0), w_2(0))\|_{\Phi_\infty}, \end{aligned} \tag{11.19}$$

*where positive constants $C$ and $K$ depend only on the norms of the initial data.*

PROOF. Let us first verify uniqueness and estimate (11.19). Indeed, let

$$(v(t), w(t)) := (v_1(t), w_1(t)) - (v_2(t), w_2(t)).$$

Then, these functions solve

$$\begin{cases} \partial_t^2 v + \varphi(v_1)\partial_t v + [\varphi(v_1) - \varphi(v_2)]\partial_t v_2 + [f(v_1) - f(v_2)] = \alpha w, \\ \partial_t w - \Delta_x w + w = v. \end{cases} \tag{11.20}$$

Now multiplying the first equation of (11.20) by $\partial_t v + \varepsilon v$, $\varepsilon > 0$ is a small positive number, using the fact that $v_i, \partial_t v_i$ are globally bounded in $L^\infty$, and applying the Gronwall inequality in a standard way (without integration by $x$!), we conclude that

(11.21)
$$\begin{aligned}\|\partial_t v(t)\|^2_{L^\infty(\Omega)} &+ \|v(t)\|^2_{L^\infty}\\ &\le Ce^{Kt}(\|\partial_t v(0)\|^2_{L^\infty(\Omega)} + \|v(0)\|^2_{L^\infty(\Omega)}) + C\int_0^t e^{K(t-s)}\|w(s)\|^2_{L^\infty(\Omega)}\,ds\end{aligned}$$

for some positive constants $C$ and $K$ depending only on the $L^\infty$-norms of $v_i$ and $\partial_t v_i$. Furthermore, due to the maximum principle for the heat equation, we have the estimate

$$\|w(t)\|_{L^\infty} \le e^{-t}\|w(0)\|_{L^\infty} + \int_0^t e^{-(t-s)}\|v(s)\|_{L^\infty}\,ds. \tag{11.22}$$

Inserting this estimate into the right-hand side of (11.21), we arrive at

$$\begin{aligned}\|\partial_t v(t)\|^2_{L^\infty(\Omega)} + \|v(t)\|^2_{L^\infty} \le C'e^{Kt}(\|\partial_t v(0)\|^2_{L^\infty(\Omega)} + \|v(0)\|^2_{L^\infty(\Omega)} + \|w(0)\|^2_{L^\infty})\\ + C'\int_0^t e^{K(t-s)}\|v(s)\|^2_{L^\infty(\Omega)}\,ds.\end{aligned} \tag{11.23}$$

Applying the Gronwall inequality again to that relation, we conclude that

(11.24)
$$\|\partial_t v(t)\|^2_{L^\infty(\Omega)} + \|v(t)\|^2_{L^\infty} \le C'e^{2Kt}(\|\partial_t v(0)\|^2_{L^\infty(\Omega)} + \|v(0)\|^2_{L^\infty(\Omega)} + \|w(0)\|^2_{L^\infty}).$$

This estimate, together with (11.22), gives the desired $L^\infty$-estimate for the triple $(v, \partial_t v, w)$. In order to finish the proof of estimate (11.19), it remains to note that the desired estimate of the $H^1$-norm of the $w$-component is immediate, since the $L^\infty$-control for the right-hand side of the heat equation for $w$ is already obtained. Thus, uniqueness and Lipschitz continuity (11.19) are proved.

So, we only need to prove the existence of a solution. It can be done in a standard way, based on a priori estimate (11.13), using the Banach fixed point theorem for proving the existence of a local solution and estimate (11.13) for extending this solution globally in time; see, e.g., [55] for the details. □

Our next goal is to establish the basic dissipative estimate in the phase space $\Phi_\infty$.

THEOREM 11.2. *Let the above assumptions hold. Then, a solution* $(v(t), w(t))$ *of problem* (11.4) *satisfies the following dissipative estimate:*

$$\|(v(t), v'(t), w(t))\|_{\Phi_\infty} \le Q(\|(v_0, v_0', w_0)\|_{\Phi_\infty})e^{-\beta t} + C_* \tag{11.25}$$

*for some positive constants* $\beta$ *and* $C_*$ *and monotone function* $Q$.

PROOF. As we can see from the proof of the previous proposition, the only problem is to obtain a *dissipative* estimate for the $L^2$-norm of $v(t)$. Indeed, if this estimate is obtained, analyzing the equation for the $w$-component analogously to (11.14), we deduce the *dissipative* estimate for the $L^\infty$-norm of $w(t)$. This, in turn, gives the dissipative estimate for the right-hand side $h(t)$ of (11.15) and the Gronwall lemma applied to inequality (11.17) will finish the derivation of estimate (11.18).

So, we only need to obtain the dissipative estimate for the $L^2$-norm. To this end, we multiply the first equation of (11.4) by $2(\partial_t v+\varepsilon v)$, $\varepsilon>0$ is a small number, which will be fixed below, and *integrate* over $x\in\Omega$. After that, we multiply the second equation of (11.4) by $2\alpha(\partial_t w+\varepsilon w)$, integrate over $x\in\Omega$, and take a sum of these two equations. Then, after the standard transformations, we end up with

(11.26)

$$6\frac{d}{dt}Z(t)+2\alpha\|\partial_t w\|^2+2((\varphi(v)-\varepsilon)\partial_t v,\partial_t v)+2\varepsilon f(v).v+2\alpha\varepsilon(\|\nabla_x w\|^2_{L^2}+\|w\|^2_{L^2})$$
$$=4\alpha\varepsilon(v,w),$$

where

$$Z(t):=\|\partial_t v(t)\|^2_{L^2}+2(F(v(t)),1)+2\varepsilon(R(v(t)),1)+2\varepsilon(v(t),\partial_t v(t))$$
$$-2\alpha(v(t),w(t))+\alpha(\|\nabla_x w(t)\|^2_{L^2}+(1+\varepsilon)\|w(t)\|^2_{L^2}). \tag{11.27}$$

We now fix $\varepsilon>0$ so small that

$$\varepsilon|(R(v(t)),1)|\le 1$$

(it is possible to do this due to estimate (11.13); of course, $\varepsilon$ will depend on the norm of the initial data). Then, due to (11.10), we have

$$\beta_2[\|\partial_t v\|^2_{L^2}+\|\nabla_x w\|^2_{L^2}+\|w\|^2_{L^2}+(|F(v)|,1)]-C_2\le Z(t)$$
$$\le\beta_1[\|\partial_t v\|^2_{L^2}+\|\nabla_x w\|^2_{L^2}+\|w\|^2_{L^2}+(|F(v)|,1)]+C_1, \tag{11.28}$$

where the positive constants $C_i$ and $\beta_i$ are independent of $\varepsilon\to 0$ and $(v,w)$. Moreover, due to the fourth assumption of (11.5),

$$F(v)\le f(v).v+Kv^2/2. \tag{11.29}$$

Inserting estimates (11.28) and (11.29) into (11.26) and again using the third assumption of (11.5), we deduce the differential inequality:

$$\partial_t Z(t)+\beta\varepsilon Z(t)\le C\varepsilon, \tag{11.30}$$

where $\varepsilon$ depends on the norm of the initial data, but the positive constants $\beta$ and $C$ are independent of $v$ and $w$. Integrating this inequality, we arrive at

$$Z(t)\le[Z(0)-\frac{C}{\beta}]e^{-\beta\varepsilon t}+\frac{C}{\beta}. \tag{11.31}$$

We see that, although the rate of convergence to the absorbing ball depends on the initial data (through the choice of $\varepsilon>0$), the radius of the absorbing ball is *independent* of $\varepsilon$ and, consequently, is independent of the norm of the initial data. This observation, together with estimate (11.28), implies that

$$\|\partial_t v(t)\|_{L^2}+\|v(t)\|_{L^2}+\|w(t)\|_{H^1}\le Q(\|(v_0,v_0',w_0)\|_{\Phi_\infty})e^{-\gamma t}+C_* \tag{11.32}$$

for some positive $\gamma$ and $C$ and a monotone function $Q$ which are all independent of $t$, $v$, and $w$. Thus, the desired dissipative estimate in $L^2$ is obtained and Theorem 11.2 is proved. □

We now formulate several auxiliary results on the smoothing property for the $w$-component and the existence of dissipative integrals in stronger norms, which will be essentially used in the next sections.

PROPOSITION 11.3. *Let the assumptions of Theorem* 11.1 *hold. Then,* $w(t) \in W^{2,p}(\Omega)$ *and* $\partial_t w(t) \in W^{2,p}(\Omega)$ *for any* $t > 0$ *and any* $p < \infty$ *and the following estimate is valid:*

$$\|w(t)\|_{W^{2,p}(\Omega)} + \|\partial_t w(t)\|_{W^{2,p}(\Omega)} \le \left(1 + t^{-N}\right) Q_p(\|(v_0, v_0', w_0)\|_{\Phi_\infty}) \tag{11.33}$$

*for some positive exponent* $N$ *and some monotone function* $Q_p$ *(depending only on* $p$*).*

PROOF. Indeed, due to the smoothing property of the heat equation, the solution $\theta(t)$ of

$$\partial_t \theta - \Delta_x \theta + \theta = 0, \quad \theta\big|_{t=0} = w_0 \tag{11.34}$$

satisfies the estimate

$$\|\theta(t)\|_{W^{2,p}(\Omega)} + \|\partial_t \theta(t)\|_{W^{2,p}(\Omega)} \le C_p t^{-N} \|w_0\|_{H^1} \tag{11.35}$$

for some exponent $N$ and positive constant $C_p$ depending only on $p$; see, e.g., [68]. The remaining $z(t) := w(t) - \theta(t)$ solve the heat equation with zero initial data

$$\partial_t z - \Delta_x z + z = v(t), \quad z\big|_{t=0} = 0.$$

Moreover, using estimate (11.13) together with the first equation of (11.4), we conclude that

$$\|v(t)\|_{L^\infty(\Omega)} + \|\partial_t v(t)\|_{L^\infty(\Omega)} + \|\partial_t^2 v(t)\|_{L^\infty(\Omega)} \le Q(\|(v_0, v_0', w_0)\|_{\Phi_\infty}). \tag{11.36}$$

Using that estimate together with the $W^{2,p}$-regularity estimate for the heat equation, we arrive at

$$\|\theta(t)\|_{W^{2,p}(\Omega)} + \|\partial_t w(t)\|_{W^{2,p}(\Omega)} \le Q_p(\|(v_0, v_0', w_0)\|_{\Phi_\infty}),$$

which, together with estimate (11.35), finishes the proof of the proposition. □

PROPOSITION 11.4. *Let the assumptions* (11.5), (11.6), *and* $\alpha > 0$ *hold. Then, the following stronger version of dissipative integrals exists:*

(11.37)
$$\begin{cases} (1) & \int_1^\infty \|\partial_t w(t)\|^2_{C(\Omega)}\, dt \le Q(\|(v_0, v_0', w_0)\|_{\Phi_\infty}), \\ (2) & \int_0^T |\partial_t^2 v(t, x_0)|^2 + |\partial_t v(t, x_0)|^2\, dt \le \varepsilon T + C_\varepsilon Q(\|(v_0, v_0', w_0)\|_{\Phi_\infty}), \end{cases}$$

*where* $\varepsilon > 0$ *is arbitrary,* $x_0$ *is a point in* $\Omega$, $C_\varepsilon > 0$ *depends only on* $\varepsilon$, *and* $Q$ *is some monotone function.*

PROOF. We first note that, due to Proposition 11.3, we may assume without loss of generality that $\partial_t w(0) \in H^1$. Differentiating the equation for the $w$-component by $t$ and denoting $z := \partial_t w$, we get

$$\partial_t z - \Delta_x z + z = \partial_t v, \quad z\big|_{t=0} = \partial_t v(0). \tag{11.38}$$

Applying the $L^2$-regularity theorem for that heat equation, we will have

$$\int_0^T \|\Delta_x z(s)\|^2_{H^2}\, ds \le C\|z(0)\|^2_{H^1} + C\int_0^2 \|\partial_t v(s)\|^2_{L^2}\, ds, \tag{11.39}$$

where the constant $C$ is independent of $T$. Together with estimate (11.12) and embedding $H^2 \subset L^\infty$, it gives the desired first estimate of (11.37). Let us now prove the second estimate of (11.37). To this end, we multiply equation (11.15) by $2y'$ (without integration by $t$!). This gives

$$((y')^2 + 2F(y) - 2\alpha wy)' + 2\varphi(y)(y')^2 = -2\alpha y\partial_t w(t, x_0). \tag{11.40}$$

Integrating this equality over $t \in [0, T]$, estimating

$$|2\alpha y\partial_t w(t, x_0)| \le C\|v(t)\|_{L^\infty}\|\partial_t w(t)\|_{L^\infty} \le \varepsilon + C_\varepsilon\|\partial_t w(t)\|^2_{L^\infty},$$

and using the first estimate of (11.37) together with the strict positivity of $\varphi$ and the fact that the $L^\infty$-norm of $v$ is under control, we deduce that

$$\int_0^T [y'(t)]^2\, dt \le \varepsilon T + \varepsilon^{-1}Q(\|v_0, v_0', w_0\|_{\Phi_\infty}) \tag{11.41}$$

for some (new) monotone function $Q$ which is independent of $T$. This gives the second estimate of (11.37) for the term $\partial_t v(t, x_0)$. Thus, in order to finish the proof of the proposition, we only need to estimate the term $\partial_t^2 v(t, x_0)$. To this end, we differentiate the first equation of (11.4) by $t$ and denote $q(t) := \partial_t v(t, x_0)$. Then, we get

$$q'' + \varphi(y)q' + \varphi'(v)q^2 + f'(y)q = \alpha\partial_t w(t, x_0).$$

Multiplying this equation by $2q'$, integrating by time, and using that the $L^\infty$-norms of $v$, $\partial_t v$, and $\partial_t^2 v$ are under control, we arrive at

$$\int_0^T [q'(t)]^2\, dt \le Q(\|v_0, v_0', w_0\|_{\Phi_\infty})\left(1 + \int_0^T [y'(t)]^2 + \|\partial_t w(t)\|^2_{L^\infty}\, dt\right),$$

for some monotone function $Q$ which is independent of $T$. Inserting estimate (11.41) and the first estimate of (11.37) into that inequality (and scaling the parameter $\varepsilon$ if necessary), we obtain the desired control for the integral of $\partial_t^2 v$ and we finish the proof of the proposition. □

We conclude this section by showing that if the initial data $(v_0, v_0', w_0)$ is smooth, the solution $(v(t), w(t))$ remains smooth for all $t$.

PROPOSITION 11.5. *Let the assumptions of Theorem* 11.1 *hold. Assume, in addition, that*

$$(v_0, v_0', w_0) \in W^{1,\infty}(\Omega). \tag{11.42}$$

*Then, the solution* $(v(t), \partial_t v(t), w(t)) \in W^{1,\infty}(\Omega)$ *for any* $t \ge 0$ *and the following estimate holds:*

$$\|v(t)\|_{W^{1,\infty}} + \|\partial_t v(t)\|_{W^{1,\infty}} + \|w(t)\|_{W^{1,\infty}} \le C\|(v_0, v_0', w_0)\|_{[W^{1,\infty}]^3}e^{Kt} \tag{11.43}$$

*for some positive constants* $C$ *and* $K$ *(which depend on the* $L^\infty$*-norms of the initial data).*

PROOF. The desired estimate for the $w$-component is factually obtained in Proposition 11.3; thus, we only need to estimate the $v$ component. Let $x_1$ and $x_2$ be two arbitrary points of $\Omega$ and let $z(t) := v(t, x_1) - v(t, x_2)$. Then, this function satisfies the ODE

$$\begin{aligned} &z''(t) + \varphi(v(t, x_1))z'(t) + [\varphi(v(t, x_1)) - \varphi(v(t, x_2))]\partial_t v(t, x_2) \\ &+ [f(v(t, x_1)) - f(v(t, x_2))] \\ &\quad = h_{x_1,x_2}(t) := \alpha(w(t, x_1) - w(t, x_2)). \end{aligned} \tag{11.44}$$

Multiplying this equation by $\partial_t z(t)$ and arguing exactly as in (11.21), we arrive at

$$|z(t)|^2 + |z'(t)|^2 \le C(|z(0)|^2 + |z'(0)|^2)e^{Kt} + C\int_0^t e^{K(t-s)}|h_{x_1,x_2}(s)|^2\,ds, \tag{11.45}$$

where the constants $C$ and $K$ depend on the $L^\infty$-norm of the solution. Furthermore, since the $W^{1,\infty}$-estimate for the $w$-component is already obtained, we have

$$\sup_{x_1,x_2\in\Omega}\frac{1}{|x_1-x_2|^2}\int_0^t e^{K(t-s)}|h_{x_1,x_2}(s)|^2\,ds$$
$$\le C_1\int_0^t e^{K(t-s)}\|w(s)\|^2_{W^{1,\infty}}\,ds \le Ce^{Kt}.$$

Finally, dividing inequality (11.45) by $|x_1-x_2|^2$ and taking the supremum over $x_1, x_2 \in \Omega$ from both parts of the inequality obtained, we get the desired estimate for the $W^{1,\infty}$-norms of $v$ and $\partial_t v$ and we finish the proof of the proposition. □

REMARK 11.1. Arguing analogously, one can show that if the initial data belong to $C^k$, the solution will be of class $C^k$ for every $t \ge 0$. Thus, the blowup in finite time of the higher norms cannot occur. However, there is a principal difference between estimate (11.25) for the $L^\infty$-norm and estimate (11.43) for the $W^{1,\infty}$-norm of the solution. Indeed, the first estimate is dissipative and shows that the $L^\infty$-norm of the solution cannot grow and even gives the absorbing ball in that norm. In contrast to that, the $W^{1,\infty}$-norm, a priori, may grow exponentially and, in this sense, the solution may become "less and less regular" as $t \to \infty$ (i.e., it may tend to a discontinuous limit). As we will see in the next sections, the answer to the question of whether or not it really happens depends in a crucial way on the monotonicity of the nonlinearity $f$.

## 11.2. The monotone case: Asymptotic compactness and regular attractor

According to the results of the previous section, equation (11.4) is uniquely solvable in the phase space $\Phi_\infty$ and the solution operators

$$S(t)(v_0, v_0', w_0) := (v(t), \partial_t v(t), w(t)) \tag{11.46}$$

generate a dissipative semigroup in $\Phi_\infty$. The aim of this section is to study the long-time behavior of solutions as $t \to \infty$ in the particular case where the nonlinearity $f$ is strictly monotone:

$$f'(v) \ge \kappa_0 > 0. \tag{11.47}$$

As we will see, in that case, the associated semigroup is asymptotically compact and possesses a smooth global attractor $\mathcal{A}$ in $\Phi_\infty$. Moreover, due to the Lyapunov functional, this attractor can be described as a finite union of finite-dimensional unstable manifolds. Our proof of the asymptotic compactness is based on the following lemma, which can be considered as a refinement of estimate (11.43).

LEMMA 11.1. *Let the assumptions of Theorem* 11.4 *hold and let, in addition,* (11.47) *be satisfied. Let us also introduce, for any $h > 0$, the following (semi)norm on the space $L^\infty(\Omega)$:*

$$\|v\|_{W_h^{1,\infty}} := \sup_{x_1,x_2\in\Omega,\,|x_1-x_2|\ge h}\frac{|v(x_1)-v(x_2)|}{|x_1-x_2|} \tag{11.48}$$

*(as mathematicians, we would write* ess sup *instead of* sup*). Then, every solution* $(v(t), w(t))$ *of problem* (11.4) *satisfies*

$$\|v(t)\|_{W_h^{1,\infty}} + \|\partial_t v(t)\|_{W_h^{1,\infty}} \le \frac{C_1}{h} e^{-\beta t} + C_2, \tag{11.49}$$

*where the positive constants* $\beta$, $C_i$ *depend on the* $L^\infty$*-norms of the initial data but are independent of* $t$ *and* $h \to 0$.

PROOF. Analogously to the proof of Proposition 11.5, we introduce a function $z(t) := v(t, x_1) - v(t, x_2)$ which solves equation (11.44). But, using the monotonicity assumption (11.47) and the dissipation integrals (11.37), we are now able to suppress the exponential divergence in estimate (11.45). To this end, we multiply equation (11.44) by $z'$ and transform the term containing the nonlinearity $f$ as follows:

$$[f(v(t, x_1)) - f(v(t, x_2))]z' = \frac{1}{2}[l(t)z^2]' - \frac{1}{2}l'(t)z^2, \tag{11.50}$$

where $l(t) := \int_0^1 f'(sv(t, x_1) + (1 - s)v(t, x_2))\, ds \ge \kappa_0 > 0$ and its derivative can be estimated as

$$|l'(t)| \le C(|\partial_t v(t, x_1)| + |\partial_t v(t, x_2)|), \tag{11.51}$$

where the constant $C$ depends on the $L^\infty$-norm of the initial data but is independent of $t$ and $x_i$. Then, using the positivity of $\varphi$ and the $L^\infty$-bounds for $v$, we get

$$\frac{1}{2}((z')^2 + lz^2)' + \gamma(z')^2 \le C|h_{x_1,x_2}|^2 + C(|\partial_t v(t, x_1)| + |\partial_t v(t, x_2)|)z^2$$

for some positive constants $\gamma$ and $C$. Now multiplying equation (11.44) by $\varepsilon z$ (where $\varepsilon > 0$ is a sufficiently small positive number) and taking a sum with the above inequality, we infer

$$\begin{aligned}\frac{1}{2}&((z')^2 + lz^2 + 2\varepsilon zz')' + (\gamma - \varepsilon)(z')^2 + \varepsilon lz^2 \\ &\le C|h_{x_1,x_2}|^2 + C(|\partial_t v(t, x_1)| + |\partial_t v(t, x_2)|)(z^2 + (z')^2).\end{aligned} \tag{11.52}$$

Let now

$$\mathcal{L}_z(t) := (z')^2 + lz^2 + 2\varepsilon zz'.$$

Then, since $l(t) \ge \kappa_0 > 0$, we may fix $\varepsilon > 0$ to be small enough that

$$\kappa(z^2 + (z')^2) \le \mathcal{L}_z(t) \le \kappa_1(z^2 + (z')^2) \tag{11.53}$$

for some positive $\kappa$ and $\kappa_1$. This inequality together with the evident estimate $|x| \le \beta + \beta^{-1}x^2$ allows us to transform (11.52) to

$$\frac{d}{dt}\mathcal{L}_z(t) + (\gamma - C(|\partial_t v(t, x_1)|^2 + |\partial_t v(t, x_2)|^2))\mathcal{L}_z(t) \le C|h_{x_1,x_2}(t)|^2. \tag{11.54}$$

Applying the Gronwall inequality to this relation, we arrive at

$$\mathcal{L}_z(t) \le \mathcal{L}_z(0)e^{-\int_0^T K(s)\, ds} + C\int_0^T e^{-\int_t^T K(s)\, ds}|h_{x_1,x_2}(t)|^2\, dt$$

with $K(s) := \gamma - C(|\partial_t v(t, x_1)|^2 + |\partial_t v(t, x_2)|^2)$. Using the dissipation integral (11.37) with $\varepsilon := \gamma/2$, we see that

$$\int_t^T K(s)\, ds \ge \gamma(T - t)/2 - C,$$

where the constant $C$ depends on the $\Phi_\infty$-norm of the initial data. This estimate, together with the bounds (11.53), gives the nondivergent analogue of estimate (11.45):

$$|z(t)|^2 + |z'(t)|^2 \le C(|z(0)|^2 + |z'(0)|^2)e^{-\gamma t} + C\int_0^t e^{-\gamma(t-s)}|h_{x_1,x_2}(s)|^2\,ds \tag{11.55}$$

for some positive $C$ and $\gamma$ depending only on the $L^\infty$-norm of the solution.

In order to deduce the desired estimate (11.48) from (11.55), we note that, due to Proposition 11.3, we may assume without loss of generality that $\|w(t)\|_{W^{1,\infty}} \le C$ for all $t \ge 0$ and, consequently,

$$\sup_{|x_1-x_2|\ge h} \frac{1}{|x_1-x_2|^2}\int_0^t e^{-\gamma(t-s)}|h_{x_1,x_2}(s)|^2\,ds \le \alpha\int_0^t e^{-\gamma(t-s)}\|w(s)\|^2_{W^{1,\infty}}\,ds \le C_1.$$

Moreover, obviously,

$$\|v\|_{W^{1,\infty}_h} \le \frac{2\|v\|_{L^\infty}}{h}.$$

Now dividing inequality (11.55) by $|x_1-x_2|^2$ and taking the supremum over all $x_i \in \Omega$, $|x_1-x_2| \ge h$, we deduce the desired estimate (11.48). Lemma 11.1 is proved. □

Our next step is to verify the existence of the global attractor $\mathcal{A}$ for semigroup (11.46) associated with problem (11.4).

THEOREM 11.3. *Let the assumptions of Lemma* 11.1 *hold. Then, the semigroup* $S(t)$ *associated with problem* (11.4) *possesses a global attractor* $\mathcal{A}$ *in the phase space* $\Phi_\infty$. *This attractor is bounded in* $[W^{1,\infty}(\Omega)]^3$ *and has the following structure:*

$$\mathcal{A} = \mathcal{K}\big|_{t=0}, \tag{11.56}$$

*where* $\mathcal{K} \subset C_b(\mathbb{R}, \Phi_\infty)$ *is a set of all solutions of problem* (11.4) *which are defined for all* $t \in \mathbb{R}$ *and are globally bounded.*

PROOF. In order to deduce the existence of a global attractor from Lemma 11.1, we will use the so-called Kuratowski measure of noncompactness. Recall that, by definition, the Kuratowski measure of noncompactness $\alpha(B)$ of a set $B$ is an infimum of all $r > 0$ for which it can be covered by a finite number of $r$-balls; see, e.g., [53] for details. To be more precise, we need the following lemma.

LEMMA 11.2. *Let*

$$B := \{v \in L^\infty(\Omega),\quad \|v\|_{L^\infty} + \|v\|_{W^{1,\infty}_h} \le R\}$$

*for some* $R$ *and* $h > 0$. *Then, its Kuratowski measure of noncompactness of the set* $B$ *can be estimated as follows:*

$$\alpha(B) \le Rh. \tag{11.57}$$

PROOF. Let $\mathcal{S}_h$ be the standard averaging operator

$$(\mathcal{S}_h v)(x) := \int_{\mathbb{R}^3} D_h(x,y)v(y)\,ds,$$

where the smooth nonnegative kernels $D_h(x,z)$ are such that

$$\begin{cases} (1) & \operatorname{supp} D_h(x,\cdot) \subset \{z \in \Omega,\ |z-x| \le h\}, \\ (2) & \int_{\mathbb{R}^3} D_h(x,y)\,dy \equiv 1, \\ (3) & |D_h(x,y)| + |\nabla_x D_h(x,y)| \le C_h, \quad x,y \in \mathbb{R}^3 \end{cases} \tag{11.58}$$

(since $\Omega$ is assumed to be smooth, such kernels exist).

Also let $B_h := \mathcal{S}_h(B)$. Then, on the one hand, the set $B_h$ consists of smooth functions and, in particular, is bounded in $C^1(\Omega)$. By the Arzela-Ascoli theorem, this means that $B_h$ is compact in $L^\infty(\Omega)$.

On the other hand,

$$|(\mathcal{S}_h v)(x) - v(x)| \le \int_{\mathbb{R}^3} D_h(x,y)|v(y)-v(x)|\,dy \le \|v\|_{W^{1,\infty}_h} h \int_{\mathbb{R}^3} D_h(x,y)\,dy \le Rh.$$

Thus, $B \subset B_h + Rh$ and $B_h$ is compact. This gives estimate (11.57) and finishes the proof of the lemma. □

We are now ready to finish the proof of the theorem. Indeed, due to Proposition 11.3, we know that the $w$-component is bounded in $W^{2,p}(\Omega)$ for every $t > 0$ and, consequently, the $w$-component of $S(t)B$ is precompact in $L^\infty \cap H^1$ for any bounded set $B$. So, the Kuratowski measure of noncompactness for $S(t)B$ is determined by the $v$-component only. Moreover, Lemma 11.1 guarantees that

$$v(t), \partial_t v(t) \subset \{u \in L^\infty(\Omega),\ \|u\|_{L^\infty} + \|u\|_{W^{1,\infty}_h} \le R\}$$

if $t \ge T(h)$ is large enough (but $R$ is independent of $h$). This gives that

$$\lim_{t\to\infty} \alpha(S(t)B) = 0 \tag{11.59}$$

for any bounded set $B$.

Since the semigroup $S(t)$ is Lipschitz continuous with respect to the initial data (see Theorem 11.1) and dissipative (see Theorem 11.2), the convergence of the Kuratowski measure (11.59) to zero implies the asymptotic compactness of the semigroup $S(t)$ and the existence of a global attractor $\mathcal{A}$; see [53]. The structure (11.56) of the attractor is also a corollary of that abstract theorem, and the fact that $\mathcal{A}$ is bounded in $W^{1,\infty}$ follows from estimate (11.48) (together with the fact that the constant $C_2$ is independent of $h$). Thus, Theorem 11.3 is proved. □

Our next task is to establish the regular structure of the attractor $\mathcal{A}$ provided by the Lyapunov functional. To this end, we need to make some preparations. As a first step, we establish the differentiability of the semigroup $S(t)$ with respect to the initial data.

PROPOSITION 11.6. *Let the assumptions of Theorem* 11.1 *hold. Then, the associated semigroup $S(t)$ is Fréchet differentiable with respect to the initial data for every fixed $t$ and its Fréchet derivative $D_\xi[S(t)\xi] \in \mathcal{L}(\Phi_\infty, \Phi_\infty)$ is Lipschitz continuous with respect to the initial data $\xi \in \Phi_\infty$ and the following estimate holds for every bounded set $B \subset \Phi_\infty$:*

$$\|S(t)\|_{C^{1,1}(B,\Phi_\infty)} \le Ce^{2Kt}, \tag{11.60}$$

*where the constants $C$ and $K$ depend only on the radius of $B$.*

The proof of this proposition is straightforward and standard, so, in order to avoid the technicalities, we leave it to the reader (see [9]).

At the next step, we need to study the equilibria of problem (11.4).

PROPOSITION 11.7. *Let the assumptions of Theorem* 11.3 *hold. Then, any equilibrium* $(v_0, w_0) \in \mathcal{R}$ *(the set of all equilibria) of problem* (11.4) *solves the following semilinear elliptic equation:*

$$-\Delta_x w + w = f^{-1}(\alpha w), \quad \partial_n w\big|_{\partial\Omega} = 0, \quad v = (-\Delta_x + 1)w \tag{11.61}$$

*($f^{-1}$ exists since $f$ is now assumed to be monotone). Moreover, the equilibrium $(v_0, w_0)$ is hyperbolic if and only if $w_0$ is hyperbolic as a solution of* (11.61), *i.e., if the equation*

$$-\Delta_x\theta + \theta = [f^{-1}]'(\alpha w_0)\alpha\theta \tag{11.62}$$

*has only trivial solution $\theta = 0$. In particular, for generic $f$, all of the equilibria $(v_0, w_0) \in \mathcal{R}$ are hyperbolic and $\mathcal{R}$ is finite.*

PROOF. Indeed, the equations on equilibria for problem (11.4)

$$f(v) = \alpha w, \quad -\Delta_x w + w = v$$

are equivalent to (11.61). Let us verify the assertion on hyperbolicity. Indeed, the asymptotic compactness of the semigroup $S(t)$ implies in a standard way that the essential spectrum (see [9], [53]) of the operator $D_\xi S(1)$ lies strictly inside of the unit circle. Thus, only eigenvalues of finite multiplicity are possible on the unit circle. Any such eigenvalue generates a *time-periodic* solution $(z, \theta)$ of the associated equation of variations

$$\begin{cases} \partial_t^2 z + \varphi(v_0)\partial_t z + f'(v_0)z = \alpha\theta, \\ \partial_t\theta - \Delta_x\theta + \theta = z. \end{cases} \tag{11.63}$$

However, analogously to the nonlinear problem (11.4), the linearized problem (11.63) possesses a global Lyapunov function (in order to find it, one needs to multiply the first and the second equations by $\partial_t z$ and $\alpha\partial_t\theta$, respectively, take a sum, and integrate over $\Omega$). Thus, every periodic solution of that linearized problem must be an equilibrium:

$$f'(v_0)z = \alpha\theta, \quad z = -\Delta_x\theta + \theta,$$

and, consequently, $z$ must solve (11.62). Vice versa, any nontrivial solution $z$ of (11.62) generates a nontrivial equilibrium of (11.63) by setting $z = -\Delta_x\theta + \theta$.

Finally, the last assertion that generically $\mathcal{R}$ is finite and all of the equilibria are hyperbolic is a standard corollary of the Sard theorem; see, e.g., [9]. Proposition 11.7 is proved. □

Thus, we will assume from now on that all of the equilibria $(v_0, w_0) \in \mathcal{R}$ are hyperbolic (which automatically implies that $\mathcal{R}$ is finite). Furthermore, for any $\xi_0 := (v_0, w_0) \in \mathcal{R}$, we define the associated unstable set $\mathcal{M}^+_{\xi_0}$ by the usual expression

(11.64)

$$\mathcal{M}^+_{\xi_0} := \{(v_0, v_0', w_0) \in \Phi_\infty,\ \exists (v(t), w(t)) \text{ which solves (11.4) for } t \le 0 \text{ such that } (v(0), \partial_t v(0), w(0)) = (v_0, v_0', w_0) \text{ and } \lim_{t\to-\infty}(v(t), \partial_t v(t), w(t)) = (v_0, v_0', w_0)\}.$$

In other words, the unstable set $\mathcal{M}^+_{\xi_0}$ consists of all complete trajectories of (11.4) which stabilize to $\xi_0$ as $t \to -\infty$.

It is well known (see, e.g., [9]) that, for hyperbolic equilibrium $\xi_0 \in \mathcal{R}$, the set $\mathcal{M}^+_{\xi_0}$ is *locally* (near $\xi_0$) a finite-dimensional submanifold of $\Phi_\infty$ and its dimension equals the instability index of $\xi_0$. But, in order to prove that the whole of $\mathcal{M}^+_{\xi_0}$ is a submanifold of $\Phi_\infty$, one needs the semigroup $S(t)$ to be injective (in other words, problem (11.4) should possess the so-called backward uniqueness property; see [9] again).

PROPOSITION 11.8. *Let the assumptions of Theorem* 11.1 *hold. Then, the semigroup $S(t)$ associated with equation* (11.4) *is injective; i.e., the equality $S(T)\xi_1 = S(T)\xi_2$, for some $T > 0$, implies that $\xi_1 = \xi_2$.*

PROOF. Indeed, let $(v_1(t), w_1(t))$ and $(v_2(t), w_2(t))$ be two solutions of problem (11.4) and let $(z(t), w(t))$ be their difference. Then, these functions solve (11.20). Let us rewrite this equation in the form

$$\partial_t \xi + \mathcal{B}\xi = \mathcal{P}(t)\xi, \tag{11.65}$$

where $\xi(t) := (z(t), \partial_t z(t), \theta(t))$,

$$\mathcal{B} = \begin{pmatrix} 1 & 0 & 0 \\ 0 & 1 & 0 \\ 0 & 0 & -\Delta_x + 1 \end{pmatrix}, \quad \mathcal{P}(t) := \begin{pmatrix} 1 & 1 & 0 \\ -[l_\varphi(t)\partial_t v_2(t) + l_f(t)] & 1 - \varphi(v_1(t)) & \alpha \\ 1 & 0 & 0 \end{pmatrix},$$

and

$$l_\varphi(t) := \int_0^1 \varphi'(sv_1(t) + (1-s)v_2(t))\, ds, \quad l_f(t) := \int_0^1 \varphi'(sv_1(t) + (1-s)v_2(t))\, ds.$$

Let us consider equation (11.65) in a Hilbert space $H := [L^2(\Omega)]^3$. Then, obviously, $\mathcal{B}$ is a positive selfadjoint (unbounded) operator in $H$ and the operator $\mathcal{P}(t)$ is uniformly bounded for all $t \geq 0$. Then, the classical backward uniqueness theorem of Agmon and Nirenberg (see [5]) is applicable here and, consequently, $\xi(T) = 0$ implies that $\xi(0) = 0$. Proposition 11.8 is proved. □

COROLLARY 11.3. *Let the assumptions of Theorem* 11.3 *hold. Then, for any hyperbolic equilibrium $\xi_0 := (v_0, w_0)$ of problem* (11.4), *the associated unstable manifold $\mathcal{M}^+_{\xi_0}$ is a finite-dimensional $C^1$-submanifold of $\Phi_\infty$ diffeomorphic to $\mathbb{R}^N$, where $N$ is the instability index of the equilibrium $\xi_0$.*

This result is a standard corollary of the existence of a Lyapunov function, injectivity, and smoothness of the semigroup $S(t)$; see [9].

We are now ready to formulate a theorem on the regular structure of the attractor $\mathcal{A}$ which can be considered as the main result of the section.

THEOREM 11.4. *Let the assumptions of Theorem* 11.3 *hold and let, in addition, all of the equilibria $\xi_0 \in \mathcal{R}$ be hyperbolic. Then:*

(1) *Any nonequilibrium complete trajectory $\xi(t)$, $t \in \mathbb{R}$, of the semigroup $S(t)$ belonging to the attractor is a heteroclinic orbit between two different equilibria $\xi_-$ and $\xi_+$:*

$$\lim_{t \pm \infty} \xi(t) = \xi_\pm, \tag{11.66}$$

*where $\xi_- \neq \xi_+$.*

(2) *The attractor $\mathcal{A}$ is a finite union of finite-dimensional submanifolds of $\Phi_\infty$:*

$$\mathcal{A} = \bigcup_{\xi_0 \in \mathcal{R}} \mathcal{M}^+_{\xi_0}, \tag{11.67}$$

*where $\mathcal{M}^+_{\xi_0}$ is an unstable manifold of $\xi_0 \in \mathcal{R}$.*

(3) *The rate of attraction to $\mathcal{A}$ is exponential; i.e., for any bounded set $B \subset \Phi_\infty$,*

$$\operatorname{dist}_{\Phi_\infty}(S(t)B, \mathcal{A}) \le Q(\|B\|_{\Phi_\infty})e^{-\gamma t} \tag{11.68}$$

*for some monotone function $Q$ and positive constant $\gamma$.*

Indeed, this theorem follows from the abstract result of Babin and Vishik on regular attractors; see [9] and Propositions 11.1, 11.6, and 11.8.

REMARK 11.2. Theorem 11.4 shows that the long-time behavior of solutions of problem (11.4) is "extremely good" if the monotonicity assumption (11.47) holds. As we will see in the next section, this behavior is "extremely bad" if the monotonicity assumption is essentially violated.

## 11.3. The nonmonotone $f$: Stabilization for the case of a weak coupling

The aim of this section is to understand what the asymptotic behavior of (11.4) may look like when the monotonicity assumption (11.47) is violated. To this end, we will consider below the case of a small coupling constant $\alpha$, where the dynamics is, in a sense, determined by the limit ODE

$$y''(t) + \varphi(y(t))y'(t) + f(y(t)) = 0. \tag{11.69}$$

In particular, in that limit case, the value of $v(t,x)$ at $x = x_0$ is determined by the value of $(v_0(x), v_0'(x))$ at $x = x_0$ only and, consequently, $v(t,x)$ and $v(t,y)$ evolve independently if $x \ne y$. Thus, if (11.69) has more than one equilibrium, then most of the trajectories will tend to a *discontinuous* equilibria, no matter whether or not the initial data $(v_0, v_0')$ is continuous. As we will see, the same property is preserved for the case of a small positive coupling constant $\alpha$.

To be more precise, we assume that the limit equation (11.69) possesses a regular attractor in $\mathbb{R}^2$, i.e., that

$$f'(u_0) \ne 0, \quad \text{for all } u_0 \text{ such that } f(u_0) = 0 \tag{11.70}$$

(since the existence of a global Lyapunov function and dissipativity are immediate, only the hyperbolicity assumption on the equilibria should be postulated).

We start our exposition by verifying that the dissipative estimate (11.25) is uniform with respect to $\alpha \to 0$.

PROPOSITION 11.9. *Let the assumptions of Theorem 11.2 hold. Then the positive constants $\beta$ and $C_*$ and the monotone function $Q$ in the dissipative estimate (11.25) are independent of $\alpha \to 0$.*

PROOF. In order to verify this assertion, we need to check that most of the estimates of Section 11.1 are uniform with respect to $\alpha \to 0$. We start with estimate (11.9). From the first point of view (based on the form of the global Lyapunov function (11.8)), one may guess that it is nonuniform and only $\alpha\|w(t)\|^2_{L^2}$ is uniformly bounded. However, the Lyapunov function gives the uniform bound for the $L^2$-norm of the $v$-component. The standard $L^2$-estimate for the second equation

$$\partial_t w - \Delta_x w + w = v(t) \tag{11.71}$$

of (11.4) gives after that the uniform $L^2$- and $L^\infty$-bounds for the $w$-component.

Thus, the uniformity as $\alpha \to 0$ is verified for estimates (11.9) and (11.14). The uniformity of the $L^\infty$-bound (11.13) follows from (11.14) exactly as in Proposition 11.2.

So, it only remains to check the dissipative estimate (11.25) and, following the proof of Theorem 11.2, we see that only the uniformity of the $L^2$-estimate (11.32) is necessary.

Analogously to (11.9) the function $Z(t)$ (defined by (11.27)) can be estimated as follows:

$$\begin{aligned} &\beta_2[\|\partial_t v\|_{L^2}^2 + \alpha(\|\nabla_x w\|_{L^2}^2 + \|w\|_{L^2}^2) + (|F(v)|, 1)] - C_2 \le Z(t) \\ &\le \beta_1[\|\partial_t v\|_{L^2}^2 + \alpha(\|\nabla_x w\|_{L^2}^2 + \|w\|_{L^2}^2) + (|F(v)|, 1)] + C_1, \end{aligned} \tag{11.72}$$

where $\beta_i$ and $C_i$ are now uniform with respect to $\alpha \to 0$. For this reason, estimates (11.30) and (11.31) do not immediately give the uniform analogue of (11.32), but only the uniform dissipative estimate for the $L^2$-norm of $v(t)$. Combining after that this estimate with the usual $L^2$-estimate for the heat equation (11.71), we verify that estimate (11.32) is indeed uniform as $\alpha \to 0$. Exactly as in Theorem 11.2, this gives the uniformity of estimate (11.25) and finishes the proof of the proposition. □

Thus, due to the previous proposition, the radius of the absorbing ball in $\Phi_\infty$ for problem (11.4) is uniform with respect to $\alpha \to 0$. In particular, the $\Phi_\infty$-norm of any equilibria of that problem is uniformly bounded. Denoting the set of equilibria for problem (11.4) by $\mathcal{R}_\alpha$, we may conclude that

$$\|\mathcal{R}_\alpha\|_{\Phi_\infty} \le C, \tag{11.73}$$

where the constant $C$ is independent of $\alpha$.

This observation together with the hyperbolicity assumption (11.70) allows us to give a complete description of the equilibria set $\mathcal{R}_\alpha$ if $\alpha > 0$ is small enough.

PROPOSITION 11.10. *Let the assumptions of Theorem 11.1 hold and let, in addition, the limit hyperbolicity assumption (11.70) be valid. Denote these hyperbolic equilibria by $\{u_1, \dots, u_N\}$. Then, there exists $\alpha_0 > 0$ such that, for every $\alpha \le \alpha_0$ and every partition*

$$\Omega = \Omega_1 \cup \Omega_2 \cup \dots \cup \Omega_N \tag{11.74}$$

*on disjoint measurable sets, $\Omega_i \cap \Omega_j = \emptyset$ for $i \ne j$. There exists a unique equilibrium $(v_0, 0, w_0) \in \Phi_\infty$ of problem (11.4) such that*

$$v_0 = \tilde v_0 + \theta, \quad \tilde v_0(x) := \sum_{i=1}^N u_i \chi_{\Omega_i}(x), \quad \|\theta\|_{L^\infty} \le C\alpha, \tag{11.75}$$

*where $\chi_V(x)$ is a characteristic function of the set $V$ and the constant $C$ is independent of $\alpha$. Moreover, every equilibrium $(v_0, 0, w_0) \in \Phi_\infty$ can be presented in such a form.*

PROOF. Indeed, in order to find the equilibrium, we need to solve

$$f(v_0) = \alpha w_0, \quad \Delta_x w_0 - w_0 = v_0,$$

which we rewrite in the form of a single equation on $v_0$ in $L^\infty(\Omega)$:

$$f(v_0) = \alpha(-\Delta_x + 1)^{-1} v_0. \tag{11.76}$$

We note that the function

$$F(v,\alpha):=f(v)-\alpha(-\Delta_x+1)^{-1}v$$

belongs to $C^1(L^\infty(\Omega)\times\mathbb{R},L^\infty(\Omega))$ (see [6]). Moreover, its derivative,

$$D_vF(\tilde{v}_0,0):=f'(\tilde{v}_0),$$

is invertible in $L^\infty$ (due to the hyperbolicity assumption (11.70)) and the norm of the inverse operator is uniformly bounded with respect to the choice of a partition. In addition, $F(\tilde{v}_0,0)=0$. Thus, the existence and uniqueness of the equilibrium $v_0$ in a small neighborhood of $\tilde{v}_0$ if $\alpha$ is small follows from the implicit function theorem.

Let us now verify that any equilibrium $(v_0,w_0)$ can be presented in that form. Indeed, let $(v_0,0,w_0)\in\Phi_\infty$ be an arbitrary equilibrium. Then, according to (11.73), $\|w_0\|_{L^\infty}\le C$ where $C$ is independent of $\alpha$. Therefore,

$$|f(v_0(x))|\le C\alpha,\quad x\in\Omega.$$

Since all of the roots $f(z)=0$ are hyperbolic, for sufficiently small $\alpha$, we conclude from this that

$$v_0(x)\in\mathcal{O}_{C\alpha}(u_{k(x)}),\quad x\in\Omega, \tag{11.77}$$

for some root $u_{k(x)}$ of $f(z)=0$. Now fixing

$$\Omega_i:=\{x\in\Omega,\ k(x)=i\}, \tag{11.78}$$

we see that the equilibrium $(v_0,w_0)$ indeed has the form of (11.75) and the proposition is proved. □

REMARK 11.3. We see that, in contrast to the case of monotone $f$, we now have an *uncountable* number of different equilibria (all of them are hyperbolic in $\Phi_\infty$) most of which are *discontinuous* (we have only a finite number of continuous equilibria associated with trivial partitions of $\Omega$). Moreover, using the explicit description given in the previous proposition, it is not difficult to show that the set $\mathcal{R}_\alpha$ is not compact in the strong topology of the space $\Phi_\infty$ and that it is not closed in the weak-$*$ topology of that space. For this reason, the possibility of applying the strong/weak global attractor theory to that problem seems very problematic. However, as the next theorem shows, any trajectory $(v(t),w(t))$ still converges to one of the equilibrium from $\mathcal{R}_\alpha$ as $t\to\infty$.

THEOREM 11.5. *Let the assumptions of Proposition* 11.10 *hold. Then, there exists* $\alpha_0>0$ *such that for every* $\alpha\le\alpha_0$ *every trajectory* $(v(t),\partial_tv(t),w(t))$ *of problem* (11.4) *stabilizes as* $t\to\infty$ *to some equilibrium* $(\bar{v},0,\bar{w})\in\mathcal{R}_\alpha$ *in the topology of* $L^p(\Omega)$:

$$\lim_{t\to\infty}\|(v(t),\partial_tv(t),w(t))-(\bar{v},0,\bar{w})\|_{[L^p(\Omega)]^3}=0 \tag{11.79}$$

*for any* $1\le p<\infty$.

PROOF. The proof of that convergence is strongly based on the perturbation theory of regular attractors and Proposition 2.1. Indeed, due to Propositions 11.9 and 11.3, we may assume without loss of generality that $(v(0),\partial_tv(0),w(0))$ belongs to the absorbing ball $B_R$ in $\Phi_\infty$ with the radius $R$ independent of $\alpha$ and that

$$\|w\|_{C_b(\mathbb{R}_+\times\Omega)}+\|\partial_tw\|_{C_b(\mathbb{R}_+\times\Omega)}\le C, \tag{11.80}$$

where the constant $C$ is also independent of $\alpha$. Thus, the first equation of (11.4)

$$\partial_t^2 v(t,x) + \varphi(v(t,x))\partial_t v(t,x) + f(v(t,x)) = \alpha w(t,x) \tag{11.81}$$

can be treated as an ODE for every fixed $x \in \Omega$. Moreover, due to the hyperbolicity assumption (11.70) and uniform estimate (11.80), the right-hand side of (11.81) can be treated as a small nonautonomous perturbation of the ODE

$$u'' + \varphi(u)u' + f(u) = 0. \tag{11.82}$$

Thus, the assumptions of Proposition 2.1 hold for problem (11.81) for every fixed $x \in \Omega$ if $\alpha \le \alpha_0$ for sufficiently small positive $\alpha_0$. Due to this proposition, we have the estimate

$$\int_0^T |\partial_t^2 v(t,x)| + |\partial_t v(t,x)|\, dt \le C_1 + C_2\alpha \int_0^T |\partial_t w(t,x)|\, dt, \tag{11.83}$$

where the positive constants $C_1$ and $C_2$ are independent of $T$, $\alpha$, and $x \in \Omega$. Integrating this inequality by $x \in \Omega$, we arrive at

$$\int_0^T \|\partial_t^2 v(t)\|_{L^1} + \|\partial_t v(t)\|_{L^1}\, dt \le C_1|\Omega| + C_2\alpha \int_0^T \|\partial_t w(t)\|_{L^1}\, dt. \tag{11.84}$$

In order to estimate the integral on the right-hand side of (11.84), we differentiate the second equation of (11.4) by $t$, denote $\theta := \partial_t w$, multiply it by $\operatorname{sgn}\theta(t)$, and integrate over $\Omega$. Then, due to the Kato inequality, we arrive at

$$\partial_t \|\theta(t)\|_{L^1} + \|\theta\|_{L^1} \le \|\partial_t v(t)\|_{L^1}. \tag{11.85}$$

Integrating this inequality, we have

$$\|\partial_t w(t)\|_{L^1} \le \|\partial_t w(0)\|_{L^1} e^{-t} + \int_0^t e^{-(t-s)}\|\partial_t v(s)\|_{L^1}\, ds.$$

Integrating the obtained inequality once more over $t \in [0,T]$ and using that $\partial_t w(0)$ is uniformly bounded, we arrive at

$$\int_0^T \|\partial_t w(t)\|_{L^1}\, dt \le C + \int_0^T \|\partial_t v(t)\|_{L^1}\, dt, \tag{11.86}$$

where $C$ is again independent of $\alpha$ and $T$ and the trajectory.

Inserting (11.86) into the right-hand side of (11.84) and assuming that $\alpha$ is small enough, we finally deduce the following $L^1$-dissipation integral

$$\int_0^T \|\partial_t^2 v(t)\|_{L^1} + \|\partial_t v(t)\|_{L^1} + \|\partial_t w(t)\|_{L^1}\, dt \le C, \tag{11.87}$$

where the constant $C$ is independent of $T$.

Thus, we have proved that $(v(t), \partial_t v(t), w(t))$ converges to some $\xi \in \Phi_\infty$ in the $L^1(\Omega)$-norm. Moreover, since we have control of the $L^\infty$-norm, the interpolation inequality gives convergence in $L^p$ for any $p < \infty$:

$$\lim_{t\to\infty} \|(v(t), \partial_t v(t), w(t)) - \xi\|_{[L^p(\Omega)]^3} = 0. \tag{11.88}$$

Thus, we only need to verify that $\xi \in \mathcal{R}_\alpha$ is an equilibrium. To this end, we will use the so-called trajectory approach (see [19] for the details) and consider positive semitrajectories instead of points in the phase spaces. Indeed, arguing exactly as in the proof of estimate (11.19) but taking the $L^p$-norm instead of the $L^\infty$-norm, we see that

$$\begin{aligned}&\|(v_1(t),\partial_t v_1(t),w_1(t))-(v_2(t),\partial_t v_2(t),w_2(t))\|_{[L^p(\Omega)]^3}\\ &\le Ce^{kt}\|(v_1(0),\partial_t v_1(0),w_1(0))-(v_2(0),\partial_t v_2(0),w_2(0))\|_{[L^p(\Omega)]^3}.\end{aligned}\tag{11.89}$$

Now define the map $\mathbb{S}:\Phi_\infty\to L^\infty(\mathbb{R}_+,\Phi_\infty)$ via the expression

$$\mathbb{S}:(v_0,v_0',w_0)\to(v(\cdot),\partial_t v(\cdot),w(\cdot))\tag{11.90}$$

and let $\mathcal{K}_+:=\mathbb{S}(\Phi_\infty)$. Then, estimate (11.89) (together with the obvious fact that $\mathcal{K}_+\subset L^\infty(\mathbb{R}_+,\Phi_\infty)$) shows that the map $\mathbb{S}$ realizes a Lipschitz continuous homeomorphism between spaces $\Phi_\infty$ and $\mathcal{K}_+$ endowed by the topology of $[L^p(\Omega)]^3$ and $\Phi_{tr}:=L^\infty(\mathbb{R}_+,[L^p(\Omega)]^3)$, respectively. The solution semigroup $S(t)$ is conjugated via that homeomorphism to the semigroup $T(t)$ of temporal shifts on $\Phi_{tr}$:

$$S(t)=\mathbb{S}^{-1}\circ T(t)\circ\mathbb{S},\quad (T(t)\xi)(s):=\xi(t+s),\ \xi\in\mathcal{K}_+,\quad t,s\ge 0.\tag{11.91}$$

Thus, the convergence (11.88) implies that

$$T(t)\mathbb{S}\xi_0:=T(t)\mathbb{S}(v(0),\partial_t v(0),w(0))\to\mathbb{S}\xi$$

in the space $\Phi_{tr}$.

Let us first check that $\mathbb{S}\xi\in\mathcal{K}_+$, i.e., that the limit trajectory $\xi(t)$ solves equation (11.4). In other words, we need to show that $\mathcal{K}_+$ is closed in $\Phi_{tr}$.

To this end, we need to show that it is possible to pass to the $\Phi_{tr}$-limit in equations (11.4) in the sense of distributions for any sequence $\xi_n(t):=(v_n(t),\partial_t v_n(t),w_n(t))$ converging in $\Phi_{tr}$ to some $\xi(t):=(\bar v(t),\partial_t\bar v(t),\bar w(t))$ and bounded in $L^\infty(\mathbb{R}_+,\Phi_\infty)$. Indeed, the passage to the limit in all linear terms is evident and only the passage to the limit in the nonlinear terms $\varphi(v)$ and $f(v)$ may a priori be problematic. But this is not the case, since convergence in $\Phi_{tr}$ implies convergence almost everywhere (up to extracting a subsequence) and this allows us to conclude in a standard way that $f(v_n)\rightharpoondown f(\bar v)$ and $\varphi(v_n)\partial_t v_n\rightharpoondown\varphi(\bar v)\partial_t\bar v$ (here we have implicitly used that $v_n$ is uniformly bounded in $L^\infty$). Thus, the limit function $\xi(t)$ indeed solves problem (11.4).

We are now ready to verify that $\xi(t)$ is an equilibrium, which will finish the proof of the theorem. Indeed, due to the dissipation integral (11.12), we see that

$$\|\partial_t v(s+\cdot)\|_{L^2(\mathbb{R}_+\times\Omega)}+\|\partial_t^2 v(s+\cdot)\|_{L^2(\mathbb{R}_+\times\Omega)}+\|\partial_t w(s+\cdot)\|_{L^2(\mathbb{R}_+\times\Omega)}\to 0$$

as $s\to\infty$. Thus, for the limit function $(\bar v(t),\partial_t\bar v(t),\bar w(t))$, we have $\partial_t\bar v\equiv\partial_t^2\bar v\equiv\partial_t\bar w\equiv 0$ and $\xi$ is indeed an equilibrium. Theorem 11.5 is proved. □

REMARK 11.4. Assume, in addition, that equation $f(z)=0$ possesses at least two solutions $v_1$ and $v_2$ such that $f'(v_i)>0$, $i=0,1$. Then, $v_i$ will be the exponential stable equilibria of equation (11.69). Now let $v_0(x)$ be a smooth function such that

$$v_0(x)=v_1,\quad x\in\Omega_1,\quad v_0(x)=v_2,\quad x\in\Omega_2,$$

for some nonempty $\Omega_i\subset\Omega$ of the nonzero measure.

Finally, let us consider the initial data for problem (11.4) of the form

$$\xi_0:=(v_0(x),0,0).$$

Then, since $\alpha>0$ is small and the equilibria $v_i$ are exponentially stable, the solution $v(t,x)$ will remain close to $v_i$ (for $x\in\Omega_i$) for all $t$. This shows that the smooth trajectory $S(t)\xi_0 = (v(t), \partial_t v(t), w(t))$ tends as $t \to \infty$ to the *discontinuous* equilibrium (in the $L^p$-topology, according to the last theorem).

This example shows that we cannot extend the assertion of the theorem to the case $p = \infty$ and obtain convergence in the topology of the phase space $\Phi_\infty$. Indeed, if the sequence of continuous functions converges in $L^\infty$ to some limit function, this function is automatically continuous. Thus, the $\omega$-limit set of the above-constructed trajectory in the topology of the phase space is empty:

$$\omega_{\Phi_\infty}(\xi_0) = \emptyset.$$

REMARK 11.5. It is clear from the proof of Proposition 11.10 that all of the equilibria $\mathcal{R}_\alpha$ are hyperbolic in the phase space $\Phi_\infty$. Thus, we may construct the infinite-dimensional stable and unstable manifolds for any equilibrium belonging to $\mathcal{R}_\alpha$ if $\alpha > 0$ is small enough. However, it does not help much for the study of the limit dynamics since, as shown in the previous proposition, generically, we do not have stabilization in the topology of $\Phi_\infty$, but only in a weaker space $[L^p(\Omega)]^3$. Also, in this weaker space the solution semigroup $S(t)$ is *not differentiable*. For this reason, we are not able to extract *exponential* convergence from the hyperbolicity of any equilibrium, and we do not know whether or not such exponential convergence takes place.

To conclude, we note that, arguing analogously to the proof of Proposition 11.10, one can extract some reasonable information about the equilibria $\mathcal{R}_\alpha$ even in the case where $\alpha$ is not small.

PROPOSITION 11.11. *Let the assumptions of Theorem* 11.1 *hold. Assume, in addition, that* $(\bar{v}, \bar{w}) \in \mathbb{R}^2$ *is a spatially homogeneous hyperbolic (in* $\Phi_\infty$*) equilibrium of problem* (11.4), *i.e., that the equation*

$$f'(\bar{v})\theta - \alpha(-\Delta_x + 1)^{-1}\theta = h \tag{11.92}$$

*is uniquely solvable for every* $h \in L^\infty(\Omega)$. *Assume, finally, that there exists another constant* $\tilde{v} \neq \bar{v}$ *such that*

$$f(\bar{v}) = f(\tilde{v}) \tag{11.93}$$

*(this, of course, may happen only in the case of nonmonotone* $f$*). Then, there exists* $\delta_0 > 0$ *such that, for any measurable partition* $\Omega = \Omega_1 \cup \Omega_2$ *on two disjoint sets where*

$$|\Omega_2| \le \delta_0, \tag{11.94}$$

*there exists a hyperbolic equilibrium* $(v, w)$ *such that* $v$ *is close (in the* $L^\infty$*-metric) to*

$$v_{12} := \bar{v}\chi_{\Omega_1}(x) + \tilde{v}\chi_{\Omega_2}(x). \tag{11.95}$$

PROOF. We first check that the equation of variations

$$f'(v_{12})\theta - \alpha(-\Delta_x + 1)^{-1}\theta = h \tag{11.96}$$

is uniquely solvable if the measure of $\Omega_2$ is small. To this end, we construct the approximative solution of this equation in the form $\tilde{\theta} := \theta_0 + \hat{\theta}$ where $\theta_0$ solves equation (11.92) and

$$\hat{\theta}(x) := [f'(\bar{v}) - f'(\tilde{v})]\theta_0(x)\chi_{\Omega_2}(x). \tag{11.97}$$

Then, since $\bar{v}$ is a hyperbolic equilibrium, we have

$$\|\theta_0\|_{L^\infty(\Omega)} \le C\|h\|_{L^\infty(\Omega)}. \tag{11.98}$$

Moreover, the approximate solution $\tilde{\theta}$ thus constructed solves

$$f'(v_{12})\tilde{\theta} - \alpha(-\Delta_x + 1)^{-1}\tilde{\theta} = \tilde{h} := -\alpha(-\Delta_x + 1)^{-1}\hat{\theta}. \tag{11.99}$$

Finally, since the measure of $\Omega_2$ is small, we have

$$\|\hat{\theta}\|_{L^2} \le C|\Omega_2|^{1/2}\|\theta_0\|_{L^\infty} \le C_1\delta_0^{1/2}\|h\|_{L^\infty}$$

and, consequently,

$$\|\tilde{h}\|_{L^\infty} \le C\|\tilde{h}\|_{H^2} \le C'\|\hat{\theta}\|_{L^2} \le C_2\delta_0^{1/2}\|h\|_{L^\infty}, \tag{11.100}$$

where the constant $C_2$ is independent of $h$ and of the concrete form of the partition $\Omega = \Omega_1 \cup \Omega_2$.

Thus, if $\delta_0 > 0$ is so small that $C_2\delta_0^{1/2} := \kappa < 1$, the norm of the reminder $\|\tilde{h}\|_{L^\infty}$ is estimated to be $\kappa\|h\|_{L^\infty}$ with $\kappa < 1$. Then, the standard iteration process gives the desired solution $\theta$ of equation (11.96) together with estimate

$$\|\theta\|_{L^\infty} \le C\|h\|_{L^\infty} \tag{11.101}$$

with the constant $C$ independent of $\delta_0 \to 0$. The uniqueness of a solution can be obtained in a standard way using the observation that the operator $f'(v_{12}) - \alpha(-\Delta_x + 1)^{-1}$ is selfadjoint in $L^2$.

It is now not difficult to finish the proof of the proposition. Indeed, we seek for the desired equilibrium $(v, w)$ in the form

$$v(x) = v_{12}(x) + \theta(x),$$

where $\theta$ is a small corrector which should satisfy the equation

$$f(v_{12} + \theta) - f(v_{12}) - \alpha(-\Delta_x + 1)^{-1}\theta = \hat{h} := \alpha(-\Delta_x + 1)[(\bar{v} - \tilde{v})\chi_{\Omega_2}] \tag{11.102}$$

and, arguing as before, we see that

$$\|\hat{h}\|_{L^\infty} \le C|\Omega_2|^{1/2}.$$

Now applying the implicit function theorem to equation (11.102), we establish the existence of a unique solution $\theta$,

$$\|\theta\|_{L^\infty} \le C_1|\Omega_2|^{1/2}$$

where the measure of $\Omega_2$ is small enough. Proposition 11.11 is proved. □

REMARK 11.6. Although we are not able either to give a complete description of the equilibria set $\mathcal{R}_\alpha$ or to verify stabilization if $\alpha$ is not small, we see that, under the assumptions of the last proposition (which are, in a sense, natural for the nonmonotone case), the set of equilibria is not compact in $\Phi_\infty$ and is not closed in the weak-$*$ topology of the phase space. These facts do not allow us to extend global attractor theory to the nonmonotone case.

Note also that, although we formulate (for simplicity) Proposition 11.11 for the case of *spatially homogeneous* hyperbolic equilibrium $(\bar{v}, \bar{w})$, it can be easily extended to the case of nonhomogeneous equilibria. This shows that the conclusion of Proposition 11.11 is somehow "generic" for the nonmonotone case.

# Bibliography


[1] G. Akagi and M. Otani, *Evolution inclusions governed by subdifferentials in reflexive Banach spaces*, Journal of Evolution Equations 4 (2004), 519–541.

[2] R. A. Adams, *Sobolev spaces*, Academic Press, New York (1975).

[3] M. Aida, M. A. Efendiev, and A. Yagi, *Quasilinear abstract parabolic evolution equations and exponential attractor*, Osaka Journal of Mathematics 42 (2005), 101–132.

[4] M. Aida, T. Tsujikawa, M. A. Efendiev, A. Yagi, and M. Mimura, *Lower estimate of the attractor dimension for a chemotaxis growth system*, Journal London Mathematical Society 74 (2006), 453–474.

[5] S. Agmon and L. Nirenberg, *Lower bounds and uniqueness theorems for solutions of differential equations in a Hilbert space*, Communications on Pure and Applied Mathematics 20 (1967), 207–229.

[6] H. Amann and J. Escher, *Analysis II*, Grundstudium Mathematik, Birkhäuser, (1999).

[7] F. Andreu, J. Mazon, F. Simondon, and J. Toledo, *Attractor for a degenerate nonlinear diffusion problem with nonlinear boundary condition*, Journal of Dynamics and Differential Equations 10 (1998), no. 3, 347–377.

[8] G. Andrews and J. M. Ball, *Asymptotic behavior and changes of phase in one-dimensional nonlinear viscoelasticity*, Journal of Differential Equations 44 (1982), 306–341.

[9] A. Babin and M. I. Vishik, *Attractors of Evolutionary Equations*, Nauka, Moscow, 1989; English translation in Studies in Applied Mathematics 25, North-Holland, Amsterdam (1992).

[10] V. Belleri and V. Pata, *Attractors for semilinear strongly damped wave equations on $\mathbb{R}^3$*, Discrete and Continuous Dynamical Systems 7 (2001), no. 4, 719–735.

[11] A. Biryuk, *Spectral properties of solutions of Burgers equation with small dissipation*, Functional Analysis and Its Applications 35 (2001), 1–12.

[12] H. Brezis, *Operateurs Maximaux Monotone et Semi-Groupes de Contractions dans les Espaces Hilbert*, North-Holland Mathematics Studies 5 (1973).

[13] H. Brezis and M. Crandall, *Uniqueness of solutions of the initial-value problem for $u_t - \Delta\varphi(u) = 0$*, Journal de Mathématiques Pures et Appliquées 58(9) (1979), 2, 153–163.

[14] T. Caraballo, P. Kloeden, and J. Real, *Pullback and forward attractors for a damped wave equation with delays*, Stochastics and Dynamics 4 (2004), 405–423.

[15] A. N. Carvalho, J. A. Langa, J. C. Robinson, and A. Surez, *Characterization of nonautonomous attractors of a perturbed infinite-dimensional gradient system*, Journal of Differential Equations 236 (2007), 570–603.

[16] D. Cheban, P. Kloeden, and B. Schmalfuss, *The relationship between pullback, forward and global attractors of nonautonomous dynamical systems*, Nonlinear Dynamics and Systems Theory 2 (2002), 9–28.

[17] Y. Chen, *Hölder estimates for solutions of uniformly degenerate quasilinear parabolic equations*, Chinese Annals of Mathematics B 5 (4) (1984), 661–678.

[18] V. V. Chepyzhov and M. A. Efendiev, *Hausdorff dimension estimation for attractors of nonautonomous dynamical systems in unbounded domains: An example*, Communications on Pure and Applied Mathematics 53 (2000), 647–665.

[19] V. V. Chepyzhov and M. I. Vishik, *Attractors for Equations of Mathematical Physics*, American Mathematical Society, Providence, Rhode Island (2002).

[20] V. V. Chepyzhov and M. I. Vishik, *Attractors of nonautonomous dynamical systems and their dimension*, Journal de Mathématiques Pures et Appliquées 73 (1994), 279–333.

[21] V. V. Chepyzhov and M. I. Vishik, *Kolmogorov's $\varepsilon$-entropy for the attractor of reaction-diffusion equations*, Sbornik: Mathematics 189 (2) (1998), 81–110.

[22] L. H. Chuan, T. Tsujikawa, and A. Yagi, *Asymptotic behavior of solutions for forest kinematic model*, Funkcialaj Ekvacioj 49 (3) (2006), 427–449.

[23] L. H. Chuan and A. Yagi, *Dynamical system for forest kinematic model*, Advances in Mathematical Sciences and Applications 16 (2006), 343–409.

[24] P. Colli, *On some doubly nonlinear evolution equations in Banach spaces*, Japan Journal of Industrial and Applied Mathematics 9 (2) (1992), 181–203.

[25] P. Colli and A. Visintin, *On a class of doubly nonlinear evolution equations*, Communications in Partial Differential Equations 15 (1990), no. 5, 737–756.

[26] J. W. Costerton, Z. Lewandowski, D. E. Caldwell, D. R. Korber, and H. M. Lappin-Scott, *Microbial Biofilms*, Annual Review Microbiology 49 (1995), 711–745.

[27] E. DiBenedetto, *Degenerate Parabolic Equations* (Universitext), Springer-Verlag, New York (1993).

[28] L. Dung and B. Nicolaenko, *Exponential attractors in Banach spaces*, Journal of Dynamics and Differential Equations 13 (2001), 791–806.

[29] A. Eden, C. Foias, B. Nicolaenko, and R. Temam, *Exponential Attractors for Dissipative Evolution Equations*, Research in Application Mathematics 37, John Wiley & Sons, New York (1994).

[30] A. Eden and J. Rakotoson, *Exponential attractors for a doubly nonlinear equation*, Journal of Mathematical Analysis and Applications 185 (2) (1994), 321–339.

[31] M. A. Efendiev, *Evolution Equations Arising in the Modeling of Life Sciences*, International Series of Numerical Mathematics, Vol. 163, Birkhäuser/Springer, 217 pages.

[32] M. A. Efendiev, *Finite and Infinite Dimensional Attractors for Evolution Equations of Mathematical Physics*, Gakkotoscho, Tokyo, 2010, 239 pages.

[33] M. A. Efendiev and L. Demaret, *On the structure of attractors for a class of degenerate reaction-diffusion systems*, Advances and Applications in Mathematical Sciences 18 (1) (2008), 105–118.

[34] M. A. Efendiev, R. Lasser, and S. Sonner, *Necessary and sufficient conditions for an infinite system of parabolic equations preserving the positive cone*, International Journal of Biomathematics and Biostatistics 1 (2010), 47–52.

[35] M. A. Efendiev, A. Miranville, and S. V. Zelik, *Exponential attractors and finite-dimensional reduction for nonautonomous dynamical systems*, Proceedings of the Royal Society of Edinburgh, Section A 135 (2005), 703–730.

[36] M. A. Efendiev, A. Miranville, and S. V. Zelik, *Exponential attractors for a nonlinear reaction-diffusion system in* $\mathbb{R}^3$, Comptes Rendus de l'Académie des Sciences, Series I, 330 (2000), 713–718.

[37] M. A. Efendiev, A. Miranville, and S. V. Zelik, *Infinite dimensional attractors for a nonautonomous reaction-diffusion system*, Mathematische Nachrichten 248–249 (2003), no. 1, 72–96.

[38] M. A. Efendiev and M. Otani, *Attractors of some class degenerate-doubly nonlinear parabolic equations*, Preprint IBB 41 (2007).

[39] M. A. Efendiev and M. Otani, *Infinite dimensional attractors for parabolic equations with p-Laplacian in heterogeneous medium*, Annales de l'Institut Henri Poincaré AN 28 (2011), 565–582.

[40] M. A. Efendiev and S. V. Zelik, *Finite and infinite dimensional attractors for porous media equations*, Proceedings London Mathematical Society 96 (2008), 51–77.

[41] M. A. Efendiev and S. Zelik, *Finite dimensional attractors and exponential attractors for degenerate doubly nonlinear equations*, Mathematical Methods in the Applied Sciences 32 (13) (2009), 1638–1668.

[42] M. A. Efendiev and S. V. Zelik, *The regular attractor for the reaction-diffusion system with a nonlinearity rapidly oscillating in time and its averaging*, Advances in Differential Equations 8 (6) (2003), 673–732.

[43] M. A. Efendiev, S. V. Zelik, and H. J. Eberl, *Existence and longtime behavior of a biofilm model*, Communications on Pure and Applied Analysis 8 (2) (2009), 509–531.

[44] M. A. Efendiev and S. V. Zelik, *Global attractor and stabilization for a coupled PDE-ODE system*, Preprint IBB (2011).

[45] M. A. Efendiev and A. Yagi, *Continuous dependence of exponential attractors for a parameter*, Journal of the Mathematical Society of Japan 57 (2005), 167–181.

[46] P. Fabrie and A. Miranville, *Exponential attractors for nonautonomous first-order evolution equations*, Discrete and Continuous Dynamical Systems 4 (1998), 225–240.

[47] E. Feireisl, Ph. Laurencot, and F. Simondon, *Global attractors for degenerate parabolic equations on unbounded domains*, Journal of Differential Equations 129 (1996), no. 2, 239–261.

[48] R. M. Ford and R. W. Harvey, *Role of chemotaxis in the transport of bacteria through saturated porous media*, Advances in Water Resources 30 (2007), 1608–1617.

[49] H. Gajewski, K. Gröger, and K. Zacharias, *Nichtlineare Operatorgleichungen und Operatordifferentialgleichungen*, Akademie-Verlag, Berlin (1974).

[50] A. Y. Goritsky and M. I. Vishik, *Integral manifolds for nonautonomous equations*, Rendiconti Academia Nazionale delle Scienze detta dei XL, Memorie di Matematica e Applicazioni. 115 (21) (1997), 109–146.

[51] M. Gurtin, *Generalized Ginzburg-Landau and Cahn-Hilliard equations based on a microforce balance*, Physica D 92 (1996), 178–192.

[52] R. Hagen, S. Roch, and B. Silbermann, *C*-algebras and Numerical Analysis*, CRC Press (2001).

[53] J. Hale, *Asymptotic behavior of dissipative systems*, Mathematical Surveys and Monographs 25, American Mathematical Society, Providence, Rhode Island (1988).

[54] A. Haraux, *Systèmes Dynamiques Dissipatifs et Applications*, Elsevier Masson (1991).

[55] D. Henry, *Geometric theory of semilinear parabolic equations*, Lecture Notes in Mathematics 840, Springer-Verlag, Berlin-New York (1981).

[56] D. Horstmann, *From 1970 until Present: The Keller-Segel Model in chemotaxis and Its Consequences, Part I*, Jahresbericht der Deutschen Mathematiker-Vereinigung 105(3) (2003), 103-165.

[57] A. V. Ivanov, *Gradient estimates for doubly nonlinear parabolic equations*, Journal of Mathematical Sciences 93 (5) (1999), 661-688.

[58] A. V. Ivanov, *Second order quasilinear degenerate and nonuniformly elliptic and parabolic equations*, Trudy Matematicheskogo Instituta imeni V. A. Steklova 160 (1982).

[59] S. Kamin and L. A. Peletier, *Large time behavior of solutions of the porous media equation with absorption*, Israel Journal of Mathematics 55 (2) (1986), 129-146.

[60] P. Kloeden, *Pullback attractors of nonautonomous semidynamical systems*, Stochastics and Dynamics 3 (2003), 101–112.

[61] A. N. Kolmogorov and V. M. Tikhomirov, *$\varepsilon$-entropy and $\varepsilon$-capacity of sets in functional space*, American Mathematical Society Translations 2, 17 (1961), 277–364.

[62] M. A. Krasnoselskii, *Topological Methods in the Theory of Nonlinear Integral Equations*, MacMilian, New York (1964).

[63] P. Krejčí and S. Zheng, *Pointwise asymptotic convergence of solutions for a phase separation model*, Discrete and Continuous Dynamical Systems 16 (1) (2006), 1–18.

[64] N. Krylov, *Nonlinear second order elliptic and parabolic equations*, Moscow: Nauka (1985).

[65] Yu. Kuznetsov, M. Antonovsky, V. Biktashev, and A. Aponina, *A cross-diffusion model of forest boundary dynamics*, Journal of Mathematical Biology 32 (1994), 219–232.

[66] O. A. Ladyzhenskaya, *Attractors for Semigroups and Evolution Equations*, Lincei Lectures, Cambridge University Press (1991).

[67] O. A. Ladyzhenskaya, *On the determination of minimal global attractors for Navier-Stokes equations and other partial differential equations*, Uspekhi Matematicheskikh Nauk 42 (1987), 25–60.

[68] O. A. Ladyzhenskaya, V. A. Solonnikov, and N. N. Uraltseva, *Linear and quasilinear equations of parabolic type*, Nauka (1967).

[69] J. L. Lions, *Quelques methodes de resolution des problemes aux limites non linearies*, Dunod, Paris (1969).

[70] J. Malek and D. Prazak, *Large time behavior by the method of $l$-trajectories*, Journal of Differential Equations 181 (2002), 243–279.

[71] A. Mielke, *Chapter 6: Evolution of Rate-independent Systems*, C. Dafermos and E. Feireisl, eds., Handbook of Differential Equations, Evolutionary Equations, 2, 461–559.

[72] A. J. Milani and N. J. Koksch, *An Introduction to Semiflows*, Chapman & Hall/CRC, Boca Raton (2005).

[73] M. Mimura and T. Tsujikawa, *Aggregating pattern dynamics in a chemotaxis model including growth*, Journal Physica A 230 (1996), 499–543.

[74] T. Nagai, *Blow-up of radially symmetric solutions to a chemotaxis system*, Advances in Mathematical Sciences and Applications 5 (1995), 581–601.

[75] T. Nagai and T. Senba, *Behavior of radially symmetric solutions of a system related to chemotaxis*, Nonlinear Analysis 30 (1997), 3837–3842.

[76] T. Nagai and T. Senba, *Global existence and blow-up of radial solutions to a parabolic-elliptic system of chemotaxis*, Advances in Mathematical Sciences and Applications 8 (1998), 145–156.

[77] H. Nakata, *Numerical simulations for forest boundary dynamics model*, Master Thesis, Osaka University (2004).

[78] S. M. Nikolskii, *Approximation of Functions of Several Variables and Embedding Theorems* (Russian), Nauka (1969).

[79] L. Nirenberg, *Topic in Nonlinear Functional Analysis*, Courant Institute Lecture Notes, New York University (1975).

[80] M. Otani, *$L^\infty$-energy method and its applications*, Nonlinear Partial Differential Equations and Their Applications 505-516, GAKUTO International Series, Mathematical Sciences and Applications, Gakkotosho, Tokyo 20 (2004).

[81] M. Otani, *$L^\infty$-Energy Method, Basic tools and usage*, Differential Equations, Chaos and Variational Problems: Progress in Nonlinear Differential Equations and Their Applications 75, Vasile Staicu (ed.), Birkhauser (2007), 357–376.

[82] M. Otani, *Nonmonotone perturbations for nonlinear parabolic equations associated with subdifferential operators*, Cauchy problems, Journal of Differential Equations 46 (12) (1982), 268-299.

[83] K. Osaki, T. Tsujikawa, A. Yagi, and M. Mimura, *Exponential attractor for a chemotaxis-growth system of equations*, Nonlinear Analysis 51 (2002), 119–144.

[84] R. Pego, *Phase transitions in one-dimensional nonlinear viscoelasticity: Admissibility and stability*, Archive for Rational Mechanics and Analysis 97 (1987), 353–394.

[85] G. Raugel, *Global Attractors in Partial Differential Equations*, Handbook of Dynamical Systems 2, North-Holland (2002), 885–982.

[86] G. R. Sell and Y. You, *Dynamics of Evolutionary Equations*, Springer-Verlag, New York (2002).

[87] T. Senba, *Blowup behavior of radial solutions to Jäger-Luckhaus system in high dimensional domains*, Funkcialaj Ekvaciog (Fako de l'Funkcialaj Ekvacioj Japana Matematika Societo, Funkcialaj Ekvaciog, Serio Internacia) 48 (2) (2005), 247–271.

[88] T. Senba, *Type II Blowup of Solutions to a Simplified Keller-Segel System in Two Dimensional Domains*, Nonlinear Analysis Series A: Theory, Methods and Applications 66 (2007), 1817–1839.

[89] T. Senba and T. Suzuki, *Chemotactic collapse in a parabolic-elliptic system of mathematical biology*, Advances in Differential Equations 6 (1) (2001), 21–50.

[90] I. V. Skrypnik, *Nonlinear Elliptic Boundary Value Problems*, Teubner-Texte zur Mathematik 91, Leipzig (1986).

[91] A. Stevens and D. Horstmann, *A constructive approach to traveling waves in chemotaxis*, Journal of Nonlinear Science 14 (1) (2004), 1–25.

[92] Y. Sugiyama, *Global existence in subcritical cases and finite time blow-up in super-critical cases to degenerate Keller-Segel system*, Differential and Integral Equations 20 (2006), 841–876.

[93] R. Temam, *Infinite-dimensional dynamical systems in mechanics and physics*, Springer-Verlag, New York (1988).

[94] H. Triebel, *Interpolation theory, function spaces, differential operators*, North-Holland, Amsterdam-New York (1978).

[95] A. Yagi, *Abstract Parabolic Evolution Equations and Their Applications*, Springer Monographs in Mathematics, Springer-Verlag, Berlin, Heidelberg (2010).

[96] J. L Vazquez, *The porous medium equations*, Clarendon Press, Oxford (2007).

[97] E. Zeidler, *Nonlinear functional analysis and its applications*, Springer-Verlag (1985).

# Index